CIVIL ENGINEERING CALCULATIONS REFERENCE GUIDE

The McGraw-Hill
Engineering Reference Guide Series

This series makes available to professionals and students a wide variety of engineering information and data available in McGraw-Hill's library of highly acclaimed books and publications. The books in the series are drawn directly from this vast resource of titles. Each one is either a condensation of a single title or a collection of sections culled from several titles. The Project Editors responsible for the books in the series are highly respected professionals in the engineering areas covered. Each Editor selected only the most relevant and current information available in the McGraw-Hill library, adding further details and commentary where necessary.

Hicks · CIVIL ENGINEERING CALCULATIONS REFERENCE GUIDE

Hicks · MACHINE DESIGN CALCULATIONS REFERENCE GUIDE

Hicks · PLUMBING DESIGN AND INSTALLATION REFERENCE GUIDE

Hicks · POWER GENERATION CALCULATIONS REFERENCE GUIDE

Hicks · POWER PLANT EVALUATION AND DESIGN REFERENCE GUIDE

Johnson & Jasik · ANTENNA APPLICATIONS REFERENCE GUIDE

Markus and Weston · CLASSIC CIRCUITS REFERENCE GUIDE

Merritt · CIVIL ENGINEERING REFERENCE GUIDE

Woodson · HUMAN FACTORS REFERENCE GUIDE FOR ELECTRONICS AND
COMPUTER PROFESSIONALS

Woodson · HUMAN FACTORS REFERENCE GUIDE FOR PROCESS PLANTS

CIVIL ENGINEERING CALCULATIONS REFERENCE GUIDE

TYLER G. HICKS, P.E., EDITOR
International Engineering Associates

Contributor

MAX KURTZ, P.E.
Consulting Engineer

Metricated by
GERALD M. EISENBERG
Project Engineering Administrator
American Society of Mechanical Engineers

McGRAW-HILL BOOK COMPANY

New York   St. Louis   San Francisco   Auckland   Bogotá   Hamburg
London   Madrid   Mexico   Milan   Montreal   New Delhi
Panama   Paris   São Paulo   Singapore   Sydney   Tokyo   Toronto

Library of Congress Cataloging-in-Publication Data

Civil engineering calculations reference guide.

  (The McGraw-Hill engineering reference guide series)
  Rev. ed. of: Standard Handbook of engineering
calculations. 2nd ed. c1985.
  Includes index.
  1. Engineering mathematics—Handbooks, manuals, etc.
I. Hicks, Tyler Gregory.   II. Kurtz, Max.
III. Standard handbooks of engineering
calculations.   IV. Series.
TA332.C58   1987   624′.0212   86-33825
ISBN 0-07-028798-8

CIVIL ENGINEERING CALCULATIONS REFERENCE GUIDE

  234567890   DOC/DOC   893210987

ISBN 0-07-028798-8

Printed and bound by R.R. Donnelley and Sons Company

CONTENTS

PREFACE

This reference guide is a concise coverage of the key areas of civil engineering, including fluid mechanics, surveying, aerial photogrammetry, and soil mechanics. The guide is a condensation of the *Standard Handbook of Engineering Calculations,* 2nd Edition.

Fully metricated, the guide contains hundreds of step-by-step calculation procedures for making quick and accurate analyses of many common and uncommon design situations. Specific topics covered include: statics, stress and strain, stresses in flexural members, deflection of beams, statically indeterminate structures, moving loads and influence lines, riveted and welded connections, steel beams and steel plate girders, steel columns and tension members, plastic design of steel structures, timber engineering, reinforced concrete, prestressed concrete, design of highway bridges, fluid mechanics, surveying and route design, aerial photogrammetry, and soil mechanics.

Most procedures given have related calculations—that is, other items that can be computed using the same general technique. This expands the coverage of the guide enormously. The result is that civil engineers, designers, and drafters have a powerful guide for quick and accurate solutions of hundreds of calculation procedures, along with worked-out real-life applications.

Both USCS and SI units are used throughout the guide. This permits easier and faster use of the guide in both the United States and overseas, where SI is widely used. Thus, a designer in the United States can easily do civil-engineering work overseas. And an overseas designer can easily do civil-engineering work for applications in the United States.

The guide is thoroughly up-to-date in its coverage, with calculations devoted to the design of a variety of structural members and structures using modern materials and methods. Latest code methods and usages are also given. Such coverage helps users of the guide cope with new design jobs they meet in their daily work.

Users who will find the guide most helpful include civil engineers, building and structural designers, drafters, cost estimators, schedulers, mechanical engineers, machine designers, architects, and students. Further, the algorithms given are ideal for use in micro, mini, and mainframe computers to effect a greater time saving for users.

The editor thanks the contributors whose work is cited in the guide. And if readers find any errors or deficiencies in the guide, the editor asks that they be pointed out to him. He will be grateful to every reader detecting such flaws who writes him in care of the publisher. Errors will be corrected in the next reprinting of the guide.

TYLER G. HICKS, P.E.

Principles of Statics; Geometric Properties of Areas

If a body remains in equilibrium under a system of forces, the following conditions obtain:

1. The algebraic sum of the components of the forces in any given direction is zero.

2. The algebraic sum of the moments of the forces with respect to any given axis is zero.

The above statements are verbal expressions of the *equations of equilibrium*. In the absence of any notes to the contrary, a clockwise moment is considered positive; a counterclockwise moment, negative.

GRAPHICAL ANALYSIS OF A FORCE SYSTEM

The body in Fig. 1*a* is acted on by forces *A*, *B*, and *C*, as shown. Draw the vector representing the equilibrant of this system.

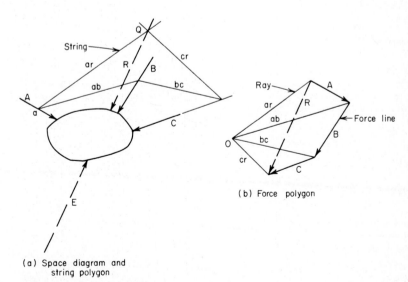

(a) Space diagram and
    string polygon

(b) Force polygon

FIG. 1 Equilibrant of force system.

1

Calculation Procedure:

1. Construct the system force line

In Fig. 1b, draw the vector chain A-B-C, which is termed the *force line*. The vector extending from the initial point to the terminal point of the force line represents the resultant R. In any force system, the resultant R is equal to and collinear with the equilibrant E, but acts in the opposite direction. The equilibrant of a force system is a single force that will balance the system.

2. Construct the system rays

Selecting an arbitrary point O as the pole, draw the rays from O to the ends of the vectors and label them as shown in Fig. 1b.

3. Construct the string polygon

In Fig. 1a, construct the string polygon as follows: At an arbitrary point a on the action line of force A, draw strings parallel to rays ar and ab. At the point where the string ab intersects the action line of force B, draw a string parallel to ray bc. At the point where string bc intersects the action line of force C, draw a string parallel to cr. The intersection point Q of ar and cr lies on the action line of R.

4. Draw the vector for the resultant and equilibrant

In Fig. 1a, draw the vector representing R. Establish the magnitude and direction of this vector from the force polygon. The action line of R passes through Q.

Last, draw a vector equal to and collinear with that representing R but opposite in direction. This vector represents the equilibrant E.

Related Calculations: Use this general method for any force system acting in a single plane. With a large number of forces, the resultant of a smaller number of forces can be combined with the remaining forces to simplify the construction.

ANALYSIS OF STATIC FRICTION

The bar in Fig. 2a weighs 100 lb (444.8 N) and is acted on by a force P that makes an angle of 55° with the horizontal. The coefficient of friction between the bar and the inclined plane is 0.20. Compute the minimum value of P required (a) to prevent the bar from sliding down the plane; (b) to cause the bar to move upward along the plane.

Calculation Procedure:

1. Select coordinate axes

Establish coordinate axes x and y through the center of the bar, parallel and perpendicular to the plane, respectively.

2. Draw a free-body diagram of the system

In Fig. 2b, draw a free-body diagram of the bar. The bar is acted on by its weight W, the force P, and the reaction R of the plane on the bar. Show R resolved into its x and y components, the former being directed upward.

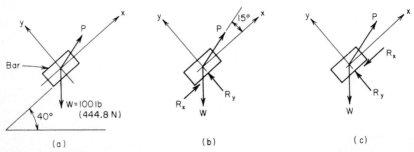

FIG. 2   Body on inclined plane.

3. Resolve the forces into their components

The forces W and P are the important ones in this step, and they must be resolved into their x and y components. Thus

$W_x = -100 \sin 40° = -64.3$ lb $(-286.0$ N$)$ $W_y = -100 \cos 40° = -76.6$ lb $(-340.7$ N$)$
$P_x = P \cos 15° = 0.966P$ $P_y = P \sin 15° = 0.259P$

4. Apply the equations of equilibrium

Consider that the bar remains at rest and apply the equations of equilibrium. Thus

$$\Sigma F_x = R_x + 0.966P - 64.3 = 0 \qquad R_x = 64.3 - 0.966P$$
$$\Sigma F_y = R_y + 0.259P - 76.6 = 0 \qquad R_y = 76.6 - 0.259P$$

5. Assume maximum friction exists and solve for the applied force

Assume that R_x, which represents the frictional resistance to motion, has its maximum potential value. Apply $R_x = \mu R_y$, where μ = coefficient of friction. Then $R_x = 0.20R_y = 0.20(76.6 - 0.259P) = 15.32 - 0.052P$. Substituting for R_x from step 4 yields $64.3 - 0.966P = 15.32 - 0.052P$; so $P = 53.6$ lb $(238.4$ N$)$.

6. Draw a second free-body diagram

In Fig. 2c, draw a free-body diagram of the bar, with R_x being directed downward.

7. Solve as in steps 1 through 5

As before, $R_y = 76.6 - 0.259P$. Also the absolute value of $R_x = 0.966P - 64.3$. But $R_x = 0.20R_y = 15.32 \times 0.052P$. Then $0.966P - 64.3 = 15.32 - 0.052P$; so $P = 78.2$ lb $(347.8$ N$)$.

ANALYSIS OF A STRUCTURAL FRAME

The frame in Fig. 3a consists of two inclined members and a tie rod. What is the tension in the rod when a load of 1000 lb (4448.0 N) is applied at the hinged apex? Neglect the weight of the frame and consider the supports to be smooth.

Calculation Procedure:

1. Draw a free-body diagram of the frame

Since friction is absent in this frame, the reactions at the supports are vertical. Draw a free-body diagram as in Fig. 3b.

With the free-body diagram shown, compute the distances x_1 and x_2. Since the frame forms a 3-4-5 right triangle, $x_1 = 16(4/5) = 12.8$ ft $(3.9$ m$)$ and $x_2 = 12(3/5) = 7.2$ ft $(2.2$ m$)$.

2. Determine the reactions on the frame

Take moments with respect to A and B to obtain the reactions:

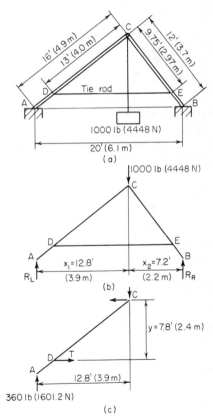

FIG. 3

$$\Sigma M_B = 20R_L - 1000(7.2) = 0 \qquad R_L = 360 \text{ lb } (1601.2 \text{ N})$$
$$\Sigma M_A = 1000(12.8) - 20R_R = 0 \qquad R_R = 640 \text{ lb } (2846.7 \text{ N})$$

3. Determine the distance y in Fig. 3c

Draw a free-body diagram of member AC in Fig. 3c. Compute $y = 13(3/5) = 7.8$ ft (2.4 m).

4. Compute the tension in the tie rod

Take moments with respect to C to find the tension T in the tie rod;

$$\Sigma M_C = 360(12.8) - 7.8T = 0 \qquad T = 591 \text{ lb } (2628.8 \text{ N})$$

5. Verify the computed result

Draw a free-body diagram of member BC, and take moments with respect to C. The result verifies that computed above.

GRAPHICAL ANALYSIS OF A PLANE TRUSS

Apply a graphical analysis to the cantilever truss in **Fig. 4a** to evaluate the forces induced in the truss members.

Calculation Procedure:

1. Label the truss for analysis

Divide the space around the truss into regions bounded by the action lines of the external and internal forces. Assign an uppercase letter to each region (Fig. 4).

2. Determine the reaction force

Take moments with respect to joint 8 (Fig. 4) to determine the horizontal component of the reaction force R_U. Then compute R_U. Thus $\Sigma M_8 = 12R_{UH} - 3(8 + 16 + 24) - 5(6 + 12 + 18) = 0$; so $R_{UH} = 27$ kips (120.1 kN) to the right.

Since R_U is collinear with the force DE, $R_{UV}/R_{UH} = {}^{13}\!/_{24}$, so $R_{UV} = 13.5$ kips (60.0 kN) upward, and $R_U = 30.2$ kips (134.3 kN).

3. Apply the equations of equilibrium

Use the equations of equilibrium to find R_L. Thus $R_{LH} = 27$ kips (120.1 kN) to the left, $R_{LV} = 10.5$ kips (46.7 kN) upward, and $R_L = 29.0$ kips (129.0 kN).

4. Construct the force polygon

Draw the force polygon in Fig. 4b by using a suitable scale and drawing vector **fg** to represent force FG. Next, draw vector **gh** to represent force GH, and so forth. Omit the arrowheads on the vectors.

5. Determine the forces in the truss members

Starting at joint 1, Fig. 4b, draw a line through a in the force polygon parallel to member AJ in the truss, and one through h parallel to member HJ. Designate the point of intersection of these lines as j. Now, vector **aj** represents the force in AJ, and vector **hj** represents the force in HJ.

6. Analyze the next joint in the truss

Proceed to joint 2, where there are now only two unknown forces—BK and JK. Draw a line through b in the force polygon parallel to BK and one through j parallel to JK. Designate the point of intersection as k. The forces BK and JK are thus determined.

7. Analyze the remaining joints

Proceed to joints 3, 4, 5, and 6, in that order, and complete the force polygon by continuing the process. If the construction is accurately performed, the vector **pe** will parallel the member PE in the truss.

8. Determine the magnitude of the internal forces

Scale the vector lengths to obtain the magnitude of the internal forces. Tabulate the results as in Table 1.

9. Establish the character of the internal forces

To determine whether an internal force is one of tension or compression, proceed in this way: Select a particular joint and proceed around the joint in a clockwise direction, listing the letters

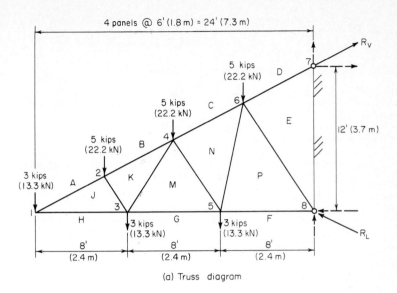

4 panels @ 6' (1.8 m) = 24' (7.3 m)

R_V

5 kips
(22.2 kN)

5 kips
(22.2 kN)

5 kips
(22.2 kN)

3 kips
(13.3 kN)

3 kips
(13.3 kN)

3 kips
(13.3 kN)

12' (3.7 m)

R_L

8'
(2.4 m)

8'
(2.4 m)

8'
(2.4 m)

(a) Truss diagram

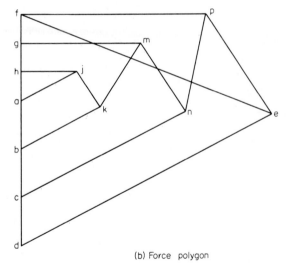

(b) Force polygon

FIG. 4

in the order in which they appear. Then refer to the force polygon pertaining to that joint, and proceed along the polygon in the same order. This procedure shows the direction in which the force is acting at that joint.

For instance, by proceeding around joint 4, *CNMKB* is obtained. By tracing a path along the force polygon in the order in which the letters appear, force *CN* is found to act upward to the right; *NM* acts upward to the left; *MK* and *KB* act downward to the left. Therefore, *CN*, *MK*, and *KB* are directed away from the joint (Fig. 4); this condition discloses that they are tensile forces. Force *NM* is directed toward the joint; therefore, it is compressive.

The validity of this procedure lies in the drawing of the vectors representing external forces while proceeding around the truss in a clockwise direction. Tensile forces are shown with a positive sign in Table 1; compressive forces are shown with a negative sign.

Related Calculations: Use this general method for any type of truss.

TRUSS ANALYSIS BY THE METHOD OF JOINTS

Applying the method of joints, determine the forces in the truss in Fig. 5a. The load at joint 4 has a horizontal component of 4 kips (17.8 kN) and a vertical component of 3 kips (13.3 kN).

TABLE 1 Forces in Truss Members (Fig. 4)

Member	Force	
	kips	kN
AJ	+6.7	+29.8
BK	+9.5	+42.2
CN	+19.8	+88.0
DE	+30.2	+134.2
HJ	−6.0	−26.7
GM	−13.0	−57.8
FP	−20.0	−88.9
JK	−4.5	−20.0
KM	+8.1	+36.0
MN	−8.6	−38.2
NP	+10.4	+46.2
PE	−12.6	−56.0

Calculation Procedure:

1. Compute the reactions at the supports

Using the usual analysis techniques, we find $R_{LV} = 19$ kips (84.5 kN); $R_{LH} = 4$ kips (17.8 kN); $R_R = 21$ kips (93.4 kN).

2. List each truss member and its slope

Table 2 shows each truss member and its slope.

3. Determine the forces at a principal joint

Draw a free-body diagram, Fig. 5b, of the pin at joint 1. For the free-body diagram, assume that the unknown internal forces AJ and HJ are tensile. Apply the equations of equilibrium to evaluate these forces, using the subscripts H and V, respectively, to identify the horizontal and vertical components. Thus $\Sigma F_H = 4.0 + AJ_H + HJ = 0$ and $\Sigma F_V = 19.0 + AJ_V = 0$; $\therefore AJ_V = -19.0$ kips (-84.5 kN); $AJ_H = -19.0/0.75 = -25.3$ kips (-112.5 kN). Substituting in the first equation gives $HJ = 21.3$ kips (94.7 kN).

The algebraic signs disclose that AJ is compressive and HJ is tensile. Record these results in Table 2, showing the tensile forces as positive and compressive forces as negative.

4. Determine the forces at another joint

Draw a free-body diagram of the pin at joint 2 (Fig. 5c). Show the known force AJ as compressive, and assume that the unknown forces BK and JK are tensile. Apply the equations of equilibrium, expressing the vertical components of BK and JK in terms of their horizontal components. Thus $\Sigma F_H = 25.3 + BK_H + JK_H = 0$; $\Sigma F_V = -6.0 + 19.0 + 0.75BK_H - 0.75JK_H = 0$.

Solve these simultaneous equations, to obtain $BK_H = -21.3$ kips (-94.7 kN); $JK_H = -4.0$ kips (-17.8 kN); $BK_V = -16.0$ kips (-71.2 kN); $JK_V = -3.0$ kips (-13.3 kN). Record these results in Table 2.

5. Continue the analysis at the next joint

Proceed to joint 3. Since there are no external horizontal forces at this joint, $CL_H = BK_H = 21.3$ kips (94.7 kN) of compression. Also, $KL = 6$ kips (26.7 kN) of compression.

6. Proceed to the remaining joints in their numbered order

Thus, for joint 4: $\Sigma F_H = -4.0 - 21.3 + 4.0 + LM_H + GM = 0$; $\Sigma F_V = -3.0 - 3.0 - 6.0 + LM_V = 0$; $LM_V = 12.0$ kips (53.4 kN); $LM_H = 12.0/2.25 = 5.3$ kips (23.6 kN). Substituting in the first equation gives $GM = 16.0$ kips (71.2 kN).

Joint 5: $\Sigma F_H = 21.3 - 5.3 + DN_H + MN_H = 0$; $\Sigma F_V = -6.0 + 16.0 - 12.0 - 0.75DN_H - 2.25MN_H = 0$; $DN_H = -22.7$ kips (-101.0 kN); $MN_H = 6.7$ kips (29.8 kN); $DN_V = -17.0$ kips (-75.6 kN); $MN_V = 15.0$ kips (66.7 kN).

Joint 6: $EP_H = DN_H = 22.7$ kips (101.0 kN) of compression; $NP = 11.0$ kips (48.9 kN) of compression.

Joint 7: $\Sigma F_H = 22.7 - PQ_H + FQ_H = 0$; $\Sigma F_V = -8.0 - 17.0 - 0.75PQ_H - 0.75FQ_H = 0$; $PQ_H = -5.3$ kips (-23.6 kN); $FQ_H = -28.0$ kips (-124.5 kN); $PQ_V = -4.0$ kips (-17.8 kN); $FQ_V = -21.0$ kips (-93.4 kN).

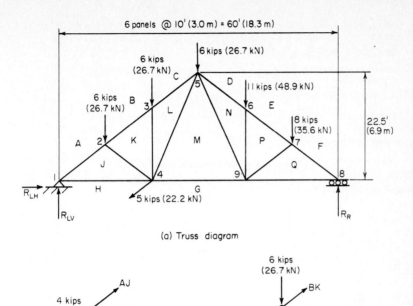

(a) Truss diagram

(b) Free-body diagram of joint I

(c) Free-body diagram of joint 2

FIG. 5

TABLE 2 Forces in Truss Members (Fig. 5)

Member	Slope	Horizontal component	Vertical component	Force kips	Force kN
AJ	0.75	25.3	19.0	−31.7	−141.0
BK	0.75	21.3	16.0	−26.7	−118.8
CL	0.75	21.3	16.0	−26.7	−118.8
DN	0.75	22.7	17.0	−28.3	−125.9
EP	0.75	22.7	17.0	−28.3	−125.9
FQ	0.75	28.0	21.0	−35.0	−155.7
HJ	0.0	21.3	0.0	+21.3	+94.7
GM	0.0	16.0	0.0	+16.0	+71.2
GQ	0.0	28.0	0.0	+28.0	+124.5
JK	0.75	4.0	3.0	−5.0	−22.2
KL	∞	0.0	6.0	−6.0	−26.7
LM	2.25	5.3	12.0	+13.1	+58.3
MN	2.25	6.7	15.0	+16.4	+72.9
NP	∞	0.0	11.0	−11.0	−48.9
PQ	0.75	5.3	4.0	−6.7	−29.8

Joint 8: $\Sigma F_H = 28.0 - GQ = 0$; $GQ = 28.0$ kips (124.5 kN); $\Sigma F_V = 21.0 - 21.0 = 0$.
Joint 9: $\Sigma F_H = -16.0 - 6.7 - 5.3 + 28.0 = 0$; $\Sigma F_V = 15.0 - 11.0 - 4.0 = 0$.

7. Complete the computation

Compute the values in the last column of Table 2 and enter them as shown.

TRUSS ANALYSIS BY THE METHOD OF SECTIONS

Using the method of sections, determine the forces in members BK and LM in Fig. 5a.

Calculation Procedure:

1. Draw a free-body diagram of one portion of the truss

Cut the truss at the plane aa (Fig. 6a), and draw a free-body diagram of the left part of the truss. Assume that BK is tensile.

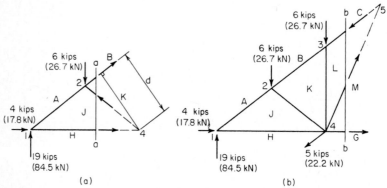

(a)   (b)

FIG. 6

2. Determine the magnitude and character of the first force

Take moments with respect to joint 4. Since each half of the truss forms a 3-4-5 right triangle, $d = 20(3/5) = 12$ ft (3.7 m), $\Sigma M_4 = 19(20) - 6(10) + 12BK = 0$, and $BK = -26.7$ kips (-118.8 kN).

The negative result signifies that the assumed direction of BK is incorrect; the force is, therefore, compressive.

3. Use an alternative solution

Alternatively, resolve BK (again assumed tensile) into its horizontal and vertical components at joint 1. Take moments with respect to joint 4. (A force may be resolved into its components at any point on its action line.) Then $\Sigma M_4 = 19(20) + 20BK_V - 6(10) = 0$; $BK_V = -16.0$ kips (-71.2 kN); $BK = -16.0(5/3) = -26.7$ kips (-118.8 kN).

4. Draw a second free-body diagram of the truss

Cut the truss at plane bb (Fig. 6b), and draw a free-body diagram of the left part. Assume LM is tensile.

5. Determine the magnitude and character of the second force

Resolve LM into its horizontal and vertical components at joint 4. Take moments with respect to joint 1: $\Sigma M_1 = 6(10 + 20) + 3(20) - 20LM_V = 0$; $LM_V = 12.0$ kips (53.4 kN); $LM_H = 12.0/2.25 = 5.3$ kips (23.6 kN); $LM = 13.1$ kips (58.3 kN).

REACTIONS OF A THREE-HINGED ARCH

The parabolic arch in Fig. 7 is hinged at A, B, and C. Determine the magnitude and direction of the reactions at the supports.

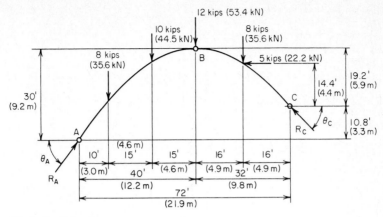

FIG. 7 Parabolic arch.

Calculation Procedure:

1. Consider the entire arch as a free body and take moments

Since a moment cannot be transmitted across a hinge, the bending moments at A, B, and C are zero. Resolve the reactions R_A and R_C (Fig. 7) into their horizontal and vertical components.

Considering the entire arch ABC as a free body, take moments with respect to A and C. Thus $\Sigma M_A = 8(10) + 10(25) + 12(40) + 8(56) - 5(25.2) - 72R_{CV} - 10.8R_{CH} = 0$, or $72R_{CV} + 10.8R_{CH} = 1132$, Eq. a. Also, $\Sigma M_C = 72R_{AV} - 10.8R_{AH} - 8(62) - 10(47) - 12(32) - 8(16) - 5(14.4) = 0$, or $72R_{AV} - 10.8R_{AH} = 1550$, Eq. b.

2. Consider a segment of the arch and take moments

Considering the segment BC as a free body, take moments with respect to B. Then $\Sigma M_B = 8(16) + 5(4.8) - 32R_{CV} + 19.2R_{CH} = 0$, or $32R_{CV} - 19.2R_{CH} = 152$, Eq. c.

3. Consider another segment and take moments

Considering segment AB as a free body, take moments with respect to B: $\Sigma M_B = 40R_{AV} - 30R_{AH} - 8(30) - 10(15) = 0$, or $40R_{AV} - 30R_{AH} = 390$, Eq. d.

4. Solve the simultaneous moment equations

Solve Eqs. b and d to determine R_A; solve Eqs. a and c to determine R_C. Thus $R_{AV} = 24.4$ kips (108.5 kN); $R_{AH} = 19.6$ kips (87.2 kN); $R_{CV} = 13.6$ kips (60.5 kN); $R_{CH} = 14.6$ kips (64.9 kN). Then $R_A = [(24.4)^2 + (19.6)^2]^{0.5} = 31.3$ kips (139.2 kN). Also $R_C = [(13.6)^2 + (14.6)^2]^{0.5} = 20.0$ kips (89.0 kN). And $\theta_A = \arctan (24.4/19.6) = 51°14'$; $\theta_C = \arctan (13.6/14.6) = 42°58'$.

LENGTH OF CABLE CARRYING KNOWN LOADS

A cable is supported at points P and Q (Fig. 8a) and carries two vertical loads, as shown. If the tension in the cable is restricted to 1800 lb (8006 N), determine the minimum length of cable required to carry the loads.

Calculation Procedure:

1. Sketch the loaded cable

Assume a position of the cable, such as $PRSQ$ (Fig. 8a). In Fig. 8b, locate points P' and Q', corresponding to P and Q, respectively, in Fig. 8a.

2. Take moments with respect to an assumed point

Assume that the maximum tension of 1800 lb (8006 N) occurs in segment PR (Fig. 8). The reaction at P, which is collinear with PR, is therefore 1800 lb (8006 N). Compute the true perpendicular

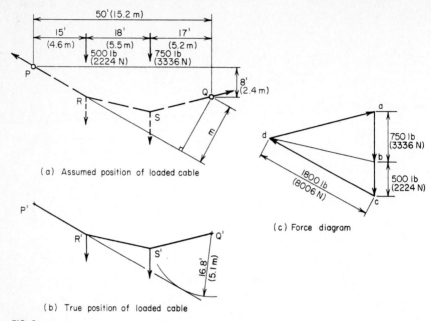

(a) Assumed position of loaded cable

(b) True position of loaded cable

(c) Force diagram

FIG. 8

distance m from Q to PR by taking moments with respect to Q. Or $\Sigma M_Q = 1800m - 500(35) - 750(17) = 0$; $m = 16.8$ ft (5.1 m). This dimension establishes the true position of PR.

3. Start the graphical solution of the problem

In Fig. 8b, draw a circular arc having Q' as center and a radius of 16.8 ft (5.1 m). Draw a line through P' tangent to this arc. Locate R' on this tangent at a horizontal distance of 15 ft (4.6 m) from P'.

4. Draw the force vectors

In Fig. 8c, draw vectors **ab**, **bc**, and **cd** to represent the 750-lb (3336-N) load, the 500-lb (2224-N) load, and the 1800-lb (8006-N) reaction at P, respectively. Complete the triangle by drawing vector **da**, which represents the reaction at Q.

5. Check the tension assumption

Scale da to ascertain whether it is less than 1800 lb (8006 N). This is found to be so, and the assumption that the maximum tension exists in PR is validated.

6. Continue the construction

Draw a line through Q' in Fig. 8b parallel to da in Fig. 8c. Locate S' on this line at a horizontal distance of 17 ft (5.2 m) from Q.

7. Complete the construction

Draw $R'S'$ and db. Test the accuracy of the construction by determining whether these lines are parallel.

8. Determine the required length of the cable

Obtain the required length of the cable by scaling the lengths of the segments in Fig. 8b. Thus $P'R' = 17.1$ ft (5.2 m); $R'S' = 18.4$ ft (5.6 m); $S'Q' = 17.6$ ft (5.4 m); and length of cable = 53.1 ft (16.2 m).

PARABOLIC CABLE TENSION AND LENGTH

A suspension bridge has a span of 960 ft (292.61 m) and a sag of 50 ft (15.2 m). Each cable carries a load of 1.2 kips per linear foot (kips/lin ft) (17,512.68 N/m) uniformly distributed along the horizontal. Compute the tension in the cable at midspan and at the supports, and determine the length of the cable.

Calculation Procedure:

1. Compute the tension at midspan

A cable carrying a load uniformly distributed along the horizontal assumes the form of a parabolic arc. In Fig. 9, which shows such a cable having supports at the same level, the tension at midspan

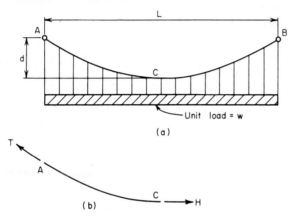

FIG. 9 Cable supporting load uniformly distributed along horizontal.

is $H = wL^2/(8d)$, where H = midspan tension, kips (kN); w = load on a unit horizontal distance, kips/lin ft (kN/m); L = span, ft (m); d = sag, ft (m). Substituting yields $H = 1.2(960)^2/[8(50)]$ = 2765 kips (12,229 kN).

2. Compute the tension at the supports

Use the relation $T = [H^2 + (wL/2)^2]^{0.5}$, where T = tension at supports, kips (kN), and the other symbols are as before. Thus, $T = [(2765)^2 + (1.2 \times 480)^2]^{0.5}$ = 2824 kips (12,561 kN).

3. Compute the length of the cable

When d/L is 1/20 or less, the cable length can be approximated from $S = L + 8d^2/(3L)$, where S = cable length, ft (m). Thus, $S = 960 + 8(50)^2/[3(960)]$ = 966.94 ft (294.72 m).

CATENARY CABLE SAG AND DISTANCE BETWEEN SUPPORTS

A cable 500 ft (152.4 m) long and weighing 3 pounds per linear foot (lb/lin ft) (43.8 N/m) is supported at two points lying in the same horizontal plane. If the tension at the supports is 1800 lb (8006 N), find the sag of the cable and the distance between the supports.

Calculation Procedure:

1. Compute the catenary parameter

A cable of uniform cross section carrying only its own weight assumes the form of a catenary. Using the notation of the previous procedure, we find the catenary parameter c from $d + c = T/w = 1800/3 = 600$ ft (182.9 m). Then $c = [(d + c)^2 - (S/2)^2]^{0.5} = [(600)^2 - (250)^2]^{0.5} = 545.4$ ft (166.2 m).

2. Compute the cable sag

Since $d + c = 600$ ft (182.9 m) and $c = 545.4$ ft (166.2 m), we know $d = 600 - 545.4 = 54.6$ ft (16.6 m).

3. Compute the span length

Use the relation $L = 2c \ln (d + c + 0.5S)/c$, or $L = 2(545.5) \ln (600 + 250)/545.4 = 484.3$ ft (147.6 m).

STABILITY OF A RETAINING WALL

Determine the factor of safety (FS) against sliding and overturning of the concrete retaining wall in Fig. 10. The concrete weighs 150 lb/ft³ (23.56 kN/m³), the earth weighs 100 lb/ft³ (15.71 kN/m³), the coefficient of friction is 0.6, and the coefficient of active earth pressure is 0.333.

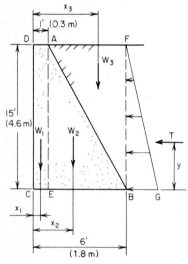

FIG. 10

Calculation Procedure:

1. Compute the vertical loads on the wall

Select a 1-ft (304.8-mm) length of wall as typical of the entire structure. The horizontal pressure of the confined soil varies linearly with the depth and is represented by the triangle BGF in Fig. 10.

Resolve the wall into the elements $AECD$ and AEB; pass the vertical plane BF through the soil. Calculate the vertical loads, and locate their resultants with respect to the toe C. Thus $W_1 = 15(1)(150) = 2250$ lb (10,008 N); $W_2 = 0.5(15)(5)(150) = 5625$; $W_3 = 0.5(15)(5)(100) = 3750$. Then $\Sigma W = 11,625$ lb (51,708 N). Also, $x_1 = 0.5$ ft; $x_2 = 1 + 0.333(5) = 2.67$ ft (0.81 m); $x_3 = 1 + 0.667(5) = 4.33$ ft (1.32 m).

2. Compute the resultant horizontal soil thrust

Compute the resultant horizontal thrust T lb of the soil by applying the coefficient of active earth pressure. Determine the location of T. Thus $BG = 0.333(15)(100) = 500$ lb/lin ft (7295 N/m); $T = 0.5(15)(500) = 3750$ lb (16,680 N); $y = 0.333(15) = 5$ ft (1.5 m).

3. Compute the maximum frictional force preventing sliding

The maximum frictional force $F_m = \mu(\Sigma W)$, where μ = coefficient of friction. Or $F_m = 0.6(11,625) = 6975$ lb (31,024.8 N).

4. Determine the factor of safety against sliding

The factor of safety against sliding is $FSS = F_m/T = 6975/3750 = 1.86$.

5. Compute the moment of the overturning and stabilizing forces

Taking moments with respect to C, we find the overturning moment $= 3750(5) = 18,750$ lb·ft (25,406.3 N·m). Likewise, the stabilizing moment $= 2250(0.5) + 5625(2.67) + 3750(4.33) = 32,375$ lb·ft (43,868.1 N·m).

6. Compute the factor of safety against overturning

The factor of safety against overturning is FSO = stabilizing moment, lb·ft (N·m)/overturning moment, lb·ft (N·m) $= 32,375/18,750 = 1.73$.

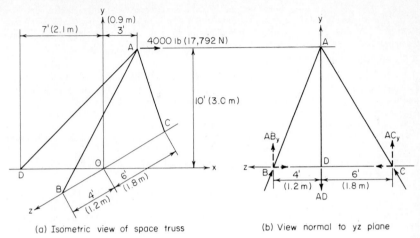

(a) Isometric view of space truss   (b) View normal to yz plane

FIG. 11

ANALYSIS OF A SIMPLE SPACE TRUSS

In the space truss shown in Fig. 11a, A lies in the xy plane, B and C lie on the z axis, and D lies on the x axis. A horizontal load of 4000 lb (17,792 N) lying in the xy plane is applied at A. Determine the force induced in each member by applying the method of joints, and verify the results by taking moments with respect to convenient axes.

Calculation Procedure:

1. Determine the projected length of members

Let d_x, d_y, and d_z denote the length of a member as projected on the x, y, and z axes, respectively. Record in Table 3 the projected lengths of each member. Record the remaining values as they are obtained.

2. Compute the true length of each member

Use the equation $d = (d_x^2 + d_y^2 + d_z^2)^{0.5}$, where d = the true length of a member.

3. Compute the ratio of the projected length to the true length

For each member, compute the ratios of the three projected lengths to the true length. For example, for member AC, $d_z/d = 6/12.04 = 0.498$.

These ratios are termed *direction cosines* because each represents the cosine of the angle between the member and the designated axis, or an axis parallel thereto.

TABLE 3 Data for Space Truss (Fig. 11)

Member	AB		AC		AD	
d_x, ft (m)	3	(0.91)	3	(0.91)	10	(3.05)
d_y, ft (m)	10	(3.0)	10	(3.0)	10	(3.0)
d_z, ft (m)	4	(1.2)	6	(1.8)	0	(0)
d, ft (m)	11.18	(3.4)	12.04	(3.7)	14.14	(4.3)
d_x/d	0.268		0.249		0.707	
d_y/d	0.894		0.831		0.707	
d_z/d	0.358		0.498		0	
Force, lb (N)	-3830	$(-17,036)$	-2750	$(-12,232)$	$+8080$	$(+35,940)$

Since the axial force in each member has the same direction as the member itself, a direction cosine also represents the ratio of the component of a force along the designated axis to the total force in the member. For instance, let AC denote the force in member AC, and let AC_x denote its component along the x axis. Then $AC_x/AC = d_x/d = 0.249$.

4. Determine the component forces

Consider joint A as a free body, and assume that the forces in the three truss members are tensile. Equate the sum of the forces along each axis to zero. For instance, if the truss members are in tension, the x components of these forces are directed to the left, and $\Sigma F_x = 4000 - AB_x - AC_x - AD_x = 0$.

Express each component in terms of the total force to obtain $\Sigma F_x = 4000 - 0.268AB - 0.249AC - 0.707AD = 0$; $\Sigma F_y = -0.894AB - 0.831AC - 0.707AD = 0$; $\Sigma F_z = 0.358AB - 0.498AC = 0$.

5. Solve the simultaneous equations in step 4 to evaluate the forces in the truss members

A positive result in the solution signifies tension; a negative result, compression. Thus, $AB = 3830$-lb (17,036-N) compression; $AC = 2750$-lb (12,232-N) compression; and $AD = 8080$-lb (35,940-N) tension. To verify these results, it is necessary to select moment axes yielding equations independent of those previously developed.

6. Resolve the reactions into their components

In Fig. 11b, show the reactions at the supports B, C, and D, each reaction being numerically equal to and collinear with the force in the member at that support. Resolve these reactions into their components.

7. Take moments about a selected axis

Take moments with respect to the axis through C parallel to the x axis. (Since the x components of the forces are parallel to this axis, their moments are zero.) Then $\Sigma M_{Cx} = 10AB_y - 6AD_y = 10(0.894)(3830) - 6(0.707)(8080) = 0$.

8. Take moments about another axis

Take moments with respect to the axis through D parallel to the x axis. So $\Sigma M_{Dx} = 4AB_y - 6AC_y = 4(0.894)(3830) - 6(0.831)(2750) = 0$.

The computed results are thus substantiated.

ANALYSIS OF A COMPOUND SPACE TRUSS

The compound space truss in Fig. 12a has the dimensions shown in the orthographic projections, Fig. 12b and c. A load of 5000 lb (22,240 N), which lies in the xy plane and makes an angle of 30° with the vertical, is applied at A. Determine the force induced in each member, and verify the results.

Calculation Procedure:

1. Compute the true length of each truss member

Since the truss and load system are symmetric with respect to the xy plane, the internal forces are also symmetric. As one component of an internal force becomes known, it will be convenient to calculate the other components at once, as well as the total force.

Record in Table 4 the length of each member as projected on the coordinate axes. Calculate the true length of each member, using geometric relations.

2. Resolve the applied load into its x and y components

Use only the absolute values of the forces. Thus $P_x = 5000 \sin 30° = 2500$ lb (11,120 N); $P_y = 5000 \cos 30° = 4330$ lb (19,260 N).

3. Compute the horizontal reactions

Compute the horizontal reactions at D and at line CC' (Fig. 12b). Thus $\Sigma M_{CC'} = 4330(12) - 2500(7) - 10H_1 = 0$; $H_1 = 3446$ lb (15,328 N); $H_2 = 3446 - 2500 = 946$ lb (4208 N).

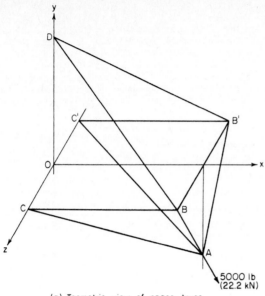

(a) Isometric view of space truss

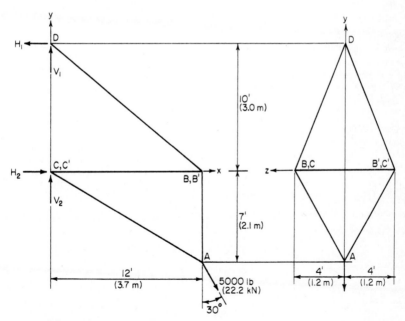

(b) View normal to xy plane

(c) View normal to yz plane

FIG. 12

4. Compute the vertical reactions

Consider the equilibrium of joint D and the entire truss when you are computing the vertical reactions. In all instances, assume that an unknown internal force is tensile. Thus, at joint D: $\Sigma F_x = -H_1 + 2BD_x = 0$; $BD_x = 1723$-lb (7664-N) tension; $BD_y = 1723(10/12) = 1436$ lb (6387 N); likewise, $\Sigma F_y = V_1 - 2BD_y = V_1 - 2(1436) = 0$; $V_1 = 2872$ lb (12,275 N).

For the entire truss, $\Sigma F_y = V_1 + V_2 - 4330 = 0$; $V_2 = 1458$ lb (6485 N).

The z components of the reactions are not required in this solution. Thus, the remaining calculations for BD are $BD_z = 1723(4/12) = 574$ lb (2553 N); $BD = 1723(16.12/12) = 2315$ lb (10,297 N).

5. Compute the unknown forces by using the equilibrium of a joint

Calculate the forces AC and BC by considering the equilibrium of joint C. Thus $\Sigma F_x = 0.5H_2 + AC_x + BC = 0$, Eq. a; $\Sigma F_y = 0.5V_2 - AC_y = 0$, Eq. b. From Eq. b, $AC_y = 729$-lb (3243-N) tension. Then $AC_x = 729(12/7) = 1250$ lb (5660 N). From Eq. a, $BC = 1723$-lb (7664-N) compression. Then $AC_z = 729(4/7) = 417$ lb (1855 N); $AC = 729(14.46/7) = 1506$ lb (6699 N).

6. Compute another set of forces by considering joint equilibrium

Calculate the forces AB and BB' by considering the equilibrium of joint B. Thus $\Sigma F_y = BD_y - AB_y = 0$; $AB_y = 1436$-lb (6387-N) tension; $AB_z = 1436(4/7) = 821$ lb (3652 N); $AB = 1436 (8.06/7) = 1653$ lb (7353 N); $\Sigma F_z = -AB_z - BD_z - BB' = 0$; $BB' = 1395$-lb (6205-N) compression.

All the internal forces are now determined. Show in Table 4 the tensile forces as positive, and the compressive forces as negative.

7. Check the equilibrium of the first joint considered

The first joint considered was A. Thus $\Sigma F_x = -2AC_x + 2500 = -2(1250) + 2500 = 0$, and $\Sigma F_y = 2AB_y + 2AC_y - 4330 = 2(1436) + 2(729) - 4330 = 0$. Since the summation of forces for both axes is zero, the joint is in equilibrium.

8. Check the equilibrium of the second joint

Check the equilibrium of joint B by taking moments of the forces acting on this joint with respect to the axis through A parallel to the x axis (Fig. 12c). Thus $\Sigma M_{Ax} = -7BB' + 7BD_z + 4BD_y = -7(1395) + 7(574) + 4(1436) = 0$.

9. Check the equilibrium of the right-hand part of the structure

Cut the truss along a plane parallel to the yz plane. Check the equilibrium of the right-hand part of the structure. Now $\Sigma F_x = -2BD_x + 2BC - 2AC_x + 2500 = -2(1723) + 2(1723) - 2(1250) + 2500 = 0$, and $\Sigma F_y = 2BD_y + 2AC_y - 4330 = 2(1436) + 2(729) - 4330 = 0$. The calculated results are thus substantiated in these equations.

GEOMETRIC PROPERTIES OF AN AREA

Calculate the polar moment of inertia of the area in Fig. 13: (a) with respect to its centroid, and (b) with respect to point A.

TABLE 4 Data for Space Truss (Fig. 12)

Member	AB	AC	BC	BD	BB'
d_x, ft (m)	0 (0)	12 (3.7)	12 (3.7)	12 (3.7)	0 (0)
d_y, ft (m)	7 (2.1)	7 (2.1)	0 (0)	10 (3.0)	0 (0)
d_z, ft (m)	4 (1.2)	4 (1.2)	0 (0)	4 (1.2)	8 (2.4)
d, ft (m)	8.06 (2.5)	14.46 (4.4)	12.00 (3.7)	16.12 (4.9)	8 (2.4)
F_x, lb (N)	0 (0)	1,250(5,560)	1,723 (7,664)	1,723 (7,664)	0 (0)
F_y, lb (N)	1,436 (6,387)	729 (3,243)	0 (0)	1,436 (6,387)	0 (0)
F_z, lb (N)	821 (3,652)	417 (1,855)	0 (0)	574 (2,553)	1,395 (6,205)
F, lb (N)	+1,653 (+7,353)	+1,506 (+6,699)	−1,723 (−7,664)	+2,315 (+10,297)	−1,395 (−6,205)

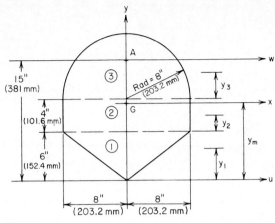

FIG. 13

Calculation Procedure:

1. Establish the area axes

Set up the horizontal and vertical coordinate axes u and y, respectively.

2. Divide the area into suitable elements

Using the American Institute of Steel Construction (AISC) *Manual*, obtain the properties of elements 1, 2, and 3 (Fig. 13) after locating the horizontal centroidal axis of each element. Thus y_1 = ⅔(6) = 4 in (101.6 mm); y_2 = 2 in (50.8 mm); y_3 = 0.424(8) = 3.4 in (86.4 mm).

3. Locate the horizontal centroidal axis of the entire area

Let x denote the horizontal centroidal axis of the entire area. Locate this axis by computing the statical moment of the area with respect to the u axis. Thus

Element	Area, in² (cm²)			×	Arm, in (cm)	=	Moment, in³ (cm³)
1	0.5(6)(16) =	48	(309.7)		4 (10.2)	=	192 (3,158.9)
2	4(16) =	64	(412.9)		8 (20.3)	=	512 (8,381.9)
3	1.57(8)² =	100.5	(648.4)		13.4 (34.0)	=	1,347 (22,045.6)
Total		212.5 (1,371.0)					2,051 (33,586.4)

Then y_m = 2051/212.5 = 9.7 in (246.4 mm). Since the area is symmetric with respect to the y axis, this is also a centroidal axis. The intersection point G of the x and y axes is, therefore, the centroid of the area.

4. Compute the distance between the centroidal axis and the reference axis

Compute k, the distance between the horizontal centroidal axis of each element and the x axis. Only absolute values are required. Thus k_1 = 9.7 − 4.0 = 5.7 in (144.8 mm); k_2 = 9.7 − 8.0 = 1.7 in (43.2 mm); k_3 = 13.4 − 9.7 = 3.7 in (94.0 mm).

5. Compute the moment of inertia of the entire area—x axis

Let I_0 denote the moment of inertia of an element with respect to its horizontal centroidal axis and A its area. Compute the moment of inertia I_x of the entire area with respect to the x axis by applying the transfer equation $I_x = \Sigma I_0 + \Sigma Ak^2$. Thus

Element	I_0, in^4 (dm^4)	Ak^2, in^4 (dm^4)
1	$\frac{1}{36}(16)(6)^3 = 96 \ (0.40)$	$48(5.7)^2 = 1560 \ (6.49)$
2	$\frac{1}{12}(16)(4)^3 = 85 \ (0.35)$	$64(1.7)^2 = 185 \ (0.77)$
3	$0.110(8)^4 = \underline{451 \ (1.88)}$	$100.5(3.7)^2 = \underline{1376 \ (5.73)}$
Total	632 (2.63)	3121 (12.99)

Then, $I_x = 632 + 3121 = 3753$ in^4 (15.62 dm^4).

6. Determine the moment of inertia of the entire area—y axis

For this computation, subdivide element 1 into two triangles having the y axis as a base. Thus

Element	I about y axis, in^4 (dm^4)
1'	$2(\frac{1}{12})(6)(8)^3 = 512 \ (2.13)$
2	$\frac{1}{12}(4)(16)^3 = 1365 \ (5.68)$
3	$\frac{1}{2}(0.785)(8)^4 = \underline{1607 \ (6.69)}$
	$I_y = 3484 \ (14.5)$

7. Compute the polar moment of inertia of the area

Apply the equation for the polar moment of inertia J_G with respect to G: $J_G = I_x + I_y = 3753 + 3484 = 7237$ in^4 (30.12 dm^4).

8. Determine the moment of inertia of the entire area—w axis

Apply the equation in step 5 to determine the moment of inertia I_w of the entire area with respect to the horizontal axis w through A. Thus $k = 15.0 - 9.7 = 5.3$ in (134.6 mm); $I_w = I_x + Ak^2 = 3753 + 212.5(5.3)^2 = 9722$ in^4 (40.46 dm^4).

9. Compute the polar moment of inertia

Compute the polar moment of inertia of the entire area with respect to A. Then $J_A = I_w + I_y = 9722 + 3484 = 13,206$ in^4 (54.97 dm^4).

PRODUCT OF INERTIA OF AN AREA

Calculate the product of inertia of the isosceles trapezoid in Fig. 14 with respect to the rectangular axes u and v.

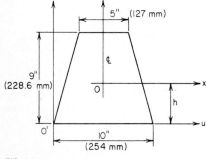

FIG. 14

Calculation Procedure:

1. Locate the centroid of the trapezoid

Using the AISC *Manual* or another suitable reference, we find h = centroid distance from the axis (Fig. 14) = $(9/3)[(2 \times 5 + 10)/(5 + 10)] = 4$ in (101.6 mm).

2. Compute the area and product of inertia P_{xy}

The area of the trapezoid is $A = \frac{1}{2}(9)(5 + 10) = 67.5$ in^2 (435.5 cm^2). Since the area is symmetrically disposed with respect to the y axis, the product of inertia with respect to the x and y axes is $P_{xy} = 0$.

3. Compute the product of inertia by applying the transfer equation

The transfer equation for the product of inertia is $P_{uv} = P_{xy} + Ax_m y_m$, where x_m and y_m are the coordinates of O' with respect to the centroidal x and y axes, respectively. Thus $P_{uv} = 0 + 67.5(-5)(-4) = 1350$ in^4 (5.6 dm^4).

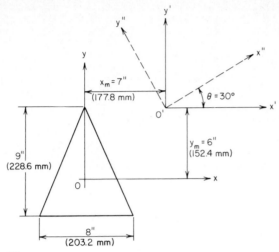

FIG. 15

PROPERTIES OF AN AREA WITH RESPECT TO ROTATED AXES

In Fig. 15, x and y are rectangular axes through the centroid of the isosceles triangle; x' and y' are axes parallel to x and y, respectively; x'' and y'' are axes making an angle of 30° with x' and y', respectively. Compute the moments of inertia and the product of inertia of the triangle with respect to the x'' and y'' axes.

Calculation Procedure:

1. Compute the area of the figure

The area of this triangle = 0.5(base)(altitude) = 0.5(8)(9) = 36 in² (232.3 cm²).

2. Compute the properties of the area with respect to the x and y axes

Using conventional moment-of-inertia relations, we find $I_x = bd^3/36 = 8(9)^3/36 = 162$ in⁴ (0.67 dm⁴); $I_y = b^3d/48 = (8)^3(9)/48 = 96$ in⁴ (0.39 dm⁴). By symmetry, the product of inertia with respect to the x and y axes is $P_{xy} = 0$.

3. Compute the properties of the area with respect to the x' and y' axes

Using the usual moment-of-inertia relations, we find $I_{x'} = I_x + Ay_m^2 = 162 + 36(6)^2 = 1458$ in⁴ (6.06 dm⁴); $I_{y'} = I_y + Ax_m^2 = 96 + 36(7)^2 = 1860$ in⁴ (7.74 dm⁴); $P_{x'y'} = P_{xy} + Ax_my_m = 0 + 36(7)(6) = 1512$ in⁴ (6.29 dm⁴).

4. Compute the properties of the area with respect to the x'' and y'' axes

For the x'' axis, $I_{x''} = I_{x'} \cos^2\theta + I_{y'} \sin^2\theta - P_{x'y'} \sin 2\theta = 1458(0.75) + 1860(0.25) - 1512(0.866) = 249$ in⁴ (1.03 dm⁴).

For the y'' axis, $I_{y''} = I_{x'} \sin^2\theta + I_{y'} \cos^2\theta + P_{x'y'} \sin 2\theta = 1458(0.25) + 1860(0.75) + 1512(0.866) = 3069$ in⁴ (12.77 dm⁴).

The product of inertia is $P_{x''y''} = P_{x'y'} \cos 2\theta + [(I_{x'} - I_{y'})/2] \sin 2\theta = 1512(0.5) + [(1458 - 1860)/2]0.866 = 582$ in⁴ (2.42 dm⁴).

Analysis of Stress and Strain

The notational system for axial stress and strain used in this section is as follows: A = cross-sectional area of a member; L = original length of the member; Δl = increase in length; P = axial force; s = axial stress; ϵ = axial strain = $\Delta l/L$; E = modulus of elasticity of material =

s/ϵ. The units used for each of these factors are given in the calculation procedure. In all instances, it is assumed that the induced stress is below the proportional limit. The basic stress and elongation equations used are $s = P/A$; $\Delta l = sL/E = PL/(AE)$. For steel, $E = 30 \times 10^6$ lb/in² (206 GPa).

STRESS CAUSED BY AN AXIAL LOAD

A concentric load of 20,000 lb (88,960 N) is applied to a hanger having a cross-sectional area of 1.6 in² (1032.3 mm²). What is the axial stress in the hanger?

Calculation Procedure:

1. Compute the axial stress

Use the general stress relation $s = P/A = 20,000/1.6 = 12,500$ lb/in² (86,187.5 kPa).
   Related Calculations: Use this general stress relation for a member of any cross-sectional shape, provided the area of the member can be computed and the member is made of only one material.

DEFORMATION CAUSED BY AN AXIAL LOAD

A member having a length of 16 ft (4.9 m) and a cross-sectional area of 2.4 in² (1548.4 mm²) is subjected to a tensile force of 30,000 lb (133.4 kN). If $E = 15 \times 10^6$ lb/in² (103 GPa), how much does this member elongate?

Calculation Procedure:

1. Apply the general deformation equation

The general deformation equation is $\Delta l = PL/(AE) = 30,000(16)(12)/[2.4(15 \times 10^6)] = 0.16$ in (4.06 mm).
   Related Calculations: Use this general deformation equation for any material whose modulus of elasticity is known. For composite materials, this equation must be altered before it can be used.

DEFORMATION OF A BUILT-UP MEMBER

A member is built up of three bars placed end to end, the bars having the lengths and cross-sectional areas shown in Fig. 16. The member is placed between two rigid surfaces and axial loads

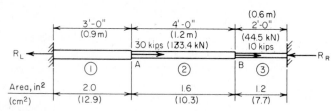

FIG. 16

of 30 kips (133 kN) and 10 kips (44 kN) are applied at A and B, respectively. If $E = 2000$ kips/in² (13,788 MPa), determine the horizontal displacement of A and B.

Calculation Procedure:

1. Express the axial force in terms of one reaction

Let R_L and R_R denote the reactions at the left and right ends, respectively. Assume that both reactions are directed to the left. Consider a tensile force as positive and a compressive force as negative. Consider a deformation positive if the body elongates and negtive if the body contracts.

Express the axial force P in each bar in terms of R_L because both reactions are assumed to be directed toward the left. Use subscripts corresponding to the bar numbers (Fig. 16). Thus, $P_1 = R_L$; $P_2 = R_L - 30$; $P_3 = R_L - 40$.

2. Express the deformation of each bar in terms of the reaction and modulus of elasticity

Thus, $\Delta l_1 = R_L(36)/(2.0E) = 18R_L/E$; $\Delta l_2 = (R_L - 30)(48)/(1.6E) = (30R_L - 900)/E$; $\Delta l_3 = (R_L - 40)24/(1.2E) = (20R_L - 800)/E$.

3. Solve for the reaction

Since the ends of the member are stationary, equate the total deformation to zero, and solve for R_L. Thus $\Delta l_t = (68R_L - 1700)/E = 0$; $R_L = 25$ kips (111 kN). The positive result confirms the assumption that R_L is directed to the left.

4. Compute the displacement of the points

Substitute the computed value of R_L in the first two equations of step 2 and solve for the displacement of the points A and B. Thus $\Delta l_1 = 18(25)/2000 = 0.225$ in (5.715 mm); $\Delta l_2 = [30(25) - 900]/2000 = -0.075$ in (-1.905 mm).

Combining these results, we find the displacement of $A = 0.225$ in (5.715 mm) to the right; the displacement of $B = 0.225 - 0.075 = 0.150$ in (3.81 mm) to the right.

5. Verify the computed results

To verify this result, compute R_R and determine the deformation of bar 3. Thus $\Sigma F_H = -R_L + 30 + 10 - R_R = 0$; $R_R = 15$ kips (67 kN). Since bar 3 is in compression, $\Delta l_3 = -15(24)/[1.2(2000)] = -0.150$ in (-3.81 mm). Therefore, B is displaced 0.150 in (3.81 mm) to the right. This verifies the result obtained in step 4.

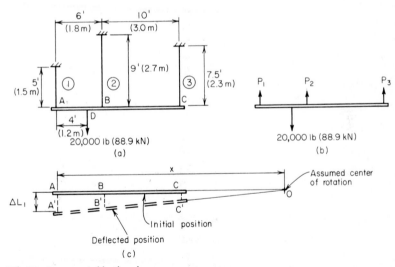

FIG. 17 Bar supported by three hangers.

REACTIONS AT ELASTIC SUPPORTS

The rigid bar in Fig. 17a is subjected to a load of 20,000 lb (88,960 N) applied at D. It is supported by three steel rods, 1, 2, and 3 (Fig. 17a). These rods have the following relative cross-sectional areas: $A_1 = 1.25$, $A_2 = 1.20$, $A_3 = 1.00$. Determine the tension in each rod caused by this load, and locate the center of rotation of the bar.

Calculation Procedure:

1. Draw a free-body diagram; apply the equations of equilibrium

Draw the free-body diagram (Fig. 17b) of the bar. Apply the equations of equilibrium: $\Sigma F_V = P_1 + P_2 + P_3 - 20,000 = 0$, or $P_1 + P_2 + P_3 = 20,000$, Eq. a; also, $\Sigma M_C = 16P_1 + 10P_2 - 20,000(12) = 0$, or $16P_1 + 10P_2 = 240,000$, Eq. b.

2. Establish the relations between the deformations

Selecting an arbitrary center of rotation O, show the bar in its deflected position (Fig. 17c). Establish the relationships among the three deformations. Thus, by similar triangles, $(\Delta l_1 - \Delta l_2)/(\Delta l_2 - \Delta l_3) = 6/10$, or $10\Delta l_1 - 16\Delta l_2 + 6\Delta l_3 = 0$, Eq. c.

3. Transform the deformation equation to an axial-force equation

By substituting axial-force relations in Eq. c, the following equation is obtained: $10P_1(5)/(1.25E) - 16P_2(9)/(1.20E) + 6P_3(7.5)/E = 0$, or $40P_1 - 120P_2 + 45P_3 = 0$, Eq. c'.

4. Solve the simultaneous equations developed

Solve the simultaneous equations a, b, and c' to obtain $P_1 = 11,810$ lb (52,530 N); $P_2 = 5100$ lb (22,684 N); $P_3 = 3090$ lb (13,744 N).

5. Locate the center of rotation

To locate the center of rotation, compute the relative deformation of rods 1 and 2. Thus $\Delta l_1 = 11,810(5)/(1.25E) = 47,240/E$; $\Delta l_2 = 5100(9)/(1.20E) = 38,250/E$.

In Fig. 17c, by similar triangles, $x/(x - 6) = \Delta l_1/\Delta l_2 = 1.235$; $x = 31.5$ ft (9.6 m).

6. Verify the computed values of the tensile forces

Calculate the moment with respect to A of the applied and resisting forces. Thus $M_{Aa} = 20,000(4) = 80,000$ lb·ft (108,400 N·m); $M_{Ar} = 5100(6) + 3090(16) = 80,000$ lb·ft (108,400 N·m). Since the moments are equal, the results are verified.

ANALYSIS OF CABLE SUPPORTING A CONCENTRATED LOAD

A cold-drawn steel wire ¼ in (6.35 mm) in diameter is stretched tightly betwen two points lying on the same horizontal plane 80 ft (24.4 m) apart. The stress in the wire is 50,000 lb/in² (344,700 kPa). A load of 200 lb (889.6 N) is suspended at the center of the cable. Determine the sag of the cable and the final stress in the cable. Verify that the results obtained are compatible.

Calculation Procedure:

1. Derive the stress and strain relations for the cable

With reference to Fig. 18, L = distance between supports, ft (m); P = load applied at center of cable span, lb (N); d = deflection of cable center, ft (m); ϵ = strain of cable caused by P; s_1 and s_2 = initial and final tensile stress in cable, respectively, lb/in² (kPa).

Refer to the geometry of the deflection diagram. Taking into account that d/L is extremely small, derive the following approximations: $s_2 = PL/(4Ad)$, Eq. a; $\epsilon = 2(d/L)^2$, Eq. b.

2. Relate stress and strain

Express the increase in stress caused by P in terms of ϵ, and apply the above two equations to derive $2E(d/L)^3 + s_1(d/L) = P/(4A)$, Eq. c.

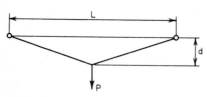

FIG. 18 Deflection of cable under concentrated load.

3. Compute the deflection at the center of the cable

Using Eq. c, we get $2(30)(10)^6(d/L)^3 + 50,000d/L = 200/[4(0.049)]$, so $d/L = 0.0157$ and ∴ $d = 0.0157(80) = 1.256$ ft (0.382 m).

4. Compute the final tensile stress

Write Eq. a as $s_2 = [P/(4A)]/(d/L) = 1020/0.0157 = 65,000$ lb/in² (448,110 kPa).

5. Verify the results computed

To demonstrate that the results are compatible, accept the computed value of d/L as correct. Then apply Eq. b to find the strain, and compute the corresponding stress. Thus $\epsilon = 2(0.0157)^2 = 4.93 \times 10^{-4}$; $s_2 = s_1 + E\epsilon = 50{,}000 + 30 \times 10^6 \times 4.93 \times 10^{-4} = 64{,}800$ lb/in^2 (446,731 kPa). This agrees closely with the previously calculated stress of 65,000 lb/in^2 (448,110 kPa).

DISPLACEMENT OF TRUSS JOINT

In Fig. 19a, the steel members AC and BC both have a cross-sectional area of 1.2 in^2 (7.7 cm^2). If a load of 20 kips (89.0 kN) is suspended at C, how much is joint C displaced?

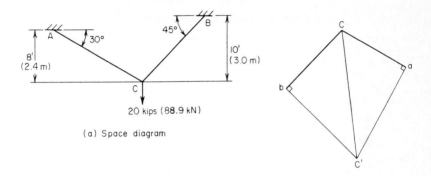

(a) Space diagram

20 kips (88.9 kN)

(b) Displacement diagram

FIG. 19

Calculation Procedure:

1. Compute the length of each member and the tensile forces

Consider joint C as a free body to find the tensile force in each member. Thus $L_{AC} = 192$ in (487.7 cm); $L_{BC} = 169.7$ in (431.0 cm); $P_{AC} = 14{,}640$ lb (65,118.7 N); $P_{BC} = 17{,}930$ lb (79,752.6 N).

2. Determine the elongation of each member

Use the relation $\Delta l = PL/(AE)$. Thus $\Delta l_{AC} = 14{,}640(192)/[1.2(30 \times 10^6)] = 0.0781$ in (1.983 mm); $\Delta l_{BC} = 17{,}930(169.7)/[1.2(30 \times 10^6)] = 0.0845$ in (2.146 mm).

3. Construct the Williott displacement diagram

Selecting a suitable scale, construct the Williott displacement diagram as follows: Draw (Fig. 19b) line Ca parallel to member AC, with $Ca = 0.0781$ in (1.98 mm). Similarly, draw Cb parallel to member BC, with $Cb = 0.0845$ in (2.146 mm).

4. Determine the displacement

Erect perpendiculars to Ca and Cb at a and b, respectively. Designate the intersection point of these perpendiculars as C'.

Line CC' represents, in both magnitude and direction, the approximate displacement of joint C under the applied load. Scaling distance CC' to obtain the displacement shows that the displacement of $C = 0.134$ in (3.4036 mm).

AXIAL STRESS CAUSED BY IMPACT LOAD

A body weighing 18 lb (80.1 N) falls 3 ft (0.9 m) before contacting the end of a vertical steel rod. The rod is 5 ft (1.5 m) long and has a cross-sectional area of 1.2 in^2 (7.74 cm^2). If the entire kinetic energy of the falling body is absorbed by the rod, determine the stress induced in the rod.

Calculation Procedure:

1. State the equation for the induced stress

Equate the energy imparted to the rod to the potential energy lost by the falling body: $s = (P/A)\{1 + [1 + 2Eh/(LP/A)]^{0.5}\}$, where h = vertical displacement of body, ft (m).

2. Substitute the numerical values

Thus, $P/A = 18/1.2 = 15$ lb/in^2 (103 kPa); $h = 3$ ft (0.9 m); $L = 5$ ft (1.5 m); $2Eh/(LP/A) = 2(30 \times 10^6)(3)/[5(15)] = 2,400,000$. Then $s = 23,250$ lb/in^2 (160,285.5 kPa).

Related Calculations: Where the deformation of the supporting member is negligible in relation to the distance h, as it is in the present instance, the following approximation is used: $s = [2PEh/(AL)]^{0.5}$.

STRESSES ON AN OBLIQUE PLANE

The prism $ABCD$ in Fig. 20a has the principal stresses of 6300- and 2400-lb/in^2 (43,438.5- and 16,548.0-kPa) tension. Applying both the analytical and graphical methods, determine the normal and shearing stress on plane AE.

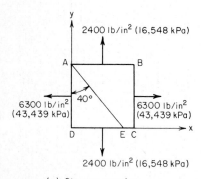

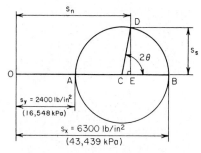

(a) Stresses on prism   (b) Mohr's circle of stress

FIG. 20

Calculation Procedure:

1. Compute the stresses, using the analytical method

A *principal stress* is a normal stress not accompanied by a shearing stress. The plane on which the principal stress exists is termed a *principal plane*. For a condition of plane stress, there are two principal planes through every point in a stressed body and these planes are mutually perpendicular. Moreover, one principal stress is the maximum normal stress existing at that point; the other is the minimum normal stress.

Let s_x and s_y = the principal stress in the x and y direction, respectively; s_n = normal stress on the plane making an angle θ with the y axis; s_s = shearing stress on this plane. All stresses are expressed in pounds per square inch (kilopascals) and all angles in degrees. Tensile stresses are positive; compressive stresses are negative.

Applying the usual stress equations yields $s_n = s_y + (s_x - s_y) \cos^2 \theta$; $s_s = \frac{1}{2}(s_x - s_y) \sin 2\theta$. Substituting gives $s_n = 2400 + (6300 - 2400)0.766^2 = 4690$-lb/in^2 (32,337.6-kPa) tension, and $s_s = \frac{1}{2}(6300 - 2400)0.985 = 1920$ lb/in^2 (13,238.4 kPa).

2. Apply the graphical method of solution

Construct, in Fig. 20b, Mohr's circle of stress thus: Using a suitable scale, draw $OA = s_y$, and $OB = s_x$. Draw a circle having AB as its diameter. Draw the radius CD making an angle of $2\theta = 80°$ with AB. Through D, drop a perpendicular DE to AB. Then $OE = s_n$ and $ED = s_s$. Scale OE and ED to obtain the normal and shearing stresses on plane AE.

Related Calculations: The normal stress may also be computed from $s_n = (s_x + s_y)0.5 + (s_x - s_y)0.5 \cos 2\theta$.

EVALUATION OF PRINCIPAL STRESSES

The prism $ABCD$ in Fig. 21a is subjected to the normal and shearing stresses shown. Construct Mohr's circle to determine the principal stresses at A, and locate the principal planes.

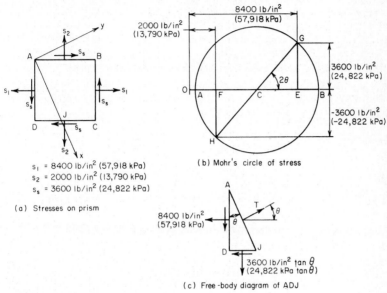

$s_1 = 8400$ lb/in² (57,918 kPa)
$s_2 = 2000$ lb/in² (13,790 kPa)
$s_s = 3600$ lb/in² (24,822 kPa)

(a) Stresses on prism

(b) Mohr's circle of stress

(c) Free-body diagram of ADJ

FIG. 21

Calculation Procedure:

1. Draw the lines representing the normal stresses (Fig. 21b)

Through the origin O, draw a horizontal base line. Locate points E and F such that $OE = 8400$ lb/in² (57,918.0 kPa) and $OF = 2000$ lb/in² (13,790.0 kPa). Since both normal stresses are tensile, E and F lie to the right of O. Note that the construction required here is the converse of that required in the previous calculation procedure.

2. Draw the lines representing the shearing stresses

Construct the vertical lines EG and FH such that $EG = 3600$ lb/in² (24,822.0 kPa), and $FH = -3600$ lb/in² ($-24,822.0$ kPa).

3. Continue the construction

Draw line GH to intersect the base line at C.

4. Construct Mohr's circle

Draw a circle having GH as diameter, intersecting the base line at A and B. Then lines OA and OB represent the principal stresses.

5. Scale the diagram

Scale OA and OB to obtain $f_{max} = 10,020$ lb/in² (69,087.9 kPa); $f_{min} = 380$ lb/in² (2620.1 kPa). Both stresses are tension.

6. Determine the stress angle

Scaie angle BCG and measure it as $48°22'$. The angle between the x axis, on which the maximum stress exists, and the side AD of the prism is one-half of BCG.

7. Construct the x and y axes

In Fig. 21a, draw the x axis, making a counterclockwise angle of $24°11'$ with AD. Draw the y axis perpendicular thereto.

8. Verify the locations of the principal planes

Consider ADJ as a free body. Set the length AD equal to unity. In Fig. 21c, since there is no shearing stress on AJ, $\Sigma F_H = T \cos \theta - 8400 - 3600 \tan \theta = 0$; $T \cos \theta = 8400 + 3600(0.45) = 10,020 \text{ lb/in}^2$ (69,087.9 kPa). The stress on $AJ = T/AJ = T \cos \theta = 10,020 \text{ lb/in}^2$ (69,087.9 kPa).

HOOP STRESS IN THIN-WALLED CYLINDER UNDER PRESSURE

A steel pipe 5 ft (1.5 m) in diameter and ⅜ in (9.53 mm) thick sustains a fluid pressure of 180 lb/in^2 (1241.1 kPa). Determine the hoop stress, the longitudinal stress, and the increase in diameter of this pipe. Use 0.25 for Poisson's ratio.

Calculation Procedure:

1. Compute the hoop stress

Use the relation $s = pD/(2t)$, where s = hoop or tangential stress, lb/in^2 (kPa); p = radial pressure, lb/in^2 (kPa); D = internal diameter of cylinder, in (mm); t = cylinder wall thickness, in (mm). Thus, for this cylinder, $s = 180(60)/[2(⅜)] = 14,400 \text{ lb/in}^2$ (99,288.0 kPa).

2. Compute the longitudinal stress

Use the relation $s' = pD/(4t)$, where s' = longitudinal stress, i.e., the stress parallel to the longitudinal axis of the cylinder, lb/in^2 (kPa), with other symbols as before. Substituting yields $s' = 7200 \text{ lb/in}^2$ (49,644.0 kPa).

3. Compute the increase in the cylinder diameter

Use the relation $\Delta D = (D/E)(s - \nu s')$, where ν = Poisson's ratio. Thus $\Delta D = 60(14,400 - 0.25 \times 7200)/(30 \times 10^6) = 0.0252 \text{ in}$ (0.6401 mm).

STRESSES IN PRESTRESSED CYLINDER

A steel ring having an internal diameter of 8.99 in (228.346 mm) and a thickness of ¼ in (6.35 mm) is heated and allowed to shrink over an aluminum cylinder having an external diameter of 9.00 in (228.6 mm) and a thickness of ½ in (12.7 mm). After the steel cools, the cylinder is subjected to an internal pressure of 800 lb/in^2 (5516 kPa). Find the stresses in the two materials. For aluminum, $E = 10 \times 10^6 \text{ lb/in}^2$ (6.895 $\times 10^7$ kPa).

Calculation Procedure:

1. Compute the radial pressure caused by prestressing

Use the relation $p = 2\Delta D/\{D^2[1/(t_a E_a) + 1/(t_s E_s)]\}$, where p = radial pressure resulting from prestressing, lb/in^2 (kPa), with other symbols the same as in the previous calculation procedure and the subscripts a and s referring to aluminum and steel, respectively. Thus, $p = 2(0.01)/\{9^2[1/(0.5 \times 10 \times 10^6) + 1/(0.25 \times 30 \times 10^6)]\} = 741 \text{ lb/in}^2$ (5109.2 kPa).

2. Compute the corresponding prestresses

Using the subscripts 1 and 2 to denote the stresses caused by prestressing and internal pressure, respectively, we find $s_{a1} = pD/(2t_a)$, where the symbols are the same as in the previous calculation procedure. Thus, $s_{a1} = 741(9)/[2(0.5)] = 6670\text{-lb/in}^2$ (45,989.7-kPa) compression. Likewise, $s_{s1} = 741(9)/[2(0.25)] = 13,340\text{-lb/in}^2$ (91,979-kPa) tension.

3. Compute the stresses caused by internal pressure

Use the relation $s_{s2}/s_{a2} = E_s/E_a$ or, for this cylinder, $s_{s2}/s_{a2} = (30 \times 10^6)/(10 \times 10^6) = 3$. Next, compute s_{a2} from $t_a s_{a2} + t_s s_{s2} = pD/2$, or $s_{a2} = 800(9)/[2(0.5 + 0.25 \times 3)] = 2880\text{-lb/in}^2$ (19,857.6-kPa) tension. Also, $s_{s2} = 3(2880) = 8640\text{-lb/in}^2$ (59,572.8-kPa) tension.

4. Compute the final stresses

Sum the results in steps 2 and 3 to obtain the final stresses: $s_{t\,i} = 6670 - 2880 = 3790\text{-lb/in}^2$ (26,132.1-kPa) compression; $s_{s3} = 13,340 + 8640 = 21,980\text{-lb}/\text{in}^2$ (151,552.1-kPa) tension.

5. Check the accuracy of the results

Ascertain whether the final diameters of the steel ring and aluminum cylinder are equal. Thus, setting $s' = 0$ in $\Delta D = (D/E)(s - vs')$, we find $\Delta D_a = -3790(9)/(10 \times 10^6) = -0.0034$ in (−0.0864 mm), $D_a = 9.0000 - 0.0034 = 8.9966$ in (228.51 mm). Likewise, $\Delta D_s = 21,980(9)/(30 \times 10^6) = 0.0066$ in (0.1676 mm), $D_s = 8.99 + 0.0066 = 8.9966$ in (228.51 mm). Since the computed diameters are equal, the results are valid.

HOOP STRESS IN THICK-WALLED CYLINDER

A cylinder having an internal diameter of 20 in (508 mm) and an external diameter of 36 in (914 mm) is subjected to an internal pressure of 10,000 lb/in² (68,950 kPa) and an external pressure of

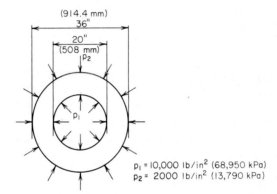

FIG. 22 Thick-walled cylinder under internal and external pressure.

2000 lb/in² (13,790 kPa) as shown in Fig. 22. Determine the hoop stress at the inner and outer surfaces of the cylinder.

Calculation Procedure:

1. Compute the hoop stress at the inner surface of the cylinder

Use the relation $s_i = [p_1(r_1^2 + r_2^2) - 2p_2 r_2^2]/(r_2^2 - r_1^2)$, where : = hoop stress at inner surface, lb/in² (kPa); p_1 = internal pressure, lb/in² (kPa); r_1 = interna radius, in (mm); r_2 = external radius, in (mm); p_2 = external pressure, lb/in² (kPa). Substituting gives $s_i = [10,000(100 + 324) - 2(2000)(324)]/(324 - 100) = 13,100\text{-lb/in}^2$ (90,324.5-kPa) tension.

2. Compute the hoop stress at the outer cylinder surface

Use the relation $s_0 = [2p_1 r_1^2 - p_2(r_1^2 + r_2^2)]/(r_2^2 - r_1^2)$, where the symbols are as before. Substituting gives $s_0 = [2(10,000)(100) - 2000(100 + 324)]/(324 - 100) = 5100\text{-lb/in}^2$ (35,164.5-kPa) tension.

3. Check the accuracy of the results

Use the relation $s_i r_1 - s_0 r_2 = [(r_2 - r_1)/(r_2 + r_1)](p_1 r_1 + p_2 r_2)$. Substituting the known values verifies the earlier calculations.

THERMAL STRESS RESULTING FROM HEATING A MEMBER

A steel member 18 ft (5.5 m) long is set snugly between two walls and heated 80°F (44.4°C). If each wall yields 0.015 in (0.381 mm), what is the compressive stress in the member? Use a coefficient of thermal expansion of $6.5 \times 10^{-6}/°F$ ($1.17 \times 10^{-5}/°C$) for steel.

Calculation Procedure:

1. Compute the thermal expansion of the member without restraint

Replace the true condition of partial restraint with the following equivalent conditions: The member is first allowed to expand freely under the temperature rise and is then compressed to its true final length.

To compute the thermal expansion without restraint, use the relation $\Delta L = cL\Delta T$, where c = coefficient of thermal expansion, $/°F$ ($/°C$); ΔT = increase in temperature, °F (°C); L = original length of member, in (mm); ΔL = increase in length of the member, in (mm). Substituting gives $\Delta L = 6.5(10^{-6})(18)(12)(80) \doteq 0.1123$ in (2.852 mm).

2. Compute the linear restraint exerted by the walls

The walls yield $2(0.015) = 0.030$ in (0.762 mm). Thus, the restraint exerted by the walls is $\Delta L_w = 0.1123 - 0.030 = 0.0823$ in (2.090 mm).

3. Determine the compressive stress

Use the relation $s = E\Delta L/L$, where the symbols are as given earlier. Thus, $s = 30(10^6)(0.0823)/[18(12)] = 11,430$ lb/in² (78,809.9 kPa).

THERMAL EFFECTS IN COMPOSITE MEMBER HAVING ELEMENTS IN PARALLEL

A ½-in (12.7-mm) diameter Copperweld bar consists of a steel core ⅜ in (9.53 mm) in diameter and a copper skin ¹⁄₁₆ in (1.6 mm) thick. What is the elongation of a 1-ft (0.3-m) length of this bar, and what is the internal force between the steel and copper arising from a temperature rise of 80°F (44.4°C)? Use the following values for thermal expansion coefficients: $c_s = 6.5 \times 10^{-6}$ and $c_c = 9.0 \times 10^{-6}$, where the subscripts s and c refer to steel and copper, respectively. Also, $E_c = 15 \times 10^6$ lb/in² (1.03×10^8 kPa).

Calculation Procedure:

1. Determine the cross-sectional areas of the metals

The total area $A = 0.1963$ in² (1.266 cm²). The area of the steel $A_s = 0.1105$ in² (0.712 cm²). By difference, the area of the copper $A_c = 0.0858$ in² (0.553 cm²).

2. Determine the coefficient of expansion of the composite member

Weight the coefficients of expansion of the two members according to their respective AE values. Thus

$A_s E_s$ (relative) = $0.1105 \times 30 \times 10^6$ =	3315
$A_c E_c$ (relative) = $0.0858 \times 15 \times 10^6$ =	1287
Total	4602

Then the coefficient of thermal expansion of the composite member is $c = (3315c_s + 1287c_c)/4602 = 7.2 \times 10^{-6}/°F$ ($1.30 \times 10^{-5}/°C$).

3. Determine the thermal expansion of the 1-ft (0.3-m) section

Using the relation $\Delta L = cL\Delta T$, we get $\Delta L = 7.2(10^{-6})(12)(80) = 0.00691$ in (0.17551 mm).

4. Determine the expansion of the first material without restraint

Using the same relation as in step 3 for copper *without* restraint yields $\Delta L_c = 9.0(10^{-6}) \times (12)(80) = 0.00864$ in (0.219456 mm).

5. Compute the restraint of the first material

The copper is restrained to the amount computed in step 3. Thus, the restraint exerted by the steel is $\Delta L_{cs} = 0.00864 - 0.00691 = 0.00173$ in (0.043942 mm).

6. Compute the restraining force exerted by the second material

Use the relation $P = (A_c E_c \Delta L_{cs})/L$, where the symbols are as given before: $P = [1,287,000(0.00173)]/12 = 185$ lb (822.9 N).

7. Verify the results obtained

Repeat steps 4, 5, and 6 with the two materials interchanged. So $\Delta L_s = 6.5(10^{-6})(12)(80) = 0.00624$ in (0.15849 mm); $\Delta L_{sc} = 0.00691 - 0.00624 = 0.00067$ in (0.01701 mm). Then $P = 3,315,000(0.00067)/12 = 185$ lb (822.9 N), as before.

THERMAL EFFECTS IN COMPOSITE MEMBER HAVING ELEMENTS IN SERIES

The aluminum and steel bars in Fig. 23 have cross-sectional areas of 1.2 and 1.0 in^2 (7.7 and 6.5 cm^2), respectively. The member is restrained against lateral deflection. A temperature rise of 100°F (55°C) causes the length of the member to increase to 42.016 in (106.720 cm). Determine the stress and deformation of each bar. For aluminum, $E = 10 \times 10^6$, $c = 13.0 \times 10^{-6}$; for steel, $c = 6.5 \times 10^{-6}$.

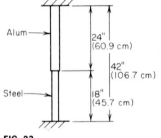

Alum →
24" (60.9 cm)
42" (106.7 cm)
Steel →
18" (45.7 cm)

FIG. 23

Calculation Procedure:

1. Express the deformation of each bar resulting from the temperature change and the compressive force

The temperature rise causes the bar to expand, whereas the compressive force resists this expansion. Thus, the net expansion is the difference between these two changes, or $\Delta L_a = cL\Delta T - PL/(AE)$, where the subscript a refers to the aluminum bar; the other symbols are the same as given earlier. Substituting gives $\Delta L_a = 13.0 \times 10^{-6}(24)(100) - P(24)/[1.2(10 \times 10^6)] = (31,200 - 2P)10^{-6}$, Eq. a. Likewise, for steel: $\Delta L_s = 6.5 \times 10^{-6}(18)(100) - P(18)/[1.0(30 \times 10^6)] = (11,700 - 0.6P)10^{-6}$, Eq. b.

2. Sum the results in step 1 to obtain the total deformation of the member

Set the result equal to 0.016 in (0.4064 mm); solve for P. Or, $\Delta L = (42,900 - 2.6P)10^{-6} = 0.016$ in (0.4064 mm); $P = (42,900 - 16,000)/2.6 = 10,350$ lb (46,037 N).

3. Determine the stresses and deformation

Substitute the computed value of P in the stress equation $s = P/A$. For aluminum $s_a = 10,350/1.2 = 8630$ lb/in^2 (59,503.9 kPa). Then $\Delta L_a = (31,200 - 2 \times 10,350)10^{-6} = 0.0105$ in (0.2667 mm). Likewise, for steel $s_s = 10,350/1.0 = 10,350$ lb/in^2 (71,363.3 kPa); and $\Delta L_s = (11,700 - 0.6 \times 10,350)10^{-6} = 0.0055$ in (0.1397 mm).

SHRINK-FIT STRESS AND RADIAL PRESSURE

An open steel cylinder having an internal diameter of 4 ft (1.2 m) and a wall thickness of ⁵⁄₁₆ in (7.9 mm) is to be heated to fit over an iron casting. The internal diameter of the cylinder before heating is ¹⁄₃₂ in (0.8 mm) less than that of the casting. How much must the temperature of the cylinder be increased to provide a clearance of ¹⁄₃₂ in (0.8 mm) all around between the cylinder and casting? If the casting is considered rigid, what stress will exist in the cylinder after it cools, and what radial pressure will it then exert on the casting?

Calculation Procedure:

1. Compute the temperature rise required

Use the relation $\Delta T = \Delta D/(cD)$, where ΔT = temperature rise required, °F (°C); ΔD = change in cylinder diameter, in (mm); c = coefficient of expansion of the cylinder = 6.5×10^{-6}/°F (1.17×10^{-5}/°C); D = cylinder internal diameter before heating, in (mm). Thus $\Delta T = (3/32)/[6.5 \times 10^{-6}(48)] = 300$°F (167°C).

2. Compute the hoop stress in the cylinder

Upon cooling, the cylinder has a diameter ¹⁄₃₂ in (0.8 mm) larger than originally. Compute the hoop stress from $s = E\Delta D/D = 30 \times 10^6 (\frac{1}{32})/48 = 19,500$ lb/in² (134,452.5 kPa).

3. Compute the associated radial pressure

Use the relation $p = 2ts/D$, where p = radial pressure, lb/in² (kPa), with the other symbols as given earlier. Thus $p = 2(5/16)(19,500)/48 = 254$ lb/in² (1751.3 kPa).

TORSION OF A CYLINDRICAL SHAFT

A torque of 8000 lb·ft (10,840 N·m) is applied at the ends of a 14-ft (4.3-m) long cylindrical shaft having an external diameter of 5 in (127 mm) and an internal diameter of 3 in (76.2 mm). What are the maximum shearing stress and the angle of twist of the shaft if the modulus of ridigity of the shaft is 6×10^6 lb/in² (4.1×10^4 MPa)?

Calculation Procedure:

1. Compute the polar moment of inertia of the shaft

For a hollow circular shaft, $J = (\pi/32)(D^4 - d^4)$, where J = polar moment of inertia of a transverse section of the shaft with respect to the longitudinal axis, in⁴ (cm⁴); D = external diameter of shaft, in (mm); d = internal diameter of shaft, in (mm). Substituting gives $J = (\pi/32)(5^4 - 3^4) = 53.4$ in⁴ (2222.6 cm⁴).

2. Compute the shearing stress in the shaft

Use the relation $s_s = TR/J$, where s_s = shearing stress, lb/in² (MPa); T = applied torque, lb·in (N·m); R = radius of shaft, in (mm). Thus $s_s = [(8000)(12)(2.5)]/53.4 = 4500$ lb/in² (31,027.5 kPa).

3. Compute the angle of twist of the shaft

Use the relation $\theta = TL/JG$, where θ = angle of twist, rad; L = shaft length, in (mm); G = modulus of ridigity, lb/in² (GPa). Thus $\theta = (8000)(12)(14)(12)/[53.4(6,000,000)] = 0.050$ rad, or 2.9°.

ANALYSIS OF A COMPOUND SHAFT

The compound shaft in Fig. 24 was formed by rigidly joining two solid segments. What torque may be applied at B if the shearing stress is not to exceed 15,000 lb/in² (103.4 MPa) in the steel and 10,000 lb/in² (69.0 MPa) in the bronze? Here $G_s = 12 \times 10^6$ lb/in² (82.7 GPa); $G_b = 6 \times 10^6$ lb/in² (41.4 GPa).

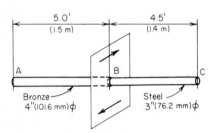

FIG. 24 Compound shaft.

Calculation Procedure:

1. Determine the relationship between the torque in the shaft segments

Since segments AB and BC (Fig. 24) are twisted through the same angle, the torque applied at the junction of these segments is distributed in proportion to their relative rigidities. Using the subscripts s and b to denote steel and bronze, respectively, we see that $\theta = T_s L_s/(J_s G_s) = T_b L_b/(J_b G_b)$, where the symbols are as given in the previous calculation procedure. Solving yields $T_s = (5/4.5)(3^4/4^4)(12/6)T_b = 0.703 T_b$.

2. Establish the relationship between the shearing stresses

For steel, $s_{ss} = 16 T_s/(\pi D^3)$, where the symbols are as given earlier. Thus $s_{ss} = 16(0.703 T_b)/(\pi\, 3^3)$. Likewise, for bronze, $s_{sb} = 16 T_b/(\pi 4^3)$, $\therefore s_{ss} = 0.703(4^3/3^3)s_{sb} = 1.67 s_{sb}$.

3. Compute the allowable torque

Ascertain which material limits the capacity of the member, and compute the allowable torque by solving the shearing-stress equation for T.

If the bronze were stressed to 10,000 lb/in^2 (69.0 MPa), inspection of the above relations shows that the steel would be stressed to 16,700 lb/in^2 (115.1 MPa), which exceeds the allowed 15,000 lb/in^2 (103.4 MPa). Hence, the steel limits the capacity. Substituting the allowed shearing stress of 15,000 lb/in^2 (103.4 MPa) gives $T_s = 15,000\pi(3^3)/[16(12)] = 6630$ lb·ft (8984.0 N·m); also, $T_b = 6630/0.703 = 9430$ lb·ft (12,777.6 N·m). Then $T = 6630 + 9430 = 16,060$ lb·ft (21,761.3 N·m).

Stresses in Flexural Members

In the analysis of beam action, the general assumption is that the beam is in a horizontal position and carries vertical loads lying in an axis of symmetry of the transverse section of the beam.

The vertical shear V at a given section of the beam is the algebraic sum of all vertical forces to the left of the section, with an upward force being considered positive.

The bending moment M at a given section of the beam is the algebraic sum of the moments of all forces to the left of the section with respect to that section, a clockwise moment being considered positive.

If the proportional limit of the beam material is not exceeded, the bending stress (also called the flexural, or fiber, stress) at a section varies linearly across the depth of the section, being zero at the neutral axis. A positive bending moment induces compressive stresses in the fibers above the neutral axis and tensile stresses in the fibers below. Consequently, the elastic curve of the beam is concave upward where the bending moment is positive.

SHEAR AND BENDING MOMENT IN A BEAM

Construct the shear and bending-moment diagrams for the beam in Fig. 25. Indicate the value of the shear and bending moment at all significant sections.

Calculation Procedure:

1. Replace the distributed load on each interval with its equivalent concentrated load

Where the load is uniformly distributed, this equivalent load acts at the center of the interval of the beam. Thus $W_{AB} = 2(4) = 8$ kips (35.6 kN); $W_{BC} = 2(6) = 12$ kips (53.3 kN); $W_{AC} = 8 + 12 = 20$ kips (89.0 kN); $W_{CD} = 3(15) = 45$ kips (200.1 kN); $W_{DE} = 1.4(5) = 7$ kips (31.1 kN).

2. Determine the reaction at each support

Take moments with respect to the other support. Thus $\Sigma M_D = 25 R_A - 6(21) - 20(20) - 45(7.5) + 7(2.5) + 4.2(5) = 0$; $\Sigma M_A = 6(4) + 20(5) + 45(17.5) + 7(27.5) + 4.2(30) - 25 R_D = 0$. Solving gives $R_A = 33$ kips (146.8 kN); $R_D = 49.2$ kips (218.84 kN).

3. Verify the computed results and determine the shears

Ascertain that the algebraic sum of the vertical forces is zero. If this is so, the computed results are correct.

Starting at A, determine the shear at every significant section, or directly to the left or right of that section if a concentrated load is present. Thus V_A at right $= 33$ kips (146.8 kN); V_B at left $= 33 - 8 = 25$ kips (111.2 kN); V_B at right $= 25 - 6 = 19$ kips (84.5 kN); $V_C = 19 - 12 = 7$ kips (31.1 kN); V_D at left $= 7 - 45 = -38$ kips (-169.0 kN); V_D at right $= -38 + 49.2 = 11.2$ kips (49.8 kN); V_E at left $= 11.2 - 7 = 4.2$ kips (18.7 kN); V_E at right $= 4.2 - 4.2 = 0$.

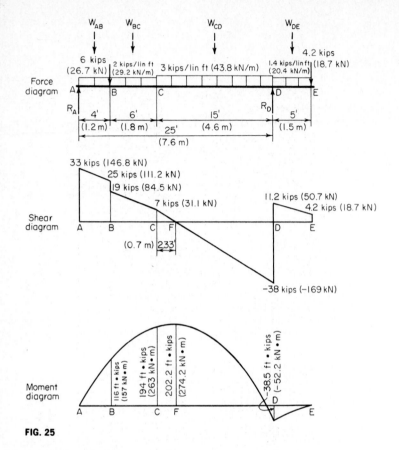

FIG. 25

4. Plot the shear diagram

Plot the points representing the forces in the previous step in the shear diagram. Since the loading betwen the significant sections is uniform, connect these points with straight lines. In general, the slope of the shear diagram is given by $dV/dx = -w$, where w = unit load at the given section and x = distance from left end to the given section.

5. Determine the bending moment at every significant section

Starting at A, determine the bending moment at every significant section. Thus $M_A = 0$; $M_B = 33(4) - 8(2) = 116$ ft·kips (157 kN·m); $M_C = 33(10) - 8(8) - 6(6) - 12(3) = 194$ ft·kips (263 kN·m). Similarly, $M_D = -38.5$ ft·kips (-52.2 kN·m); $M_E = 0$.

6. Plot the bending-moment diagram

Plot the points representing the values in step 5 in the bending-moment diagram (Fig. 25). Complete the diagram by applying the slope equation $dM/dx = V$, where V denotes the shear at the given section. Since this shear varies linearly between significant sections, the bending-moment diagram comprises a series of parabolic arcs.

7. Alternatively, apply a moment theorem

Use this theorem: If there are no externally applied moments in an interval 1-2 of the span, the difference between the bending moments is $M_2 - M_1 = \int_1^2 V \, dx$ = the area under the shear diagram across the interval.

Calculate the areas under the shear diagram to obtain the following results: $M_A = 0$; $M_B =$

M_A + ¼(4)(33 + 25) = 116 ft·kips (157.3 kN·m); M_C = 116 + ¼(6)(19 + 7) = 194 ft·kips (263 kN·m); M_D = 194 + ¼(15)(7 − 38) = −38.5 ft·kips (−52.2 kN·m); M_E = −38.5 + ¼(5)(11.2 + 4.2) = 0.

8. Locate the section at which the bending moment is maximum

As a corollary of the equation in step 6, the maximum moment occurs where the shear is zero or passes through zero under a concentrated load. Therefore, CF = 7/3 = 2.33 ft (0.710 m).

9. Compute the maximum moment

Using the computed value for CF, we find M_F = 194 + ½(2.33)(7) = 202.2 ft·kips (274.18 kN·m).

BEAM BENDING STRESSES

A beam having the trapezoidal cross section shown in Fig. 26a carries the loads indicated in Fig. 26b. What is the maximum bending stress at the top and at the bottom of this beam?

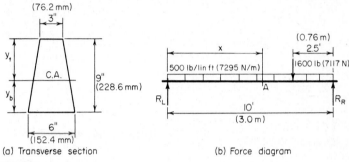

(a) Transverse section   (b) Force diagram

FIG. 26

Calculation Procedure:

1. Compute the left reaction and the section at which the shear is zero

The left reaction R_L = ½(10)(500) + 1600(2.5/10) = 2900 lb (12,899.2 N). The section A at which the shear is zero is x = 2900/500 = 5.8 ft (1.77 m).

2. Compute the maximum moment

Use the relation M_A = ½(2900)(5.8) = 8410 lb·ft (11,395.6 N·m) = 100,900 lb·in (11,399.682 N·m).

3. Locate the centroidal axis of the section

Use the AISC Manual for properties of the trapezoid. Or y_t = (9/3)[(2 × 6 + 3)/(6 + 3)] = 5 in (127 mm); y_b = 4 in (101.6 mm).

4. Compute the moment of inertia of the section

Using the AISC Manual, I = (9³/36)[(6² + 4 × 6 × 3 + 3²)/(6 + 3)] = 263.3 in⁴ (10,959.36 cm⁴).

5. Compute the stresses in the beam

Use the relation f = My/I, where f = bending stress in a given fiber, lb/in² (kPa); y = distance from neutral axis to given fiber, in. Thus f_{top} = 100,900(5)/263.3 = 1916-lb/in² (13,210.8-kPa) compression, f_{bottom} = 100,900(4)/263.3 = 1533-lb/in² (10,570.0-kPa) tension.

In general, the maximum bending stress at a section where the moment is M is given by f = Mc/I, where c = distance from the neutral axis to the outermost fiber, in (mm). For a section that is symmetric about its centroidal axis, it is convenient to use the section modulus S of the section, this being defined as S = I/c. Then f = M/S.

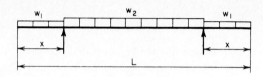

(a) Loads carried by overhanging beam

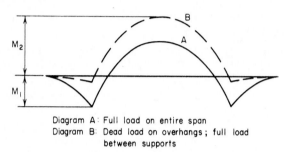

Diagram A: Full load on entire span
Diagram B: Dead load on overhangs; full load
between supports

(b) Bending-moment diagrams

FIG. 27

ANALYSIS OF A BEAM ON MOVABLE SUPPORTS

The beam in Fig. 27a rests on two movable supports. It carries a uniform live load of w lb/lin ft and a uniform dead load of $0.2w$ lb/lin ft. If the allowable bending stresses in tension and compression are identical, determine the optimal location of the supports.

Calculation Procedure:

1. Place full load on the overhangs, and compute the negative moment

Refer to the moment diagrams. For every position of the supports, there is a corresponding maximum bending stress. The position for which this stress has the smallest value must be identified.

As the supports are moved toward the interior of the beam, the bending moments between the supports diminish in algebraic value. The optimal position of the supports is that for which the maximum potential negative moment M_1 is numerically equal to the maximum potential positive moment M_2. Thus, $M_1 = -1.2w(x^2/2) = -0.6wx^2$.

2. Place only the dead load on the overhangs and the full load between the supports. Compute the positive moment.

Sum the areas under the shear diagram to compute M_2. Thus, $M_2 = \frac{1}{2}[1.2w(L/2 - x)^2 - 0.2wx^2] = w(0.15L^2 - 0.6Lx + 0.5x^2)$.

3. Equate the absolute values of M_1 and M_2 and solve for x

Substituting gives $0.6x^2 = 0.15L^2 - 0.6Lx + 0.5x^2$; $x = L(\overline{10.5}^{0.5} - 3) = 0.240L$.

FLEXURAL CAPACITY OF A COMPOUND BEAM

A 16WF45 (W16 × 45) steel beam in an existing structure was reinforced by welding an ST6WF20 (WT6 × 20) to the bottom flange, as in Fig. 28. If the allowable bending stress is 20,000 lb/in² (137,900 kPa), determine the flexural capacity of the built-up member.

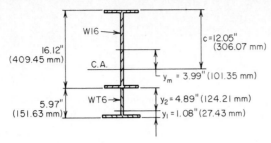

FIG. 28 Compound beam.

Calculation Procedure:

1. Obtain the properties of the elements

Using the AISC *Manual*, determine the following properties. For the 16WF45, $d = 16.12$ in (409.45 mm); $A = 13.24$ in^2 (85.424 cm^2); $I = 583$ in^4 (24,266 cm^4). For the ST6WF20, $d = 5.97$ in (151.63 mm); $A = 5.89$ in^2 (38.002 cm^2); $I = 14$ in^4 (582.7 cm^4); $y_1 = 1.08$ in (27.43 mm); $y_2 = 5.97 - 1.08 = 4.89$ in (124.21 mm).

2. Locate the centroidal axis of the section

Locate the centroidal axis of the section with respect to the centerline of the 16WF45, and compute the distance c from the centroidal axis to the outermost fiber. Thus, $y_m = 5.89[(8.06 + 4.89)]/(5.89 + 13.24) = 3.99$ in (101.346 mm). Then $c = 8.06 + 3.99 = 12.05$ in (306.07 mm).

3. Find the moment of inertia of the section with respect to its centroidal axis

Use the relation $I_0 + Ak^2$ for each member, and take the sum for the two members to find I for the built-up beam. Thus, for the 16WF45: $k = 3.99$ in (101.346 mm); $I_0 + Ak^2 = 583 + 13.24(3.99)^2 = 793$ in^4 (33,007.1 cm^4). For the ST6WF20: $k = 8.06 - 3.99 + 4.89 = 8.96$ in (227.584 mm); $I_0 + Ak^2 = 14 + 5.89(8.96)^2 = 487$ in^4 (20,270.4 cm^4). Then $I = 793 + 487 = 1280$ in^4 (53,277.5 cm^4).

4. Apply the moment equation to find the flexural capacity

Use the relation $M = fI/c = 20,000(1280)/[12.05(12)] = 177,000$ lb·ft (240,012 N·m).

ANALYSIS OF A COMPOSITE BEAM

An 8 × 12 in (203.2 × 304.8 mm) timber bean (exact size) is reinforced by the addition of a 7 × ½ in (177.8 × 12.7 mm) steel plate at the top and a 7-in (177.8-mm) 9.8-lb (43.59-N) steel channel at the bottom, as shown in Fig. 29a. The allowable bending stresses are 22,000 lb/in^2 (151,690 kPa) for steel and 1200 lb/in^2 (8274 kPa) for timber. The modulus of elasticity of the timber is 1.2×10^6 lb/in^2 (8.274×10^6 kPa). How does the flexural strength of the reinforced beam compare with that of the original timber beam?

Calculation Procedure:

1. Compute the rigidity of the steel compared with that of the timber

Let n = the relative rigidity of the steel and timber. Then $n = E_s/E_t = (30 \times 10^6)/(1.2 \times 10^6) = 25$.

2. Transform the composite beam to an equivalent homogeneous beam

To accomplish this transformation, replace the steel with timber. Sketch the cross section of the transformed beam as in Fig. 29b. Determine the sizes of the hypothetical elements by retaining the dimensions normal to the axis of bending but multiplying the dimensions parallel to this axis by n.

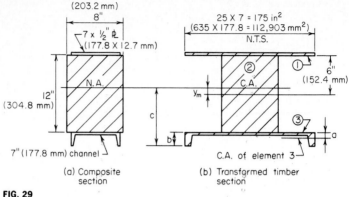

FIG. 29

(a) Composite section

(b) Transformed timber section

3. Record the properties of each element of the transformed section

Element 1: $A = 25(7)(\frac{1}{2}) = 87.5$ in^2 (564.55 cm^2); I_0 is negligible.
Element 2: $A = 8(12) = 96$ in^2 (619.4 cm^2); $I_0 = \frac{1}{12}(8)12^3 = 1152$ in^4 (4.795 dm^4).
Element 3: Refer to the AISC *Manual* for the data; $A = 25(2.85) = 71.25$ in^2 (459.71 cm^2); $I_0 = 25(0.98) = 25$ in^4 (1040.6 cm^4); $a = 0.55$ in (13.97 mm); $b = 2.09$ in (53.09 mm).

4. Locate the centroidal axis of the transformed section

Take static moments of the areas with respect to the centerline of the 8 $\times$ 12 in (203.2 $\times$ 304.8 mm) rectangle. Then $y_m = [87.5(6.25) - 71.25(6.55)]/(87.5 + 96 + 71.25) = 0.31$ in (7.87 mm). The neutral axis of the composite section is at the same location as the centroidal axis of the transformed section.

5. Compute the moment of inertia of the transformed section

Apply the relation in step 3 of the previous calculation procedure. Then compute the distance c to the outermost fiber. Thus, $I = 1152 + 25 + 87.5(6.25 - 0.31)^2 + 96(0.31)^2 + 71.25(6.55 + 0.31)^2 = 7626$ in^4 (31.74 dm^4). Also, $c = 0.31 + 6 + 2.09 = 8.40$ in (213.36 mm).

6. Determine which material limits the beam capacity

Assume that the steel is stressed to capacity, and compute the corresponding stress in the transformed beam. Thus, $f = 22,000/25 = 880$ lb/in^2 (6067.6 kPa) < 1200 lb/in^2 (8274 kPa).

In the actual beam, the maximum timber stress, which occurs at the back of the channel, is even less than 880 lb/in^2 (6067.6 kPa). Therefore, the strength of the member is controlled by the allowable stress in the steel.

7. Compare the capacity of the original and reinforced beams

Let subscripts 1 and 2 denote the original and reinforced beams, respectively. Compute the capacity of these members, and compare the results. Thus $M_1 = fI/c = 1200(1152)/6 = 230,000$ lb·in (25,985.4 N·m); $M_2 = 880(7626)/8.40 = 799,000$ lb·in (90,271.02 N·m); $M_2/M_1 = 799,000/230,000 = 3.47$. Thus, the reinforced beam is nearly 3½ times as strong as the original beam, before reinforcing.

BEAM SHEAR FLOW AND SHEARING STRESS

A timber beam is formed by securely bolting a 3 $\times$ 6 in (76.2 $\times$ 152.4 mm) member to a 6 $\times$ 8 in (152.4 $\times$ 203.2 mm) member (exact size), as shown in Fig. 30. If the beam carries a uniform load of 600 lb/lin ft (8.756 kN/m) on a simple span of 13 ft (3.9 m), determine the longitudinal shear flow and the shearing stress at the juncture of the two elements at a section 3 ft (0.91 m) from the support.

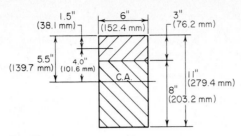

FIG. 30

Calculation Procedure:

1. Compute the vertical shear at the given section

Shear flow is the shearing force acting on a unit distance. In this instance, the shearing force on an area having the same width as the beam and a length of 1 in (25.4 mm) measured along the beam span is required.

Using dimensions and data from Fig. 30, we find $R = \frac{1}{2}(600)(13) = 3900$ lb (17,347.2 N); $V = 3900 - 3(600) = 2100$ lb (9340.8 N).

2. Compute the moment of inertia of the cross section

$$I = (\tfrac{1}{12})(bd^3) = (\tfrac{1}{12})(6)(11)^3 = 666 \text{ in}^4 \ (2.772 \text{ dm}^4)$$

3. Determine the static moment of the cross-sectional area

Calculate the static moment Q of the cross-sectional area above the plane under consideration with respect to the centroidal axis of the section. Thus, $Q = Ay = 3(6)(4) = 72$ in^3 (1180.1 cm^3).

4. Compute the shear flow

Compute the shear flow q, using $q = VQ/I = 2100(72)/666 = 227$ lb/lin in (39.75 kN/m).

5. Compute the shearing stress

Use the relation $v = q/t = VQ/(It)$, where $t =$ width of the cross section at the given plane. Then $v = 227/6 = 38$ lb/in^2 (262.0 kPa).

Note that v represents both the longitudinal and the transverse shearing stress at a particular point. This is based on the principle that the shearing stresses at a given point in two mutually perpendicular directions are equal.

LOCATING THE SHEAR CENTER OF A SECTION

A cantilever beam carries the load shown in Fig. 31a and has the transverse section shown in Fig. 31b. Locate the shear center of the section.

Calculation Procedure:

1. Construct a free-body diagram of a portion of the beam

Consider that the transverse section of a beam is symmetric solely about its horizontal centroidal axis. If bending of the beam is not to be accompanied by torsion, the vertical shearing force at any section must pass through a particular point on the centroidal axis designated as the *shear*, or *flexural, center*.

Cut the beam at section 2, and consider the left portion of the beam as a free body. In Fig. 31b, indicate the resisting shearing forces V_1, V_2, and V_3 that the right-hand portion of the beam exerts on the left-hand portion at section 2. Obtain the directions of V_1 and V_2 this way: Isolate the segment of the beam contained between sections 1 and 2; then isolate a segment $ABDC$ of the top flange, as shown in Fig. 31c. Since the bending stresses at section 2 exceed those at section 1, the resultant tensile force T_2 exceeds T_1. The resisting force on CD is therefore directed to the

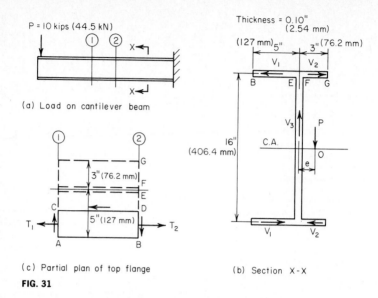

(a) Load on cantilever beam

(c) Partial plan of top flange

(b) Section X-X

FIG. 31

left. From the equation of equilibrium $\Sigma M = 0$ it follows that the resisting shears on AC and BD have the indicated directions to constitute a clockwise couple.

This analysis also reveals that the shearing stress varies linearly from zero at the edge of the flange to a maximum value at the juncture with the web.

2. Compute the shear flow

Determine the shear flow at E and F (Fig. 31) by setting Q in $q = VQ/I$ equal to the static moment of the overhanging portion of the flange. (For convenience, use the dimensions to the centerline of the web and flange.) Thus $I = \frac{1}{12}(0.10)(16)^3 + 2(8)(0.10)(8)^2 = 137$ in⁴ (5702.3 cm⁴); $Q_{BE} = 5(0.10)(8) = 4.0$ in³ (65.56 cm³); $Q_{FG} = 3(0.10)(8) = 2.4$ in³ (39.34 cm³); $q_E = VQ_{BE}/I = 10,000(4.0)/137 = 292$ lb/lin in (51,137.0 N/m); $q_F = 10,000(2.4)/137 = 175$ lb/lin in (30,647.2 N/m).

3. Compute the shearing forces on the transverse section

Since the shearing stress varies linearly across the flange, $V_1 = \frac{1}{2}(292)(5) = 730$ lb (3247.0 N); $V_2 = \frac{1}{2}(175)(3) = 263$ lb (1169.8 N); $V_3 = P = 10,000$ lb (44,480 N).

4. Locate the shear center

Take moments of all forces acting on the left-hand portion of the beam with respect to a longitudinal axis through the shear center O. Thus $V_3 e + 16(V_2 - V_1) = 0$, or $10,000e + 16(263 - 730) = 0$; $e = 0.747$ in (18.9738 mm).

5. Verify the computed values

Check the computed values of q_E and q_F by considering the bending stresses directly. Apply the equation $\Delta f = Vy/I$, where $\Delta f =$ increase in bending stress per unit distance along the span at distance y from the neutral axis. Then $\Delta f = 10,000(8)/137 = 584$ lb/(in²·in) (158.52 MPa/m).

In Fig. 31c, set $AB = 1$ in (25.4 mm). Then $q_E = 584(5)(0.10) = 292$ lb/lin in (51,137.0 N/m); $q_F = 584(3)(0.10) = 175$ lb/lin in (30,647.1 N/m).

Although a particular type of beam (cantilever) was selected here for illustrative purposes and a numeric value was assigned to the vertical shear, note that the value of e is independent of the type of beam, form of loading, or magnitude of the vertical shear. The location of the shear center is a geometric characteristic of the transverse section.

BENDING OF A CIRCULAR FLAT PLATE

A circular steel plate 2 ft (0.61 m) in diameter and ½ in (12.7 mm) thick, simply supported along its periphery, carries a uniform load of 20 lb/in² (137.9 kPa) distributed over the entire area. Determine the maximum bending stress and deflection of this plate, using 0.25 for Poisson's ratio.

Calculation Procedure:

1. Compute the maximum stress in the plate

If the maximum deflection of the plate is less than about one-half the thickness, the effects of diaphragm behavior may be disregarded.

Compute the maximum stress, using the relation $f = (\%)(3 + \nu)w(R/t)^2$, where R = plate radius, in (mm); t = plate thickness, in (mm); ν = Poisson's ratio. Thus, $f = (\%)(3.25)(20)(12/0.5)^2 = 14,000$ lb/in² (96,530.0 kPa).

2. Compute the maximum deflection of the plate

Use the relation $y = (1 - \nu)(5 + \nu)fR^2/[2(3 + \nu)Et] = 0.75(5.25)(14,000)(12)^2/[2(3.25)(30 \times 10^6)(0.5)] = 0.081$ in (2.0574 mm). Since the deflection is less than one-half the thickness, the foregoing equations are valid in this case.

BENDING OF A RECTANGULAR FLAT PLATE

A 2 × 3 ft (61.0 × 91.4 cm) rectangular plate, simply supported along its periphery, is to carry a uniform load of 8 lb/in² (55.2 kPa) distributed over the entire area. If the allowable bending stress is 15,000 lb/in² (103.4 MPa), what thickness of plate is required?

Calculation Procedure:

1. Select an equation for the stress in the plate

Use the approximation $f = a^2b^2w/[2(a^2 + b^2)t^2]$, where a and b denote the length of the plate sides, in (mm).

2. Compute the required plate thickness

Solve the equation in step 1 for t. Thus $t^2 = a^2b^2w/[2(a^2 + b^2)f] = 2^2(3)^2(144)(8)/[2(2^2 + 3^2)(15,000)] = 0.106$; $t = 0.33$ in (8.382 mm).

COMBINED BENDING AND AXIAL LOAD ANALYSIS

A post having the cross section shown in Fig. 32 carries a concentrated load of 100 kips (444.8 kN) applied at R. Determine the stress induced at each corner.

Calculation Procedure:

1. Replace the eccentric load with an equivalent system

Use a concentric load of 100 kips (444.8 kN) and two couples producing the following moments with respect to the coordinate axes:

$$M_x = 100,000(2) = 200,000 \text{ lb·in } (25,960 \text{ N·m})$$

$$M_y = 100,000(1) = 100,000 \text{ lb·in } (12,980 \text{ N·m})$$

2. Compute the section modulus

Determine the section modulus of the rectangular cross section with respect to each axis. Thus $S_x = (\%)bd^2 = (\%)(18)(24)^2 = 1728$ in³ (28,321.9 cm³); $S_y = (\%)(24)(18)^2 = 1296$ in³ (21,241 cm³).

3. Compute the stresses produced

Compute the uniform stress caused by the concentric load and the stresses at the edges caused by the bending moments. Thus $f_1 = P/A = 100,000/[18(24)] = 231$ lb/in² (1592.7 kPa); $f_x = M_x/$

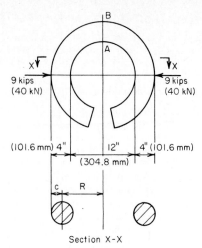

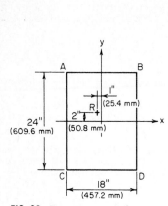

FIG. 32 Transverse section of a post.

FIG. 33 Curved member in bending.

$S_x = 200,000/1728 = 116$ lb/in^2 (799.8 kPa); $f_y = M_y/S_y = 100,000/1296 = 77$ lb/in^2 (530.9 kPa).

4. Determine the stress at each corner

Combine the results obtained in step 3 to obtain the stress at each corner. Thus $f_A = 231 + 116 + 77 = 424$ lb/in^2 (2923.4 kPa); $f_B = 231 + 116 - 77 = 270$ lb/in^2 (1861.5 kPa); $f_c = 231 - 116 + 77 = 192$ lb/in^2 (1323.8 kPa); $f_D = 231 - 116 - 77 = 38$ lb/in^2 (262.0 kPa). These stresses are all compressive because a positive stress is considered compressive, whereas a tensile stress is negative.

5. Check the computed corner stresses

Use the following equation that applies to the special case of a rectangular cross section: $f = (P/A)(1 \pm 6e_x/d_x \pm 6e_y/d_y)$, where e_x and e_y = eccentricity of load with respect to the x and y axes, respectively; d_x and d_y = side of rectangle, in (mm), normal to x and y axes, respectively. Solving for the quantities within the brackets gives $6e_x/d_x = 6(2)/24 = 0.5$; $6e_y/d_y = 6(1)/18 = 0.33$. Then $f_A = 231(1 + 0.5 + 0.33) = 424$ lb/in^2 (2923.4 kPa); $f_B = 231(1 + 0.5 - 0.33) = 270$ lb/in^2 (1861.5 kPa); $f_C = 231(1 - 0.5 + 0.33) = 192$ lb/in^2 (1323.8 kPa); $f_D = 231(1 - 0.5 - 0.33) = 38$ lb/in^2 (262.0 kPa). These results verify those computed in step 4.

FLEXURAL STRESS IN A CURVED MEMBER

The ring in Fig. 33 has an internal diameter of 12 in (304.8 mm) and a circular cross section of 4-in (101.6-mm) diameter. Determine the normal stress at A and at B (Fig. 33).

Calculation Procedure:

1. Determine the geometrical properties of the cross section

The area of the cross section is $A = 0.7854(4)^2 = 12.56$ in^2 (81.037 cm^2); the section modulus is $S = 0.7854(2)^3 = 6.28$ in^3 (102.92 cm^3). With $c = 2$ in (50.8 mm), the radius of curvature to the centroidal axis of this section is $R = 6 + 2 = 8$ in (203.2 mm).

2. Compute the R/c ratio and determine the correction factors

Refer to a table of correction factors for curved flexural members, such as Roark—*Formulas for Stress and Strain*, and extract the correction factors at the inner and outer surface associated with the R/c ratio. Thus $R/c = 8/2 = 4$; $k_i = 1.23$; $k_o = 0.84$.

3. Determine the normal stress

Find the normal stress at A and B caused by an equivalent axial load and moment. Thus $f_A = P/A + k_i(M/S) = 9000/12.56 + 1.23(9000 \times 8)/6.28 = 14{,}820\text{-lb/in}^2$ (102,183.9-kPa) compression; $f_B = 9000/12.56 - 0.84(9000 \times 8)/6.28 = 8930\text{-lb/in}^2$ (61,572.3-kPa) tension.

SOIL PRESSURE UNDER DAM

A concrete gravity dam has the profile shown in Fig. 34. Determine the soil pressure at the toe and heel of the dam when the water surface is level with the top.

Calculation Procedure:

1. Resolve the dam into suitable elements

The soil prism underlying the dam may be regarded as a structural member subjected to simultaneous axial load and bending, the cross section of the member being identical with the bearing surface of the dam. Select a 1-ft (0.3-m) length of dam as representing the entire structure. The weight of the concrete is 150 lb/ft³ (23.56 kN/m³).

Resolve the dam into the elements AED and $EBCD$. Compute the weight of each element, and locate the resultant of the weight with respect to the toe. Thus $W_1 = \frac{1}{2}(12)(20)(150) = 18{,}000$ lb (80.06 kN); $W_2 = 3(20)(150) = 9000$ lb (40.03 kN); $\Sigma W = 18{,}000 + 9000 = 27{,}000$ lb (120.10 kN). Then $x_1 = (\frac{2}{3})(12) = 8.0$ ft (2.44 m); $x_2 = 12 + 1.5 = 13.5$ ft (4.11 m).

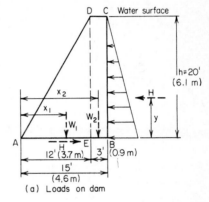

(a) Loads on dam

2. Find the magnitude and location of the resultant of the hydrostatic pressure

Calling the resultant $H = \frac{1}{2}wh^2 = \frac{1}{2}(62.4)(20)^2 = 12{,}480$ lb (55.51 kN), where w = weight of water, lb/ft³ (N/m³), and h = water height, ft (m), then $y = (\frac{1}{3})(20) = 6.67$ ft (2.03 m).

3. Compute the moment of the loads with respect to the base centerline

Thus, $M = 18{,}000(8 - 7.5) + 9000(13.5 - 7.5) - 12{,}480(6.67) = 20{,}200$ lb·ft (27,391 N·m) counterclockwise.

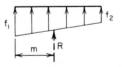

(b) Soil pressure under dam

4. Compute the section modulus of the base

Use the relation $S = (\frac{1}{6})bd^2 = (\frac{1}{6})(1)(15)^2 = 37.5$ ft³ (1.06 m³).

FIG. 34

5. Determine the soil pressure at the dam toe and heel

Compute the soil pressure caused by the combined axial load and bending. Thus $f_1 = \Sigma W/A + M/S = 27{,}000/15 + 20{,}200/37.5 = 2339$ lb/ft² (111.99 kPa); $f_2 = 1800 - 539 = 1261$ lb/ft² (60.37 kPa).

6. Verify the computed results

Locate the resultant R of the trapezoidal pressure prism, and take its moment with respect to the centerline of the base. Thus $R = 27{,}000$ lb (120.10 kN); $m = (15/3)[(2 \times 1261 + 2339)/(1261 + 2339)] = 6.75$ ft (2.05 m); $M_R = 27{,}000(7.50 - 6.75) = 20{,}200$ lb·ft (27,391 N·m). Since the applied and resisting moments are numerically equal, the computed results are correct.

LOAD DISTRIBUTION IN PILE GROUP

A continuous wall is founded on three rows of piles spaced 3 ft (0.91 m) apart. The longitudinal pile spacing is 4 ft (1.21 m) in the front and center rows and 6 ft (1.82 m) in the rear row. The resultant of vertical loads on the wall is 20,000 lb/lin ft (291.87 kN/m) and lies 3 ft 3 in (99.06 cm) from the front row. Determine the pile load in each row.

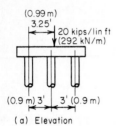

(0.99 m)
3.25'
20 kips/lin ft
(292 kN/m)

(0.9 m) 3'   3' (0.9 m)

(a) Elevation

(b) Plan

(c) Pile reactions

FIG. 35

Calculation Procedure:

1. Identify the "repeating group" of piles

The concrete footing (Fig. 35a) binds the piles, causing the surface along the top of the piles to remain a plane as bending occurs. Therefore, the pile group may be regarded as a structural member subjected to axial load and bending, the cross section of the member being the aggregate of the cross sections of the piles.

Indicate the "repeating group" as shown in Fig. 35b.

2. Determine the area of the pile group and the moment of inertia

Calculate the area of the pile group, locate its centroidal axis, and find the moment of inertia. Since all the piles have the same area, set the area of a single pile equal to unity. Then $A = 3 + 3 + 2 = 8$.

Take moments with respect to row A. Thus $8x = 3(0) + 3(3) + 2(6)$; $x = 2.625$ ft (66.675 mm). Then $I = 3(2.625)^2 + 3(0.375)^2 + 2(3.375)^2 = 43.9$.

3. Compute the axial load and bending moment on the pile group

The axial load $P = 20,000(12) = 240,000$ lb (1067.5 kN); then $M = 240,000(3.25 - 2.625) = 150,000$ lb·ft (203.4 kN·m).

4. Determine the pile load in each row

Find the pile load in each row resulting from the combined axial load and moment. Thus, $P/A = 240,000/8 = 30,000$ lb (133.4 kN) per pile; then $M/I = 150,000/43.9 = 3420$. Also, $p_a = 30,000 - 3420(2.625) = 21,020$ lb (93.50 kN) per pile; $p_b = 30,000 + 3420(0.375) = 31,280$ lb (139.13 kN) per pile; $p_c = 30,000 + 3420(3.375) = 41,540$ lb (184.76 kN) per pile.

5. Verify the above results

Compute the total pile reaction, the moment of the applied load, and the pile reaction with respect to row A. Thus, $R = 3(21,020) + 3(31,280) + 2(41,540) = 239,980$ lb (1067.43 kN); then $M_a = 240,000(3.25) = 780,000$ lb·ft (1057.68 kN·m), and $M_r = 3(31,280)(3) + 2(41,540)(6) = 780,000$ lb·ft (1057.68 kN·m). Since $M_a = M_r$, the computed results are verified.

Deflection of Beams

In this handbook the slope of the elastic curve at a given section of a beam is denoted by θ, and the deflection, in inches, by y. The slope is considered positive if the section rotates in a clockwise direction under the bending loads. A downward deflection is considered positive. In all instances, the beam is understood to be prismatic, if nothing is stated to the contrary.

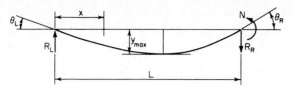

FIG. 36 Deflection of simple beam under end moment.

DOUBLE-INTEGRATION METHOD OF DETERMINING BEAM DEFLECTION

The simply supported beam in Fig. 36 is subjected to a counterclockwise moment N applied at the right-hand support. Determine the slope of the elastic curve at each support and the maximum deflection of the beam.

Calculation Procedure:

1. Evaluate the bending moment at a given section

Make this evaluation in terms of the distance x from the left-hand support to this section. Thus $R_L = N/L$; $M = Nx/L$.

2. Write the differential equation of the elastic curve; integrate twice

Thus $EI\ d^2y/dx^2 = -M = -Nx/L$; $EI\ dy/dx = EI\theta = -Nx^2/(2L) + c_1$; $EIy = -Nx^3/(6L) + c_1x + c_2$.

3. Evaluate the constants of integration

Apply the following boundary conditions: When $x = 0$, $y = 0$; $\therefore c_2 = 0$; when $x = L$, $y = 0$; $\therefore c_1 = NL/6$.

4. Write the slope and deflection equations

Substitute the constant values found in step 3 in the equations developed in step 2. Thus $\theta = [N/(6EIL)](L^2 - 3x^2)$; $y = [Nx/(6EIL)](L^2 - x^2)$.

5. Find the slope at the supports

Substitute the values $x = 0$, $x = L$ in the slope equation to determine the slope at the supports. Thus $\theta_L = NL/(6EI)$; $\theta_R = -NL/(3EI)$.

6. Solve for the section of maximum deflection

Set $\theta = 0$ and solve for x to locate the section of maximum deflection. Thus $L^2 - 3x^2 = 0$; $x = L/3^{0.5}$. Substituting in the deflection equation gives $y_{max} = NL^2/(9EI3^{0.5})$.

MOMENT-AREA METHOD OF DETERMINING BEAM DEFLECTION

Use the moment-area method to determine the slope of the elastic curve at each support and the maximum deflection of the beam shown in Fig. 36.

Calculation Procedure:

1. Sketch the elastic curve of the member and draw the M/(EI) diagram

Let A and B denote two points on the elastic curve of a beam. The moment-area method is based on the following theorems:

The difference between the slope at A and that at B is numerically equal to the area of the $M/(EI)$ diagram within the interval AB.

The deviation of A from a tangent to the elastic curve through B is numerically equal to the static moment of the area of the $M/(EI)$ diagram within the interval AB with respect to A. This tangential deviation is measured normal to the unstrained position of the beam.

Draw the elastic curve and the $M/(EI)$ diagram as shown in Fig. 37.

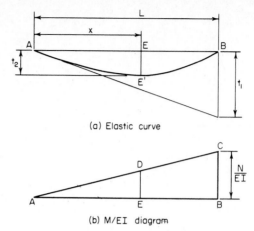

(a) Elastic curve

(b) M/EI diagram

FIG. 37

2. Calculate the deviation t_1 of B from the tangent through A

Thus, t_1 = moment of $\triangle ABC$ about $BC = [NL/(2EI)](L/3) = NL^2/(6EI)$. Also, $\theta_L = t_1/L = NL/(6EI)$.

3. Determine the right-hand slope in an analogous manner

4. Compute the distance to the section where the slope is zero

Area $\triangle AED$ = area $\triangle ABC(x/L)^2 = Nx^2/(2EIL)$; $\theta_E = \theta_L$ − area $\triangle AED = NL/(6EI) - Nx^2/(2EIL) = 0$; $x = L/3^{0.5}$.

5. Evaluate the maximum deflection

Evaluate y_{max} by calculating the deviation t_2 of A from the tangent through E' (Fig. 37). Thus area $\triangle AED = \theta_L = NL/(6EI)$; $y_{max} = t_2 = [NL/(6EI)](2x/3) = [NL/(6EI)][(2L/(3 \times 3^{0.5})] = NL^2/(9EI3^{0.5})$, as before.

CONJUGATE-BEAM METHOD OF DETERMINING BEAM DEFLECTION

The overhanging beam in Fig. 38 is loaded in the manner shown. Compute the deflection at C.

Calculation Procedure:

1. Assign supports to the conjugate beam

If a conjugate beam of identical span as the given beam is loaded with the $M/(EI)$ diagram of the latter, the shear V' and bending moment M' of the conjugate beam are equal, respectively, to the slope θ and deflection y at the corresponding section of the given beam.

Assign supports to the conjugate beam that are compatible with the end conditions of the given beam. At A, the given beam has a specific slope but zero deflection. Correspondingly, the conjugate beam has a specific shear but zero moment; i.e., it is simply supported at A.

At C, the given beam has both a specific slope and a specific deflection. Correspondingly, the conjugate beam has both a shear and a bending moment; i.e., it has a fixed support at C.

2. Construct the M/(EI) diagram of the given beam

Load the conjugate beam with this area. The moment at B is $-wd^2/2$; the moment varies linearly from A to B and parabolically from C to B.

3. Compute the resultant of the load in selected intervals

Compute the resultant W_1' of the load in interval AB and the resultant W_2' of the load in the interval BC. Locate these resultants. (Refer to the AISC *Manual* for properties of the complement of a half parabola.) Then $W_1' = (L/2)[wd^2/(2EI)] = wd^2L/(4EI)$; $x_1 = \frac{2}{3}L$; $W_2' = (d/3)[wd^2/(2EI)] = wd^3/(6EI)$; $x_2 = \frac{3}{4}d$.

4. Evaluate the conjugate-beam reaction

Since the given beam has zero deflection at B, the conjugate beam has zero moment at this section. Evaluate the reaction R_L' accordingly. Thus $M_B' = -R_L'L + W_1'L/3 = 0$; $R_L' = W_1'/3 = wd^2L/(12EI)$.

5. Determine the deflection

Determine the deflection at C by computing M_c'. Thus $y_c = M_c' = -R_L'(L + d) + W_1'(d + L/3) + W_2'(3d/4) = wd^3(4L + 3d)/(24EI)$.

UNIT-LOAD METHOD OF COMPUTING BEAM DEFLECTION

The cantilever beam in Fig. 39a carries a load that varies uniformly from w lb/lin ft at the free end to zero at the fixed end. Determine the slope and deflection of the elastic curve at the free end.

Calculation Procedure:

1. Apply a unit moment to the beam

Apply a counterclockwise unit moment at A (Fig. 39b). (This direction is selected because it is known that the end section rotates in this manner.) Let x = distance from A to given section; w_x = load intensity at the given section; M and m = bending moment at the given section induced by the actual load and by the unit moment, respectively.

2. Evaluate the moments in step 1

Evaluate M and m. By proportion, $w_x = w(L - x)/L$; $M = -(x^2/6)(2w + w_x) = -(wx^2/6)[2 + (L - x)/L] = -wx^2(3L - x)/(6L)$; $m = -1$.

3. Apply a suitable slope equation

Use the equation $\theta_A = \int_0^L [Mm/(EI)] \, dx$. Then $EI\theta_A = \int_0^L [wx^2(3L - x)/(6L)] \, dx = [w/(6L)] \int_0^L (3Lx^2 - x^3) \, dx = [w/(6L)](3Lx^3/3 - x^4/4)|_0^L = [w/(6L)](L^4 - L^4/4)$; thus, $\theta_A = \frac{1}{8}wL^3/(EI)$ counterclockwise. This is the slope at A.

4. Apply a unit load to the beam

Apply a unit downward load at A as shown in Fig. 39c. Let m' denote the bending moment at a given section induced by the unit load.

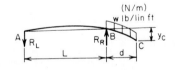

(a) Actual load on beam

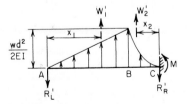

(a) Force diagram of given beam

(b) Force diagram of conjugate beam

FIG. 38 Deflection of overhanging beam.

(b) Superimposed moment to find θ_A

(c) Superimposed load to find y_A

FIG. 39

5. Evaluate the bending moment induced by the unit load; find the deflection

Apply $y_A = \int_0^L [Mm'/(EI)]\, dx$. Then $m' = -x$; $EIy_A = \int_0^L [wx^3(3L - x)/(6L)]\, dx = [w/(6L)]\int_0^L x^3(3L - x)\, dx$; $y_A = (11/120)wL^4/(EI)$.

The first equation in step 3 is a statement of the work performed by the unit moment at A as the beam deflects under the applied load. The left-hand side of this equation expresses the external work, and the right-hand side expresses the internal work. These work equations constitute a simple proof of Maxwell's theorem of reciprocal deflections, which is presented in a later calculation procedure.

DEFLECTION OF A CANTILEVER FRAME

The prismatic rigid frame $ABCD$ (Fig. 40a) carries a vertical load P at the free end. Determine the horizontal displacement of A by means of both the unit-load method and the moment-area method.

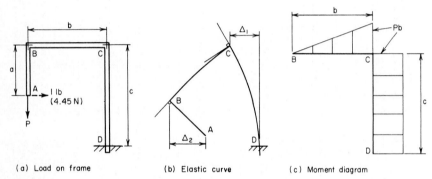

(a) Load on frame   (b) Elastic curve   (c) Moment diagram

FIG. 40

Calculation Procedure:

1. Apply a unit horizontal load

Apply the unit horizontal load at A, directed to the right.

2. Evaluate the bending moments in each member

Let M and m denote the bending moment at a given section caused by the load P and by the unit load, respectively. Evaluate these moments in each member, considering a moment positive if it induces tension in the outer fibers of the frame. Thus:

Member AB: Let x denote the vertical distance from A to a given section. Then $M = 0$; $m = x$.

Member BC: Let x denote the horizontal distance from B to a given section. Then $M = Px$; $m = a$.

Member CD: Let x denote the vertical distance from C to a given section. Then $M = Pb$; $m = a - x$.

3. Evaluate the required deflection

Calling the required deflection Δ, we apply $\Delta = \int [Mm/(EI)]\, dx$; $EI\Delta = \int_0^b Pax\, dx + \int_0^c Pb(a - x)\, dx = Pax^2/2]_0^b + Pb(ax - x^2/2)]_0^c = Pab^2/2 + Pabc - Pbc^2/2$; $\Delta = [Pb/(2EI)](ab + 2ac - c^2)$.

If this value is positive, A is displaced in the direction of the unit load, i.e., to the right. Draw the elastic curve in hyperbolic fashion (Fig. 40b). The above three steps constitute the unit-load method of solving this problem.

4. Construct the bending-moment diagram

Draw the diagram as shown in Fig. 40c.

5. Compute the rotation and horizontal displacement by the moment-area method

Determine the rotation and horizontal displacement of C. (Consider only absolute values.) Since there is no rotation at D, $EI\theta_C = Pbc$; $EI\Delta_1 = Pbc^2/2$.

6. Compute the rotation of one point relative to another and the total rotation

Thus $EI\theta_{BC} = Pb^2/2$; $EI\theta_B = Pbc + Pb^2/2 = Pb(c + b/2)$. The horizontal displacement of B relative to C is infinitesimal.

7. Compute the horizontal displacement of one point relative to another

Thus, $EI\Delta_2 = EI\theta_B a = Pb(ac + ab/2)$.

8. Combine the computed displacements to obtain the absolute displacement

Thus $EI\Delta = EI(\Delta_2 - \Delta_1) = Pb(ac + ab/2 - c^2/2)$; $\Delta = [Pb/(2EI)](2ac + ab - c^2)$.

Statically Indeterminate Structures

A structure is said to be *statically determinate* if its reactions and internal forces may be evaluated by applying solely the equations of equilibrium and *statically indeterminante* if such is not the case. The analysis of an indeterminate structure is performed by combining the equations of equilibrium with the known characteristics of the deformation of the structure.

SHEAR AND BENDING MOMENT OF A BEAM ON A YIELDING SUPPORT

The beam in Fig. 41a has an EI value of 35×10^9 lb·in^2 (100,429 kN·m^2) and bears on a spring at B that has a constant of 100 kips/in (175,126.8 kN/m); i.e., a force of 100 kips (444.8 kN) will compress the spring 1 in (25.4 mm). Neglecting the weight of the member, construct the shear and bending-moment diagrams.

Calculation Procedure:

1. Draw the free-body diagram of the beam

Draw the diagram in Fig. 41b. Consider this as a simply supported member carrying a 50-kip (222.4-kN) load at D and an upward load R_B at its center.

2. Evaluate the deflection

Evaluate the deflection at B by applying the equations presented for cases 7 and 8 in the AISC *Manual*. With respect to the 50-kip (222.4-kN) load, $b = 7$ ft (2.1 m) and $x = 14$ ft (4.3 m). If y is in inches and R_B is in pounds, $y = 50,000(7)(14)(28^2 - 7^2 - 14^2)1728/[6(35)(10)^9 28] - R_B(28)^3 1728/[48(35)(10)^9] = 0.776 - (2.26/10^5)R_B$.

3. Express the deflection in terms of the spring constant

The deflection at B is, by proportion, $y/1 = R_B/100,000$; $y = R_B/100,000$.

4. Equate the two deflection expressions, and solve for the upward load

Thus $R_B/10^5 = 0.776 - (2.26/10^5)R_B$; $R_B = 0.776(10)^5/3.26 = 23,800$ lb (105,862.4 N).

5. Calculate the reactions R_A and R_C by taking moments

We have $\Sigma M_C = 28R_A - 50,000(21) + 23,800(14) = 0$; $R_A = 25,600$ lb (113,868.8 N); $\Sigma M_A = 50,000(7) - 23,800(14) - 28R_C = 0$; $R_C = 600$ lb (2668.8 N).

6. Construct the shear and moment diagrams

Construct these diagrams as shown in Fig. 41. Then $M_D = 7(25,600) = 179,200$ lb·ft (242,960 N·m); $M_B = 179,200 - 7(24,400) = 8400$ lb·ft (11,390.4 N·m).

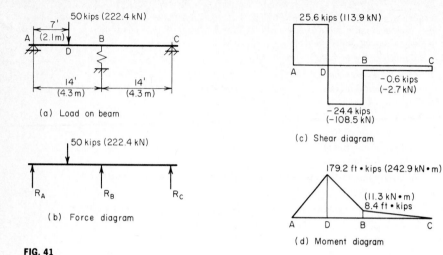

(a) Load on beam

(b) Force diagram

(c) Shear diagram

(d) Moment diagram

FIG. 41

MAXIMUM BENDING STRESS IN BEAMS JOINTLY SUPPORTING A LOAD

In Fig. 42a, a 16WF40 (W16 × 40) beam and a 12WF31 (W12 × 31) beam cross each other at the vertical line V, the bottom of the 16-in (406.4-mm) beam being ⅜ in (9.53 mm) above the top of the 12-in (304.8-mm) beam before the load is applied. Both members are simply supported. A column bearing on the 16-in (406.4-mm) beam transmits a load of 15 kips (66.72 kN) at the indicated location. Compute the maximum bending stress in the 12-in (304.8-mm) beam.

Calculation Procedure:

1. Determine whether the upper beam engages the lower beam

To ascertain whether the upper beam engages the lower one as it deflects under the 15-kip (66.72-kN) load, compute the deflection of the 16-in (406.4-mm) beam at V if the 12-in (304.8-mm)

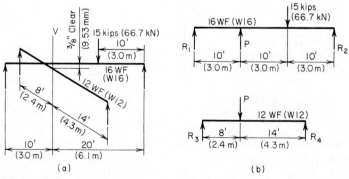

FIG. 42 Load carried by two beams.

beam were absent. This distance is 0.74 in (18.80 mm). Consequently, the gap between the members is closed, and the two beams share the load.

2. Draw a free-body diagram of each member

Let P denote the load transmitted to the 12-in (304.8-mm) beam by the 16-in (406.4-mm) beam [or the reaction of the 12-in (304.8-mm) beam on the 16-in (406.4-mm) beam]. Draw, in Fig. 42b, a free-body diagram of each member.

3. Evaluate the deflection of the beams

Evaluate, in terms of P, the deflections y_{12} and y_{16} of the 12-in (304.8-mm) and 16-in (406.4-mm) beams, respectively, at line V.

4. Express the relationship between the two deflections

Thus, $y_{12} = y_{16} - 0.375$.

5. Replace the deflections in step 4 with their values as obtained in step 3

After substituting these deflections, solve for P.

6. Compute the reactions of the lower beam

Once the reactions of the lower beam are computed, obtain the maximum bending moment. Then compute the corresponding flexural stress.

THEOREM OF THREE MOMENTS

For the two-span beam in Fig. 43a, compute the reactions at the supports. Apply the theorem of three moments to arrive at the results.

Calculation Procedure:

1. Using the bending-moment equation, determine M_B

Figure 43b represents a general case. For a prismatic beam, the bending moments at the three successive supports are related by $M_1L_1 + 2M_2(L_1 + L_2) + M_3L_2 = -\frac{1}{4}w_1L_1^3 - \frac{1}{4}w_2L_2^3 -$

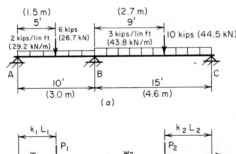

(a)

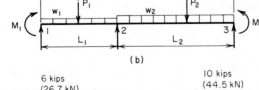

(b)

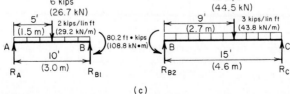

(c)

FIG. 43

$P_1L_1^2(k_1 - k_1^3) - P_2L_2^2(k_2 - k_2^3)$. Substituting in this equation gives $M_1 = M_3 = 0$; $L_1 = 10$ ft (3.0 m); $L_2 = 15$ ft (4.6 m); $w_1 = 2$ kips/lin ft (29.2 kN/m); $w_2 = 3$ kips/lin ft (43.8 kN/m); $P_1 = 6$ kips (26.7 N); $P_2 = 10$ kips (44.5 N); $k_1 = 0.5$; $k_2 = 0.4$; $2M_B(10 + 15) = -\frac{1}{4}(2)(10)^3 - \frac{1}{4}(3)(15)^3 - 6(10)^2(0.5 - 0.125) - 10(15)^2(0.4 - 0.064)$; $M_B = -80.2$ ft·kips $(-108.8$ kN·m).

2. Draw a free-body diagram of each span

Figure 43c shows the free-body diagrams.

3. Take moments with respect to each support to find the reactions

Span AB: $\Sigma M_A = 6(5) + 2(10)(5) + 80.2 - 10R_{B1} = 0$; $R_{B1} = 21.02$ kips (93.496 kN); $\Sigma M_B = 10R_A - 6(5) - 2(10)(5) + 80.2 = 0$; $R_A = 4.98$ kips (22.151 kN).

Span BC: $\Sigma M_B = -80.2 + 10(9) + 3(15)(7.5) - 15R_C = 0$; $R_C = 23.15$ kips (102.971 kN); $\Sigma M_C = 15R_{B2} - 80.2 - 10(6) - 3(15)(7.5) = 0$; $R_{B2} = 31.85$ kips (144.668 kN); $R_B = 21.02 + 31.85 = 52.87$ kips (235.165 kN).

THEOREM OF THREE MOMENTS: BEAM WITH OVERHANG AND FIXED END

Determine the reactions at the supports of the continuous beam in Fig. 44a. Use the theorem of three moments.

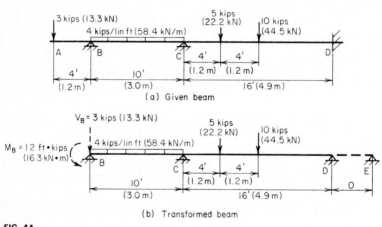

FIG. 44

Calculation Procedure:

1. Transform the given beam to one amenable to analysis by the theorem of three moments

Perform the following operations to transform the beam:

a. Remove the span AB, and introduce the shear V_B and moment M_B that the load on AB induces at B, as shown in Fig. 44b.

b. Remove the fixed support at D and add the span DE of zero length, with a hinged support at E.

For the interval BD, the transformed beam is then identical in every respect with the actual beam.

2. Apply the equation for the theorem of three moments

Consider span BC as span 1 and CD as span 2. For the 5-kip (22.2-kN) load, $k_2 = 12/16 = 0.75$; for the 10-kip (44.5-kN) load, $k_2 = 8/16 = 0.5$. Then $-12(10) + 2M_C(10 + 16) + 16M_D =$

$-\frac{1}{4}(4)(10)^3 - 5(16)^2(0.75 - 0.422) - 10(16)^2(0.5 - 0.125)$. Simplifying gives $13M_C + 4M_D = -565.0$, Eq. a.

3. Apply the moment equation again

Considering CD as span 1 and DE as span 2, apply the moment equation again. Or, for the 5-kip (22.2-kN) load, $k_1 = 0.25$; for the 10-kip (44.5-kN) load, $k_1 = 0.5$. Then $16M_C + 2M_D(16 + 0) = -5(16)^2(0.25 - 0.016) - 10(16)^2(0.50 - 0.125)$. Simplifying yields $M_C + 2M_D = -78.7$, Eq. b.

4. Solve the moment equations

Solving Eqs. a and b gives $M_C = -37.1$ ft·kips (-50.30 kN·m); $M_D = -20.8$ ft·kips (-28.20 kN·m).

5. Determine the reactions by using a free-body diagram

Find the reactions by drawing a free-body diagram of each span and taking moments with respect to each support. Thus $R_B = 20.5$ kips (91.18 kN); $R_C = 32.3$ kips (143.67 kN); $R_D = 5.2$ kips (23.12 kN).

BENDING-MOMENT DETERMINATION BY MOMENT DISTRIBUTION

Using moment distribution, determine the bending moments at the supports of the member in Fig. 45. The beams are rigidly joined at the supports and are composed of the same material.

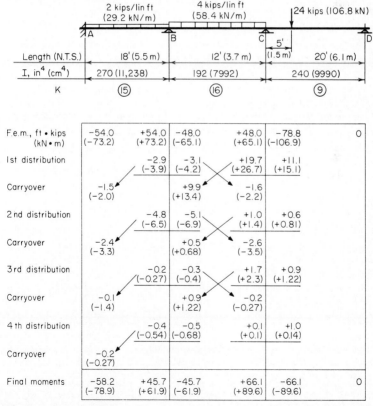

FIG. 45 Moment distribution.

Calculation Procedure:

1. Calculate the flexural stiffness of each span

Using K to denote the flexural stiffness, we see that $K = I/L$ if the far end remains fixed during moment distribution; $K = 0.75I/L$ if the far end remains hinged during moment distribution. Then $K_{AB} = 270/18 = 15$; $K_{BC} = 192/12 = 16$; $K_{CD} = 0.75(240/20) = 9$. Record all the values on the drawing as they are obtained.

2. For each span, calculate the required fixed-end moments at those supports that will be considered fixed

These are the *external* moments with respect to the span; a clockwise moment is considered positive. (For additional data, refer to cases 14 and 15 in the AISC *Manual.*) Then $M_{AB} = -wL^2/12 = -2(18)^2/12 = -54.0$ ft·kips (-73.2 kN·m); $M_{BA} = +54.0$ ft·kips (73.22 kN·m). Similarly, $M_{BC} = -48.0$ ft·kips (-65.1 kN·m); $M_{CB} = +48.0$ ft·kips (65.1 kN·m); $M_{CD} = -24(15)(5)(15 + 20)/[2(20)^2] = -78.8$ ft·kips (-106.85 kN·m).

3. Calculate the unbalanced moments

Computing the unbalanced moments at B and C yields the following: At B, $+54.0 - 48.0 = +6.0$ ft·kips (8.14 kN·m); at C, $+48.0 - 78.8 = -30.8$ ft·kips (-41.76 kN·m).

4. Apply balancing moments; distribute them in proportion to the stiffness of the adjoining spans

Apply the balancing moments at B and C, and distribute them to the two adjoining spans in proportion to their stiffness. Thus $M_{BA} = -6.0(15/31) = -2.9$ ft·kips (-3.93 kN·m); $M_{BC} = -6.0(16/31) = -3.1$ ft·kips (-4.20 kN·m); $M_{CB} = +30.8(16/25) = +19.7$ ft·kips (26.71 kN·m); $M_{CD} = +30.8(9/25) = +11.1$ ft·kips (15.05 kN·m).

5. Perform the "carry-over" operation for each span

To do this, take one-half the distributed moment applied at one end of the span, and add this to the moment at the far end if that end is considered to be fixed during moment distribution.

6. Perform the second cycle of moment balancing and distribution

Thus $M_{BA} = -9.9(15/31) = -4.8$; $M_{BC} = -9.9(16/31) = -5.1$; $M_{CB} = +1.6(16/25) = +1.0$; $M_{CD} = +1.6(9/25) = +0.6$.

7. Continue the foregoing procedure until the carry-over moments become negligible

Total the results to obtain the following bending moments: $M_A = -58.2$ ft·kips (-78.91 kN·m); $M_B = -45.7$ ft·kips (-61.96 kN·m); $M_C = -66.1$ ft·kips (-89.63 kN·m).

ANALYSIS OF A STATICALLY INDETERMINATE TRUSS

Determine the internal forces of the truss in Fig. 46a. The cross-sectional areas of the members are given in Table 5.

Calculation Procedure:

1. Test the structure for static determinateness

Apply the following criterion. Let j = number of joints; m = number of members; r = number of reactions. Then if $2j = m + r$, the truss is statically determinate; if $2j < m + r$, the truss is statically indeterminate and the deficiency represents the degree of indeterminateness.

In this truss, $j = 6$, $m = 10$, $r = 3$, consisting of a vertical reaction at A and D and a horizontal reaction at D. Thus $2j = 12$; $m + r = 13$. The truss is therefore statically indeterminate to the first degree; i.e., there is *one* redundant member.

The method of analysis comprises the following steps: Assume a value for the internal force in a particular member, and calculate the relative displacement Δ_i of the two ends of that member caused solely by this force. Now remove this member to secure a determinate truss, and calculate the relative displacement Δ_a caused solely by the applied loads. The true internal force is of such magnitude that $\Delta_i = -\Delta_a$.

2. Assume a unit force for one member

Assume for convenience that the force in BF is 1-kip (4.45-kN) tension. Remove this member, and replace it with the assumed 1-kip (4.45-kN) force that it exerts at joints B and F, as shown in Fig. 46b.

3. Calculate the force induced in each member solely by the unit force

Calling the induced force U, produced solely by the unit tension in BF, record the results in Table 5, considering tensile forces as positive and compressive forces as negative.

4. Calculate the force induced in each member solely by the applied loads

With BF eliminated, calculate the force S induced in each member solely by the applied loads.

5. Evaluate the true force in the selected member

Use the relation $BF = -[\Sigma SUL/(AE)]/[\Sigma U^2 L/(AE)]$. The numerator represents Δ_a; the denominator represents Δ_i for a 1-kip (4.45-kN) tensile force in BF. Since E is constant, it cancels. Substituting the values in Table 5 gives $BF = -(-266.5/135.5) = 1.97$ kips (8.76 kN). The positive result confirms the assumption that BF is tensile.

6. Evaluate the true force in each member

Use the relation $S' = S + 1.97U$, where $S' =$ true force. The results are shown in Table 5.

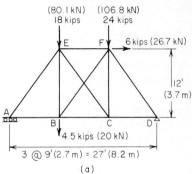

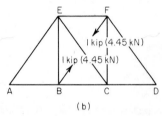

FIG. 46 Statically indeterminate truss.

Moving Loads and Influence Lines

ANALYSIS OF BEAM CARRYING MOVING CONCENTRATED LOADS

The loads shown in Fig. 47a traverse a beam of 40-ft (12.2-m) simple span while their spacing remains constant. Determine the maximum bending moment and maximum shear induced in the beam during transit of these loads. Disregard the weight of the beam.

Calculation Procedure:

1. Determine the magnitude of the resultant and its location

Since the member carries only concentrated loads, the maximum moment at any instant occurs under one of these loads. Thus, the problem is to determine the position of the load system that causes the *absolute* maximum moment.

The magnitude of the resultant R is $R = 10 + 4 + 15 = 29$ kips (129.0 kN). To determine the location of R, take moments with respect to A (Fig. 47). Thus $\Sigma M_A = 29AD = 4(5) + 15(17)$, or $AD = 9.48$ ft (2.890 m).

2. Assume several trial load positions

Assume that the maximum moment occurs under the 10-kip (44.5-kN) load. Place the system in the position shown in Fig. 47b, with the 10-kip (44.5-kN) load as far from the adjacent support as the resultant is from the other support. Repeat this procedure for the two remaining loads.

3. Determine the support reactions for the trial load positions

For these three trial positions, calculate the reaction at the support adjacent to the load under consideration. Determine whether the vertical shear is zero or changes sign at this load. Thus, for

TABLE 5 Forces in Truss Members (Fig. 46)

Member	A, in² (cm²)	L, in (mm)	U, kips (kN)	S, kips (kN)	U²L/A	SUL/A	S', kips (kN)
AB	5 (32.2)	108 (2,743.2)	0 (0)	+15.25 (+67.832)	0 (0)	0 (0)	+15.25 (+67.832)
BC	5 (32.2)	108 (2,743.2)	−0.60 (−2.668)	+15.25 (+67.832)	+7.8 (+615.54)	−197.6 (−15,417.78)	+14.07 (+62.583)
CD	5 (32.2)	108 (2,743.2)	0 (0)	+13.63 (+60.626)	0 (0)	0 (0)	+13.63 (+60.626)
EF	4 (25.8)	108 (2,743.2)	−0.60 (−2.668)	−13.63 (−60.626)	+9.7 (+756.84)	+220.8 (+17,198.18)	−14.81 (−65.874)
BE	4 (25.8)	144 (3,657.6)	−0.80 (−3.558)	+4.50 (+20.016)	+23.0 (+1,794.68)	−129.6 (−10,096.24)	+2.92 (+12.988)
CF	4 (25.8)	144 (3,657.6)	−0.80 (−3.558)	+2.17 (+9.952)	+23.0 (+1,794.68)	−62.5 (−4,868.55)	+0.59 (+2.624)
AE	6 (38.7)	180 (4,572.0)	0 (0)	−25.42 (−113.068)	0 (0)	0 (0)	−25.42 (−113.068)
BF	5 (32.2)	180 (4,572.0)	+1.00 (+4.448)	0 (0)	+36.0 (+2,809.18)	0 (0)	+1.97 (+8.762)
CE	5 (32.2)	180 (4,572.0)	+1.00 (+4.448)	−2.71 (−9.652)	+36.0 (+2,809.18)	−97.6 (−6,095.82)	−0.74 (−3.291)
DF	6 (38.7)	180 (4,572.0)	0 (0)	−32.71 (−145.494)	0 (0)	0 (0)	−32.71 (−145.494)
Total					+135.5 (+10,580.1)	−266.5 (−19,280.2)	

position 1: $R_L = 29(15.26)/40 = 11.06$ kips (49.194 kN). Since the shear does not change sign at the 10-kip (44.5-kN) load, this position lacks significance.

Position 2: $R_L = 29(17.76)/40 = 12.88$ kips (57.290 kN). The shear changes sign at the 4-kip (17.8 kN) load.

Position 3: $R_R = 29(16.24)/40 = 11.77$ kips (52.352 kN). The shear changes sign at the 15-kip (66.7-kN) load.

4. Compute the maximum bending moment associated with positions having a change in the shear sign

This applies to positions 2 and 3. The absolute maximum moment is the larger of these values. Thus, for position 2: $M = 12.88(17.76) − 10(5) = 178.7$ ft·kips (242.32 kN·m). Position 3: $M = 11.77(16.24) = 191.1$ ft·kips (259.13 kN·m). Thus, $M_{max} = 191.1$ ft·kips (259.13 kN·m).

5. Determine the absolute maximum shear

For absolute maximum shear, place the 15-kip (66.7-kN) load an infinitesimal distance to the left of the right-hand support. Then $V_{max} = 29(40 − 7.52)/40 = 23.5$ kips (104.53 kN).

When the load spacing is large in relation to the beam span, the absolute maximum moment may occur when only part of the load system is on the span. This possibility requires careful investigation.

INFLUENCE LINE FOR SHEAR IN A BRIDGE TRUSS

The Pratt truss in Fig. 48a supports a bridge at its bottom chord. Draw the influence line for shear in panel cd caused by a moving load traversing the bridge floor.

Calculation Procedure:

1. Compute the shear in the panel being considered with a unit load to the right of the panel

Cut the truss at section YY. The algebraic sum of vertical forces acting on the truss at panel points to the left of YY is termed the *shear* in panel cd.

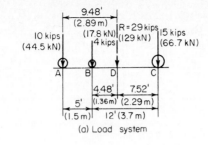

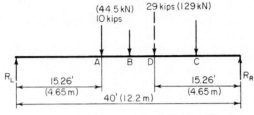

(a) Load system

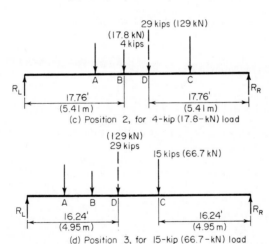

(b) Position 1, for 10-kip (44.5-kN) load

(c) Position 2, for 4-kip (17.8-kN) load

(d) Position 3, for 15-kip (66.7-kN) load

FIG. 47

Consider that a moving load traverses the bridge floor from right to left and that the portion of the load carried by the given truss is 1 kip (4.45 kN). This unit load is transmitted to the truss as concentrated loads at two adjacent bottom-chord panel points, the latter being components of the unit load. Let x denote the instantaneous distance from the right-hand support to the moving load.

Place the unit load to the right of d, as shown in Fig. 48b, and compute the shear V_{cd} in panel cd. The truss reactions may be obtained by considering the unit load itself rather than its panel-point components. Thus: $R_L = x/120$; $V_{cd} = R_L = x/120$, Eq. a.

2. Compute the panel shear with the unit load to the left of the panel considered

Placing the unit load to the left of c yields $V_{cd} = R_L - 1 = x/120 - 1$, Eq. b.

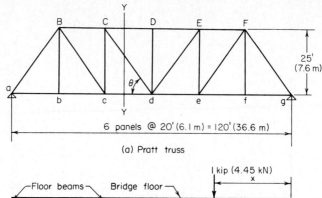

(a) Pratt truss

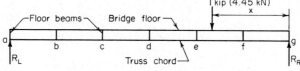

(b) Transmission of load through floor beams

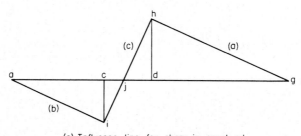

(c) Influence line for shear in panel cd

FIG. 48

3. Determine the panel shear with the unit load within the panel

Place the unit load within panel cd. Determine the panel-point load P_c at c, and compute V_{cd}. Thus $P_c = (x - 60)/20 = x/20 - 3$; $V_{cd} = R_L - P_c = x/120 - (x/20 - 3) = -x/24 + 3$, Eq. c.

4. Construct a diagram representing the shear associated with every position of the unit load

Apply the foregoing equations to represent the value of V_{cd} associated with every position of the unit load. This diagram, Fig. 48c, is termed an *influence line*. The point j at which this line intersects the base is referred to as the *neutral point*.

5. Compute the slope of each segment of the influence line

Line a, $dV_{cd}/dx = 1/120$; line b, $dV_{cd}/dx = 1/120$; line c, $dV_{cd}/dx = -1/24$. Lines a and b are therefore parallel because they have the same slope.

FORCE IN TRUSS DIAGONAL CAUSED BY A MOVING UNIFORM LOAD

The bridge floor in Fig. 48a carries a moving uniformly distributed load. The portion of the load transmitted to the given truss is 2.3 kips/lin ft (33.57 kN/m). Determine the limiting values of the force induced in member Cd by this load.

Calculation Procedure:

1. Locate the neutral point, and compute dh

The force in Cd is a function of V_{cd}. Locate the neutral point j in Fig. 48c and compute dh. From Eq. c of the previous calculation procedure, $V_{cd} = -jg/24 + 3 = 0$; $jg = 72$ ft (21.9 m). From Eq. a of the previous procedure, $dh = 60/120 = 0.5$.

2. Determine the maximum shear

To secure the maximum value of V_{cd}, apply uniform load continuously in the interval jg. Compute V_{cd} by multiplying the area under the influence line by the intensity of the applied load. Thus, $V_{cd} = \frac{1}{2}(72)(0.5)(2.3) = 41.4$ kips (184.15 kN).

3. Determine the maximum force in the member

Use the relation $Cd_{max} = V_{cd}(\csc \theta)$, where $\csc \theta = [(20^2 + 25^2)/25^2]^{0.5} = 1.28$. Then $Cd_{max} = 41.4(1.28) = 53.0$-kip (235.74-kN) tension.

4. Determine the minimum force in the member

To secure the minimum value of V_{cd}, apply uniform load continuously in the interval aj. Perform the final calculation by proportion. Thus, $Cd_{min}/Cd_{max} = $ area $aij/$area$jhg = -(2/3)^2 = -4/9$. Then $Cd_{min} = -(4/9)(53.0) = 23.6$-kip (104.97-kN) compression.

FORCE IN TRUSS DIAGONAL CAUSED BY MOVING CONCENTRATED LOADS

The truss in Fig. 49a supports a bridge that transmits the moving-load system shown in Fig. 49b to its bottom chord. Determine the maximum tensile force in De.

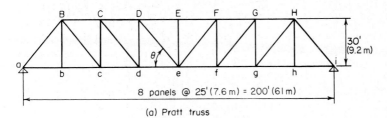

(a) Pratt truss

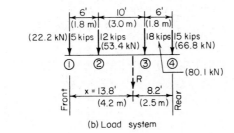

(b) Load system

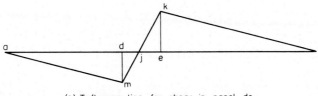

(c) Influence line for shear in panel de

FIG. 49

Calculation Procedure:

1. Locate the resultant of the load system

The force in De (Fig. 49) is a function of the shear in panel de. This shear is calculated without recourse to a set rule in order to show the principles involved in designing for moving loads.

To locate the resultant of the load system, take moments with respect to load 1. Thus, $R = 50$ kips (222.4 kN). Then $\Sigma M_1 = 12(6) + 18(16) + 15(22) = 50x$; $x = 13.8$ ft (4.21 m).

2. Construct the influence line for V_{de}

In Fig. 49c, draw the influence line for V_{de}. Assume right-to-left locomotion, and express the slope of each segment of the influence line. Thus slope of $ik =$ slope of $ma = 1/200$; slope of $km = -7/200$.

3. Assume a load position, and determine whether V_{de} increases or decreases

Consider that load 1 lies within panel de and the remaining loads lie to the right of this panel. From the slope of the influence line, ascertain whether V_{de} increases or decreases as the system is displaced to the left. Thus $dV_{de}/dx = 5(-7/200) + 45(1/200) > 0$; $\therefore$ V_{de} increases.

4. Repeat the foregoing calculation with other assumed load positions

Consider that loads 1 and 2 lie within the panel de and the remaining loads lie to the right of this panel. Repeat the foregoing calculation. Thus $dV_{de}/dx = 17(-7/200) + 33(1/200) < 0$; $\therefore$ V_{de} decreases.

From these results it is concluded that as the system moves from right to left, V_{de} is maximum at the instant that load 2 is at e.

5. Place the system in the position thus established, and compute V_{de}

Thus, $R_L = 50(100 + 6 - 13.8)/200 = 23.1$ kips (102.75 kN). The load at panel point d is $P_d = 5(6)/25 = 1.2$ kips (5.34 kN); $V_{de} = 23.1 - 1.2 = 21.9$ kips (97.41 kN).

6. Assume left-to-right locomotion; proceed as in step 3

Consider that load 4 is within panel de and the remaining loads are to the right of this panel. Proceeding as in step 3, we find $dV_{de}/dx = 15(7/200) + 35(-1/200) > 0$.

So, as the system moves from left to right, V_{de} is maximum at the instant that load 4 is at e.

7. Place the system in the position thus established, and compute V_{de}

Thus $V_{de} = R_L = [50(100 - 8.2)]/200 = 23.0$ kips (102.30 kN); $\therefore$ $V_{de,max} = 23.0$ kips (102.30 kN).

8. Compute the maximum tensile force in De

Using the same relation as in step 3 of the previous calculation procedure, we find $\csc \theta = [(25^2 + 30^2)/30^2]^{0.5} = 1.30$; then $De = 23.0(1.30) = 29.9$-kip (133.00-kN) tension.

INFLUENCE LINE FOR BENDING MOMENT IN BRIDGE TRUSS

The Warren truss in Fig. 50a supports a bridge at its top chord. Draw the influence line for the bending moment at b caused by a moving load traversing the bridge floor.

Calculation Procedure:

1. Place the unit load in position, and compute the bending moment

The moment of all forces acting on the truss at panel points to the left of b with respect to b is termed the *bending moment* at that point. Assume that the load transmitted to the given truss is 1 kip (4.45 kN), and let x denote the instantaneous distance from the right-hand support to the moving load.

Place the unit load to the right of C, and compute the bending moment M_b. Thus $R_L = x/120$; $M_b = 45R_L = 3x/8$, Eq. a.

2. Place the unit load on the other side and compute the bending moment

Placing the unit load to the left of B and computing M_b, $M_b = 45R_L - (x - 75) = -5x/8 + 75$, Eq. b.

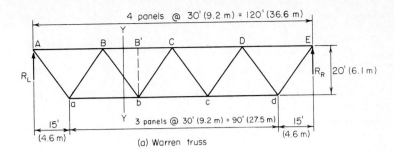

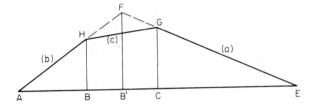

(a) Warren truss

(b) Influence line for bending moment at b

FIG. 50

3. Place the unit load within the panel; compute the panel-point load and bending moment

Place the unit load within panel BC. Determine the panel-point load P_B and compute M_b. Thus $P_B = (x - 60)/30 = x/30 - 2$; $M_b = 45R_L - 15P_B = 3x/8 - 15(x/30 - 2) = -x/8 + 30$, Eq. c.

4. Applying the foregoing equations, draw the influence line

Figure 50b shows the influence line for M_b. Computing the significant values yields $CG = (3/8)(60) = 22.50$ ft·kips (30.51 kN·m); $BH = -(5/8)(90) + 75 = 18.75$ ft·kips (25.425 kN·m).

5. Compute the slope of each segment of the influence line

This computation is made for subsequent reference. Thus, line a, $dM_b/dx = 3/8$; line b, $dM_b/dx = -5/8$; line c, $dM_b/dx = -1/8$.

FORCE IN TRUSS CHORD CAUSED BY MOVING CONCENTRATED LOADS

The truss in Fig. 50a carries the moving-load system shown in Fig. 51. Determine the maximum force induced in member BC during transit of the loads.

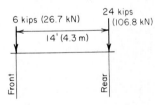

FIG. 51

Calculation Procedure:

1. Assume that locomotion proceeds from right to left, and compute the bending moment

The force in BC is a function of the bending moment M_b at b. Refer to the previous calculation procedure for the slope of each segment of the influence line. Study of these slopes shows that M_b increases as the load system moves until the rear load is at C, the front load being 14 ft (4.3 m) to the left of C. Calculate the value of M_b corresponding to this load disposition by applying the computed properties of the influence line. Thus, $M_b = 22.50(24) + (22.50 - 1/8 \times 14)(6) = 664.5$ ft·kips (901.06 kN·m).

2. Assume that locomotion proceeds from left to right, and compute the bending moment

Study shows that M_b increases as the system moves until the rear load is at C, the front load being 14 ft (4.3 m) to the right of C. Calculate the corresponding value of M_b. Thus, $M_b = 22.50(24) + (22.50 - 3/8 \times 14)(6) = 643.5$ ft·kips (872.59 kN·m). ∴ $M_{b,max} = 664.5$ ft·kips (901.06 kN·m).

3. Determine the maximum force in the member

Cut the truss at plane YY. Determine the maximum force in BC by considering the equilibrium of the left part of the structure. Thus, $\Sigma M_b = M_b - 20BC = 0$; $BC = 664.5/20 = 33.2$-kips (147.67-kN) compression.

INFLUENCE LINE FOR BENDING MOMENT IN THREE-HINGED ARCH

The arch in Fig. 52a is hinged at A, B, and C. Draw the influence line for bending moment at D, and locate the neutral point.

Calculation Procedure:

1. Start the graphical construction

Draw a line through A and C, intersecting the vertical line through B at E. Draw a line through B and C, intersecting the vertical line through A and F. Draw the vertical line GH through D.

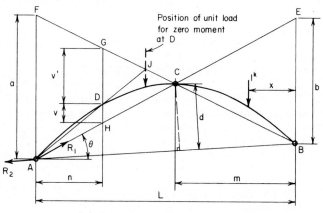

(a) Three-hinged arch

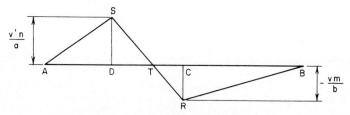

(b) Influence line for bending moment at D

FIG. 52

Let θ denote the angle between AE and the horizontal. Lines through B and D perpendicular to AE (omitted for clarity) make an angle θ with the vertical.

2.   Resolve the reaction into components

Resolve the reaction at A into the components R_1 and R_2 acting along AE and AB, respectively (Fig. 52).

3.   Determine the value of the first reaction

Let x denote the horizontal distance from the right-hand support to the unit load, where x has any value between 0 and L. Evaluate R_1 by equating the bending moment at B to zero. Thus $M_B = R_1 b \cos \theta - x = 0$; or $R_1 = x/(b \cos \theta)$.

4.   Evaluate the second reaction

Place the unit load within the interval CB. Evaluate R_2 by equating the bending moment at C to zero. Thus $M_C = R_2 d = 0$; $\therefore$ $R_2 = 0$.

5.   Calculate the bending moment at D when the unit load lies within the interval CB

Thus, $M_D = -R_1 v \cos \theta = -[(v \cos \theta)/(b \cos \theta)]x$, or $M_D = -vx/b$, Eq. a. When $x = m$, $M_D = -vm/b$.

6.   Place the unit load in a new position, and determine the bending moment

Place the unit load within the interval AD. Working from the right-hand support, proceed in an analogous manner to arrive at the following result: $M_D = v'(L - x)/a$, Eq. b. When $x = L - n$, $M_D = v'n/a$.

7.   Place the unit load within another interval, and evaluate the second reaction

Place the unit load within the interval DC, and evaluate R_2. Thus $M_C = R_2 d - (x - m) = 0$, or $R_2 = (x - m)/d$.

Since both R_1 and R_2 vary linearly with respect to x, it follows that M_D is also a linear function of x.

8.   Complete the influence line

In Fig. 52b, draw lines BR and AS to represent Eqs. a and b, respectively. Draw the straight line SR, thus completing the influence line. The point T at which this line intersects the base is termed the *neutral point*.

9.   Locate the neutral point

To locate T, draw a line through A and D in Fig. 52a intersecting BF at J. The neutral point in the influence line lies vertically below J; that is, M_D is zero when the action line of the unit load passes through J.

The proof is as follows: Since $M_D = 0$ and there are no applied loads in the interval AD, it follows that the total reaction at A is directed along AD. Similarly, since $M_C = 0$ and there are no applied loads in the interval CB, it follows that the total reaction at B is directed along BC. Because the unit load and the two reactions constitute a balanced system of forces, they are collinear. Therefore, J lies on the action line of the unit load.

Alternatively, the location of the neutral point may be established by applying the geometric properties of the influence line.

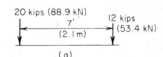

20 kips (88.9 kN)   12 kips (53.4 kN)

7' (2.1 m)

(a)

DEFLECTION OF A BEAM UNDER MOVING LOADS

The moving-load system in Fig. 53a traverses a beam on a simple span of 40 ft (12.2 m). What disposition of the system will cause the maximum deflection at midspan?

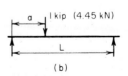

1 kip (4.45 kN)

a

L

(b)

FIG. 53

Calculation Procedure:

1.   Develop the equations for the midspan deflection under a unit load

The maximum deflection will manifestly occur when the two loads lie on opposite sides of the centerline of the span. In calculating the deflection at midspan caused by a load applied at any point on the span, it is advantageous to apply Maxwell's theorem of reciprocal deflections, which states the following: *The deflection at A caused by a load at B equals the deflection at B caused by this load at A.*

In Fig. 53b, consider the beam on a simple span L to carry a unit load applied at a distance a from the left-hand support. By referring to case 7 of the AISC *Manual* and applying the principle of reciprocal deflections, derive the following equations for the midspan deflection under the unit load: When $a < L/2$, $y = (3L^2a - 4a^3)/(48EI)$. When $a > L/2$, $y = [3L^2(L - a) - 4(L - a)^3]/(48EI)$.

2. Position the system for purposes of analysis

Position the system in such a manner that the 20-kip (89.0-kN) load lies to the left of center and the 12-kip (53.4-kN) load to the right of center. For the 20-kip (89.0-kN) load, set $a = x$. For the 12-kip (53.4-kN) load, $a = x + 7$; $L - a = 40 - (x + 7) = 33 - x$.

3. Express the total midspan deflection in terms of x

Substitute in the preceding equations. Combining all constants into a single term k, we find $ky = 20(3 \times 40^2x - 4x^3) + 12[3 \times 40^2(33 - x) - 4(33 - x)^3]$.

4. Solve for the unknown distance

Set $dy/dx = 0$ and solve for x. Thus, $x = 17.46$ ft (5.321 m).

For maximum deflection, position the load system with the 20-kip (89.0-kN) load 17.46 ft (5.321 m) from the left-hand support.

Riveted and Welded Connections

In the design of riveted and welded connections in this handbook, the American Institute of Steel Construction *Specification for the Design, Fabrication and Erection of Structural Steel for Buildings* is applied. This is presented in Part 5 of the *Manual of Steel Construction*.

The structural members considered here are made of ASTM A36 steel having a yield-point stress of 36,000 lb/in² (248,220 kPa). (The yield-point stress is denoted by F_y in the *Specification*.) All connections considered here are made with A141 hot-driven rivets or fillet welds of A233 class E60 series electrodes.

From the *Specification*, the allowable stresses are as follows: Tensile stress in connected member, 22,000 lb/in² (151,690.0 kPa); shearing stress in rivet, 15,000 lb/in² (103,425.0 kPa); bearing stress on projected area of rivet, 48,500 lb/in² (334,408.0 kPa); stress on throat of fillet weld, 13,600 lb/in² (93,772.0 kPa).

Let n denote the number of sixteenths included in the size of a fillet weld. For example, for a ⅜-in (9.53-mm) weld, $n = 6$. Then weld size = $n/16$. And throat area per linear inch of weld = $0.707n/16 = 0.0442n$ in². Also, capacity of weld = $13,600(0.0442n) = 600n$ lb/lin in (108.0n N/mm).

As shown in Fig. 54, a rivet is said to be in *single shear* if the opposing forces tend to shear the shank along one plane and in *double shear* if they tend to shear it along two planes. The symbols R_{ss}, R_{ds}, and R_b are used here to designate the shearing capacity of a rivet in single shear, the shearing capacity of a rivet in double shear, and the bearing capacity of a rivet, respectively, expressed in pounds (newtons).

CAPACITY OF A RIVET

Determine the values of R_{ss}, R_{ds}, and R_b for a ¾-in (19.05-mm) and ⅞-in (22.23-mm) rivet.

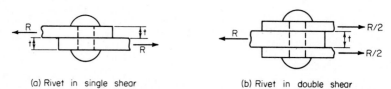

(a) Rivet in single shear   (b) Rivet in double shear

FIG. 54

Calculation Procedure:

1. Compute the cross-sectional area of the rivet

For the ¾-in (19.05-mm) rivet, area $= A = 0.785(0.75)^2 = 0.4418$ in^2 (2.8505 cm^2). Likewise, for the ⅞-in (22.23-mm) rivet, $A = 0.785(0.875)^2 = 0.6013$ in^2 (3.8796 cm^2).

2. Compute the single and double shearing capacity of the rivet

Let t denote the thickness, in inches (millimeters) of the connected member, as shown in Fig. 54. Multiply the stressed area by the allowable stress to determine the shearing capacity of the rivet. Thus, for the ¾-in (19.05-mm) rivet, $R_{ss} = 0.4418(15,000) = 6630$ lb (29,490.2 N); $R_{ds} = 2(0.4418)(15,000) = 13,250$ lb (58,936.0 N). Note that the factor of 2 is used for a rivet in double shear.

   Likewise, for the ⅞-in (22.23-mm) rivet, $R_{ss} = 0.6013(15,000) = 9020$ lb (40,121.0 N); $R_{ds} = 2(0.6013)(15,000) = 18,040$ lb (80,242.0 N).

3. Compute the rivet bearing capacity

The effective bearing area of a rivet of diameter d in (mm) $= dt$. Thus, for the ¾-in (19.05-mm) rivet, $R_b = 0.75t(48,500) = 36,380t$ lb (161,709t N). For the ⅞-in (22.23-mm) rivet, $R_b = 0.875t(48,500) = 42,440t$ lb (188,733t N). By substituting the value of t in either relation, the numerical value of the bearing capacity could be obtained.

INVESTIGATION OF A LAP SPLICE

The hanger in Fig. 55a is spliced with nine ¾-in (19.05-mm) rivets in the manner shown. Compute the load P that may be transmitted across the joint.

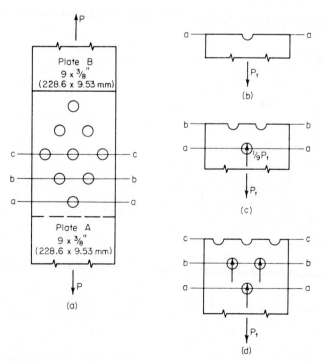

FIG. 55

Calculation Procedure:

1. Compute the capacity of the joint in shear and bearing

There are three criteria to be considered: the shearing strength of the connection, the bearing strength of the connection, and the tensile strength of the net section of the plate at each row of rivets.

Since the load is concentric, assume that the load transmitted through each rivet is $\frac{1}{9}P$. As plate A (Fig. 55) deflects, it bears against the upper half of each rivet. Consequently, the reaction of the rivet on plate A is exerted *above* the horizontal diametral plane of the rivet.

Computing the capacity of the joint in shear and in bearing yields $P_{SS} = 9(6630) = 59,700$ lb (265,545.6 N); $P_b = 9(36,380)(0.375) = 122,800$ lb (546,214.4 N).

2. Compute the tensile capacity of the plate

The tensile capacity P_t lb (N) of plate A (Fig. 55) is required. In structural fabrication, rivet holes are usually punched $\frac{1}{16}$ in (1.59 mm) larger than the rivet diameter. However, to allow for damage to the adjacent metal caused by punching, the *effective* diameter of the hole is considered to be $\frac{1}{8}$ in (3.18 mm) larger than the rivet diameter.

Refer to Fig. 55b, c, and d. Equate the tensile stress at each row of rivets to 22,000 lb/in² (151,690.0 kPa) to obtain P_t. Thus, at aa, residual tension = P_t; net area = $(9 - 0.875)(0.375)$ = 3.05 in² (19.679 cm²). The stress $s = P_t/3.05 = 22,000$ lb/in² (151,690.0 kPa); $P_t = 67,100$ lb (298,460.0 N).

At bb, residual tension = $\frac{8}{9}P_t$; net area = $(9 - 1.75)(0.375) = 2.72$ in² (17.549 cm²); $s = \frac{8}{9}P_t/2.72 = 22,000$; $P_t = 67,300$ lb (299,350.0 N).

At cc, residual tension = $\frac{6}{9}P_t$; net area = $(9 - 2.625)(0.375) = 2.39$ in² (15.420 cm²); $s = \frac{6}{9}P_t/2.39 = 22,000$; $P_t = 78,900$ lb (350,947.0 N).

3. Select the lowest of the five computed values as the allowable load

Thus, $P = 59,700$ lb (265,545.6 N).

DESIGN OF A BUTT SPLICE

A tension member in the form of a $10 \times \frac{1}{2}$ in (254.0 × 12.7 mm) steel plate is to be spliced with $\frac{7}{8}$-in (22.23-mm) rivets. Design a butt splice for the maximum load the member may carry.

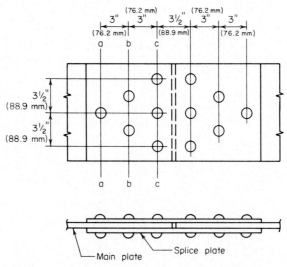

FIG. 56

Calculation Procedure:

1. Establish the design load

In a butt splice, the load is transmitted from one member to another through two auxiliary plates called *cover*, *strap*, or *splice* plates. The rivets are therefore in double shear.

Establish the design load, P lb (N), by computing the allowable load at a cross section having one rivet hole. Thus net area = $(10 - 1)(0.5) = 4.5$ in^2 (29.03 cm^2). Then $P = 4.5(22,000) = 99,000$ lb (440,352.0 N).

2. Determine the number of rivets required

Applying the values of rivet capacity found in an earlier calculation procedure in this section of the handbook, determine the number of rivets required. Thus, since the rivets are in double shear, $R_{ds} = 18,040$ lb (80,241.9 N); $R_b = 42,440(0.5) = 21,220$ lb (94,386.6 N). Then $99,000/18,040 = 5.5$ rivets; use the next largest whole number, or 6 rivets.

3. Select a trial pattern for the rivets; investigate the tensile stress

Conduct this investigation of the tensile stress in the main plate at each row of rivets.

The trial pattern is shown in Fig. 56. The rivet spacing satisfies the requirements of the AISC *Specification*. Record the calculations as shown:

Section	Residual tension in main plate, lb (N)	÷	Net area, in^2 (cm^2)	=	Stress, lb/in^2 (kPa)
aa	99,000 (440,352.0)		4.5 (29.03)		22,000 (151,690.0)
bb	82,500 (366,960.0)		4.0 (25.81)		20,600 (142,037.0)
cc	49,500 (220,176.0)		3.5 (22.58)		14,100 (97,219.5)

Study of the above computations shows that the rivet pattern is satisfactory.

4. Design the splice plates

To the left of the centerline, each splice plate bears against the *left* half of the rivet. Therefore, the entire load has been transmitted to the splice plates at cc, which is the critical section. Thus the tension in splice plate = $\frac{1}{2}(99,000) = 49,500$ lb (220,176.0 N); plate thickness required = $49,500/[22,000(7)] = 0.321$ in (8.153 mm). Make the splice plates $10 \times \frac{3}{8}$ in (254.0 × 9.53 mm).

DESIGN OF A PIPE JOINT

A steel pipe 5 ft 6 in (1676.4 mm) in diameter must withstand a fluid pressure of 225 lb/in^2 (1551.4 kPa). Design the pipe and the longitudinal lap splice, using $\frac{3}{4}$-in (19.05-mm) rivets.

Calculation Procedure:

1. Evaluate the hoop tension in the pipe

Let L denote the length (Fig. 57) of the *repeating group* of rivets. In this case, this equals the rivet pitch. In Fig. 57, let T denote the hoop tension, in pounds (newtons), in the distance L. Evaluate the tension, using $T = pDL/2$, where p = internal pressure, lb/in^2 (kPa); D = inside diameter of pipe, in (mm); L = length considered, in (mm). Thus, $T = 225(66)L/2 = 7425L$.

2. Determine the required number of rows of rivets

Adopt, tentatively, the minimum allowable pitch, which is 2 in (50.8 mm) for $\frac{3}{4}$-in (19.05-mm) rivets. Then establish a feasible rivet pitch. From an earlier calculation procedure in this section, $R_{ss} = 6630$ lb (29,490.0 N). Then $T = 7425(2) = 6630n$; $n = 2.24$. Use the next largest whole number of rows, or three rows of rivets. Also, $L_{max} = 3(6630)/7425 = 2.68$ in (68.072 mm). Use a 2½-in (63.5-mm) pitch, as shown in Fig. 57a.

3. Determine the plate thickness

Establish the thickness t in (mm) of the steel plates by equating the stress on the net section to its allowable value. Since the holes will be drilled, take $^{13}\!/_{16}$ in (20.64 mm) as their diameter. Then $T = 22,000t(2.5 - 0.81) = 7425(2.5); t = 0.50$ in (12.7 mm); use ½-in (12.7-mm) plates. Also, $R_b = 36,380(0.5) > 6630$ lb (29,490.2 N). The rivet capacity is therefore limited by shear, as assumed.

MOMENT ON RIVETED CONNECTION

The channel in Fig. 58a is connected to its supporting column with ¾-in (19.05-mm) rivets and resists the couple indicated. Compute the shearing stress in each rivet.

Calculation Procedure:

1. Compute the polar moment of inertia of the rivet group

The moment causes the channel (Fig. 58) to rotate about the centroid of the rivet group and thereby exert a tangential thrust on each rivet. This thrust is directly proportional to the radial distance to the center of the rivet.

Establish coordinate axes through the centroid of the rivet group. Compute the polar moment of inertia of the group with respect to an axis through its centroid, taking the cross-sectional area of a rivet as unity. Thus, $J = \Sigma(x^2 + y^2) = 8(2.5)^2 + 4(1.5)^2 + 4(4.5)^2 = 140$ in^2 (903.3 cm^2).

2. Compute the radial distance to each rivet

Using the right-angle relationship, we see that $r_1 = r_4 = (2.5^2 + 4.5^2)^{0.5} = 5.15$ in (130.810 mm); $r_2 = r_3 = (2.5^2 + 1.5^2)^{0.5} = 2.92$ in (74.168 mm).

3. Compute the tangential thrust on each rivet

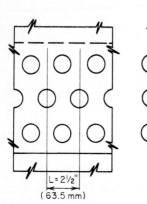

L = 2½"
(63.5 mm)

(a) Longitudinal pipe joint

(b) Free-body diagram of upper half
of pipe and contents

FIG. 57

Use the relation $f = Mr/J$. Since $M = 12,000(8) = 96,000$ lb·in (10,846.1 N·m), $f_1 = f_4 = 96,000(5.15)/140 = 3530$ lb (15,701.4 N); and $f_2 = f_3 = 96,000(2.92)/140 = 2000$ lb (8896.0 N). The directions are shown in Fig. 58b.

4. Compute the shearing stress

Using $s = P/A$, we find $s_1 = s_4 = 3530/0.442 = 7990$ lb/in^2 (55,090 kPa); also, $s_2 = s_3 = 2000/0.442 = 4520$ lb/in^2 (29,300 kPa).

5. Check the rivet forces

Check the rivet forces by summing their moments with respect to an axis through the centroid. Thus $M_1 = M_4 = 3530(5.15) = 18,180$ in·lb (2054.0 N·m); $M_2 = M_3 = 2000(2.92) = 5840$ in·lb (659.8 N·m). Then $\Sigma M = 4(18,180) + 4(5840) = 96,080$ in·lb (10,855.1 N·m).

ECCENTRIC LOAD ON RIVETED CONNECTION

Calculate the maximum force exerted on a rivet in the connection shown in Fig. 59a.

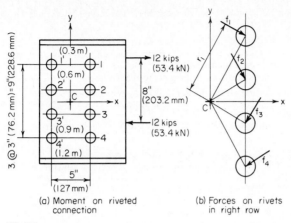

FIG. 58

(a) Moment on riveted connection

(b) Forces on rivets in right row

Calculation Procedure:

1. Compute the effective eccentricity

To account implicitly for secondary effects associated with an eccentrically loaded connection, the AISC *Manual* recommends replacing the true eccentricity with an *effective* eccentricity.

To compute the effective eccentricity, use $e_e = e_a - (1 + n)/2$, where e_e = effective eccentricity, in (mm); e_a = actual eccentricity of the load, in (mm); n = number of rivets in a vertical row. Substituting gives $e_e = 8 - (1 + 3)/2 = 6$ in (152.4 mm).

2. Replace the eccentric load with an equivalent system

The equivalent system is comprised of a concentric load P lb (N) and a clockwise moment M in·lb (N·m). Thus, $P = 15,000$ lb (66,720.0 N), $M = 15,000(6) = 90,000$ in·lb (10,168.2 N·m).

3. Compute the polar moment of inertia of the rivet group

Compute the polar moment of inertia of the rivet group with respect to an axis through its centroid. Thus, $J = \Sigma(x^2 + y^2) = 6(3)^2 + 4(4)^2 = 118$ in² (761.3 cm²).

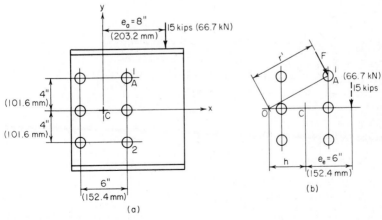

FIG. 59

4. Resolve the tangential thrust on each rivet into its horizontal and vertical components

Resolve the tangential thrust f lb (N) on each rivet caused by the moment into its horizontal and vertical components f_x and f_y, respectively. These forces are as follows: $f_x = My/J$ and $f_y = Mx/J$. Computing these forces for rivets 1 and 2 (Fig. 59) yields $f_x = 90,000(4)/118 = 3050$ lb (13,566.4 N); $f_y = 90,000(3)/118 = 2290$ lb (10,185.9 N).

5. Compute the thrust on each rivet caused by the concentric load

This thrust is $f'_y = 15,000/6 = 2500$ lb (11,120.0 N).

6. Combine the foregoing results to obtain the total force on the rivets being considered

The total force F lb (N) on rivets 1 and 2 is desired. Thus, $F_x = f_x = 3050$ lb (13,566.4 N); $F_y = f_y + f'_y = 2290 + 2500 = 4790$ lb (21,305.9 N). Then $F = [(3050)^2 + (4790)^2]^{0.5} = 5680$ lb (25,264.6 N).

The above six steps comprise method 1. A second way of solving this problem, method 2, is presented below.

The total force on each rivet may also be found by locating the *instantaneous center of rotation* associated with this eccentric load and treating the connection as if it were subjected solely to a moment (Fig. 59b).

7. Locate the instantaneous center of rotation

To locate this center, apply the relation $h = J/(e_eN)$, where N = total number of rivets and the other relations are as given earlier. Then $h = 118/[6(6)] = 3.28$ in (83.31 m).

8. Compute the force on the rivets

Considering rivets 1 and 2, use the equation $F = Mr'/J$, where r' = distance from the instantaneous center of rotation O to the center of the given rivet, in. For rivets 1 and 2, $r' = 7.45$ in (189.230 mm). Then $F = 90,000(7.45)/118 = 5680$ lb (25,264.6 N). The force on rivet 1 has an action line normal to the radius OA.

DESIGN OF A WELDED LAP JOINT

The 5-in (127.0-mm) leg of a 5 × 3 × ⅜ in (127.0 × 76.2 × 9.53 mm) angle is to be welded to a gusset plate, as shown in Fig. 60. The member will be subjected to repeated variation in stress. Design a suitable joint.

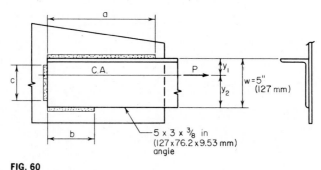

FIG. 60

Calculation Procedure:

1. Determine the properties of the angle

In accordance with the AISC *Specification*, arrange the weld to have its centroidal axis coincide with that of the member. Refer to the AISC *Manual* to obtain the properties of the angle. Thus $A = 2.86$ in² (18.453 cm²); $y_1 = 1.70$ in (43.2 mm); $y_2 = 5.00 - 1.70 = 3.30$ in (83.820 mm).

2. Compute the design load and required weld length

The design load P lb (N) $= As = 2.86(22,000) = 62,920$ lb (279,868.2 N). The AISC *Specification* restricts the weld size to $\frac{5}{16}$ in (7.94 mm). Hence, the weld capacity $= 5(600) = 3000$ lb/lin in (525,380.4 N/m); $L =$ weld length, in (mm) $= P$/capacity, lb/lin in $= 62,920/3000 = 20.97$ in (532.638 mm).

3. Compute the joint dimensions

In Fig. 60, set $c = 5$ in (127.0 mm), and compute a and b by applying the following equations: $a = Ly_2/w - c/2$; $b = Ly_1/w - c/2$. Thus, $a = (20.97 \times 3.30)/5 - \frac{5}{8} = 11.34$ in (288.036 mm); $b = (20.97 \times 1.70)/5 - \frac{5}{8} = 4.63$ in (117.602 mm). Make $a = 11.5$ in (292.10 mm) and $b = 5$ in (127.0 mm).

ECCENTRIC LOAD ON A WELDED CONNECTION

The bracket in Fig. 61 is connected to its support with a $\frac{1}{4}$-in (6.35-mm) fillet weld. Determine the maximum stress in the weld.

Calculation Procedure:

1. Locate the centroid of the weld group

Refer to the previous eccentric-load calculation procedure. This situation is analogous to that. Determine the stress by locating the instantaneous center of rotation. The maximum stress occurs at A and B (Fig. 61).

Considering the weld as concentrated along the edge of the supported member, locate the centroid of the weld group by taking moments with respect to line aa. Thus $m = 2(4)(2)/(12 + 2 \times 4) = 0.8$ in (20.32 mm).

2. Replace the eccentric load with an equivalent concentric load and moment

Thus $P = 13,500$ lb (60,048.0 N); $M = 124,200$ in·lb (14,032.1 N·m).

3. Compute the polar moment of inertia of the weld group

This moment should be computed with respect to an axis through the centroid of the weld group. Thus $I_x = (1/12)(12)^3 + 2(4)(6)^2 = 432$ in^3 (7080.5 cm^3); $I_y = 12(0.8)^2 + 2(1/12)(4)^3 + 2(4)(2 - 0.8)^2 = 29.9$ in^3 (490.06 cm^3). Then $J = I_x + I_y = 461.9$ in^3 (7570.54 cm^3).

4. Locate the instantaneous center of rotation O

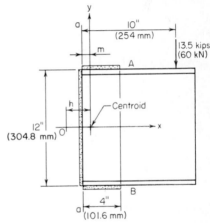

FIG. 61

This center is associated with this eccentric load by applying the equation $h = J/(eL)$, where e = eccentricity of load, in (mm), and L = total length of weld, in (mm). Thus, $e = 10 - 0.8 = 9.2$ in (233.68 mm); $L = 12 + 2(4) = 20$ in (508.0 mm); then $h = 461.9/[9.2(20)] = 2.51$ in (63.754 mm).

5. Compute the force on the weld

Use the equation $F = Mr'/J$, lb/lin in (N/m), where r' = distance from the instantaneous center of rotation to the given point, in (mm). At A and B, $r' = 8.28$ in (210.312 mm); then $F = [124,200(8.28)]/461.9 = 2230$ lb/lin in (390,532.8 N/m).

6. Calculate the corresponding stress on the throat

Thus, $s = P/A = 2230/[0.707(0.25)] = 12,600$ lb/in^2 (86,877.0 kPa), where the value 0.707 is the sine of $45°$, the throat angle.

Steel Beams and Plate Girders

In the following calculation procedures, the design of steel members is executed in accordance with the *Specification for the Design, Fabrication and Erection of Structural Steel for Buildings* of the American Institute of Steel Construction. This specification is presented in the AISC *Manual of Steel Construction.*

Most allowable stresses are functions of the yield-point stress, denoted as F_y in the *Manual*. The appendix of the *Specification* presents the allowable stresses associated with each grade of structural steel together with tables intended to expedite the design. The *Commentary* in the *Specification* explains the structural theory underlying the *Specification.*

Unless otherwise noted, the structural members considered here are understood to be made of ASTM A36 steel, having a yield-point stress of 36,000 lb/in² (248,220.0 kPa).

The notational system used conforms with that adopted earlier, but it is augmented to include the following: A_f = area of flange, in² (cm²); A_w = area of web, in² (cm²); b_f = width of flange, in (mm); d = depth of section, in (mm); d_w = depth of web, in (mm); t_f = thickness of flange, in (mm). t_w = thickness of web, in (mm); L' = unbraced length of compression flange, in (mm); f_y = yield-point stress, lb/in² (kPa).

MOST ECONOMIC SECTION FOR A BEAM WITH A CONTINUOUS LATERAL SUPPORT UNDER A UNIFORM LOAD

A beam on a simple span of 30 ft (9.2 m) carries a uniform superimposed load of 1650 lb/lin ft (24,079.9 N/m). The compression flange is laterally supported along its entire length. Select the most economic section.

Calculation Procedure:

1. Compute the maximum bending moment and the required section modulus

Assume that the beam weighs 50 lb/lin ft (729.7 N/m) and satisfies the requirements of a compact section as set forth in the *Specification.*

The maximum bending moment is $M = (1/8)wL^2 = (1/8)(1700)(30)^2(12) = 2,295,000$ in·lb (259,289.1 N·m).

Referring to the *Specification* shows that the allowable bending stress is 24,000 lb/in² (165,480.0 kPa). Then $S = M/f = 2,295,000/24,000 = 95.6$ in³ (1566.88 cm³).

2. Select the most economic section

Refer to the AISC *Manual*, and select the most economic section. Use 18WF55 (W18 × 55); $S = 98.2$ in³ (1609.50 cm³); section compact. The disparity between the assumed and actual beam weight is negligible.

A second method for making this selection is shown below.

3. Calculate the total load on the member

Thus, the total load = $W = 30(1700) = 51,000$ lb (226,848.0 N).

4. Select the most economic section

Refer to the tables of allowable uniform loads in the *Manual*, and select the most economic section. Thus use 18WF55; $W_{allow} = 52,000$ lb (231,296.0 N). The capacity of the beam is therefore slightly greater than required.

MOST ECONOMIC SECTION FOR A BEAM WITH INTERMITTENT LATERAL SUPPORT UNDER UNIFORM LOAD

A beam on a simple span of 25 ft (7.6 m) carries a uniformly distributed load, including the estimated weight of the beam, of 45 kips (200.2 kN). The member is laterally supported at 5-ft (1.5-m) intervals. Select the most economic member (a) using A36 steel; (b) using A242 steel, having a yield-point stress of 50,000 lb/in² (344,750.0 kPa) when the thickness of the metal is ¾ in (19.05 mm) or less.

Calculation Procedure:

1. Using the AISC allowable-load tables, select the most economic member made of A36 steel

After a trial section has been selected, it is necessary to compare the unbraced length L' of the compression flange with the properties L_c and L_u of that section in order to establish the allowable bending stress. The variables are defined thus: L_c = maximum unbraced length of the compression flange if the allowable bending stress = $0.66f_y$, measured in ft (m); L_u = maximum unbraced length of the compression flange, ft (m), if the allowable bending stress is to equal $0.60f_y$.

The values of L_c and L_u associated with each rolled section made of the indicated grade of steel are recorded in the allowable-uniform-load tables of the AISC *Manual*. The L_c value is established by applying the definition of a *laterally supported* member as presented in the *Specification*. The value of L_u is established by applying a formula given in the *Specification*.

There are four conditions relating to the allowable stress:

Condition	Allowable stress
Compact section; $L' \leq L_c$	$0.66f_y$
Compact section; $L_c < L' \leq L_u$	$0.60f_y$
Noncompact section; $L' \leq L_u$	$0.60f_y$
$L' > L_u$	Apply the *Specification* formula—use the larger value obtained when the two formulas given are applied.

The values of allowable uniform load given in the AISC *Manual* apply to beams of A36 steel satisfying the first or third condition above, depending on whether the section is compact or noncompact.

Referring to the table in the *Manual*, we see that the most economic section made of A36 steel is 16WF45 (W16 × 45); W_{allow} = 46 kips (204.6 kN), where W_{allow} = allowable load on the beam, kips (kN). Also, L_c = 7.6 > 5. Hence, the beam is acceptable.

2. Compute the equivalent load for a member of A242 steel

To apply the AISC *Manual* tables to choose a member of A242 steel, assume that the shape selected will be compact. Transform the actual load to an equivalent load by applying the conversion factor 1.38, that is, the ratio of the allowable stresses. The conversion factors are recorded in the *Manual* tables. Thus, equivalent load = 45/1.38 = 32.6 kips (145.0 N).

3. Determine the lightest satisfactory section

Enter the *Manual* allowable-load table with the load value computed in step 2, and select the lightest section that appears to be satisfactory. Try 16WF36 (W16 × 36); W_{allow} = 36 kips (160.1 N). However, this section is noncompact in A242 steel, and the equivalent load of 32.6 kips (145.0 N) is not valid for this section.

4. Revise the equivalent load

To determine whether the 16WF36 will suffice, revise the equivalent load. Check the L_u value of this section in A242 steel. Then equivalent load = 45/1.25 = 36 kips (160.1 N), L_u = 6.3 ft (1.92 m) > 5 ft (1.5 m); use 16WF36.

5. Verify the second part of the design

To verify the second part of the design, calculate the bending stress in the 16WF36, using S = 56.3 in³ (922.76 cm³) from the *Manual*. Thus M = $(1/8)WL$ = $(1/8)(45,000)(25)(12)$ = 1,688,000 in·lb (190,710.2 N·m); f = M/S = 1,688,000/56.3 = 30,000 lb/in² (206,850.0 kPa). This stress is acceptable.

DESIGN OF A BEAM WITH REDUCED ALLOWABLE STRESS

The compression flange of the beam in Fig. 62a will be braced only at points A, B, C, D, and E. Using AISC data, a designer has selected a 21WF55 (W21 × 55) section for the beam. Verify the design.

Calculation Procedure:

1. *Calculate the reactions; construct the shear and bending-moment diagrams*

The results of this step are shown in Fig. 62.

2. *Record the properties of the selected section*

Using the AISC *Manual*, record the following properties of the 21WF55 section: $S = 109.7$ in³ (1797.98 cm³); $I_y = 44.0$ in⁴ (1831.41 cm⁴); $b_f = 8.215$ in (208.661 mm); $t_f = 0.522$ in (13.258

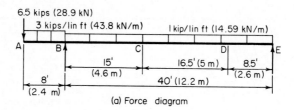

(a) Force diagram

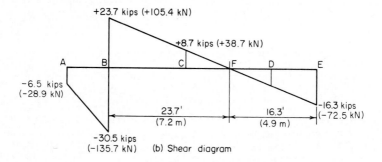

(b) Shear diagram

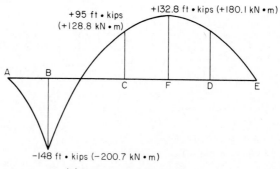

(c) Bending—moment diagram

FIG. 62

mm); d = 20.80 in (528.32 mm); t_w = 0.375 in (9.525 mm); d/A_f = 4.85/in (0.1909/mm); L_c = 8.9 ft (2.71 m); L_u = 9.4 ft (2.87 m).

Since $L' > L_u$, the allowable stress must be reduced in the manner prescribed in the *Manual*.

3. Calculate the radius of gyration

Calculate the radius of gyration with respect to the y axis of a T section comprising the compression flange and one-sixth the web, neglecting the area of the fillets. Referring to Fig. 63, we see A_f = 8.215(0.522) = 4.29 in^2 (27.679 cm^2); (1/6)A_w = (1/6)(19.76)(0.375) = 1.24; A_T = 5.53 in^2 (35.680 cm^2); I_T = 0.5I_y of the section = 22.0 in^4 (915.70 cm^4); r = (22.0/5.53)$^{0.5}$ = 1.99 in (50.546 mm).

4. Calculate the allowable stress in each interval between lateral supports

By applying the provisions of the *Manual*, calculate the allowable stress in each interval between lateral supports, and compare this with the actual stress. For A36 steel, the *Manual* formula (4) reduces to f_1 = 22,000 − 0.679$(L'/r)^2/C_b$ lb/in^2 (kPa). By *Manual* formula (5), f_2 = 12,000,000/$(L'd/A_f)$ lb/in^2 (kPa). Set the allowable stress equal to the greater of these values.

For interval AB: L' = 8 ft (2.4 m) $< L_c$; ∴ f_{allow} = 24,000 lb/in^2 (165,480.0 kPa); f_{max} = 148,000(12)/109.7 = 16,200 lb/in^2 (111,699.0 kPa)—this is acceptable.

For interval BC: L'/r = 15(12)/1.99 = 90.5; M_1/M_2 = 95/(−148) = −0.642; C_b = 1.75 − 1.05(−0.642) + 0.3(−0.642)^2 = 2.55; ∴ set C_b = 2.3; f_1 = 22,000 − 0.679(90.5)^2/2.3 = 19,600 lb/in^2 (135,142.0 kPa); f_2 = 12,000,000/[15(12)(4.85)] = 13,700 lb/in^2 (94,461.5 kPa); f_{max} = 16,200 < 19,600 lb/in^2 (135,142.0 kPa). This is acceptable.

Interval CD: Since the maximum moment occurs within the interval rather than at a boundary section, C_b = 1; L'/r = 16.5(12)/1.99 = 99.5; f_1 = 22,000 − 0.679(99.5)^2 = 15,300 lb/in^2 (105,493.5 kPa); f_2 = 12,000,000/[16.5(12)(4.85)] = 12,500 lb/in^2 (86,187.5 kPa); f_{max} = 132,800(12)/109.7 = 14,500 < 15,300 lb/in^2 (105,493.5 kPa). This stress is acceptable.

Interval DE: The allowable stress is 24,000 lb/in^2 (165,480.0 kPa), and the actual stress is considerably below this value. The 21WF55 is therefore satisfactory. Where deflection is the criterion, the member should be checked by using the *Specification*.

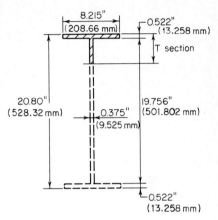

FIG. 63 Dimensions of 21WF55.

DESIGN OF A COVER-PLATED BEAM

Following the fabrication of an 18WF60 (W18 × 60) beam, a revision was made in the architectural plans, and the member must now be designed to support the loads shown in Fig. 64a. Cover plates are to be welded to both flanges to develop the required strength. Design these plates and their connection to the WF shape, using fillet welds of A233 class E60 series electrodes. The member has continuous lateral support.

Calculation Procedure:

1. Construct the shear and bending-moment diagrams

These are shown in Fig. 64. Also, M_E = 340.3 ft·kips (461.44 kN·m).

2. Calculate the required section modulus, assuming the built-up section will be compact

The section modulus S = M/f = 340.3(12)/24 = 170.2 in^3 (2789.58 cm^3).

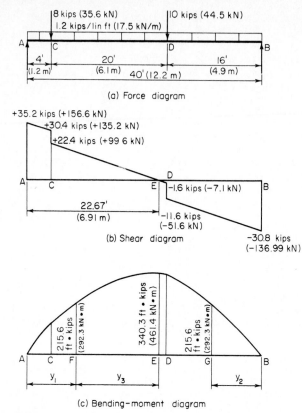

(a) Force diagram

(b) Shear diagram

(c) Bending–moment diagram

FIG. 64

3. Record the properties of the beam section

Refer to the AISC *Manual*, and record the following properties for the 18WF60; d = 18.25 in (463.550 mm); b_f = 7.56 in (192.024 mm); t_f = 0.695 in (17.653 mm); I = 984 in^4 (40.957 cm^4); S = 107.8 in^3 (1766.84 cm^3).

4. Select a trial section

Apply the approximation $A = 1.05(S - S_{WF})/d_{WF}$, where A = area of one cover plate, in^2 (cm^2); S = section modulus required, in^3 (cm^3); S_{WF} = section modulus of wide-flange shape, in^3 (cm^3); d_{WF} = depth of wide-flange shape, in (mm). Then A = [1.05(170.2 − 107.8)]/18.25 = 3.59 in^2 (23.163 cm^2).

Try 10 × ⅜ in (254.0 × 9.5 mm) plates with A = 3.75 in^2 (24.195 cm^2). Since the beam flange is 7.5 in (190.50 mm) wide, ample space is available to accommodate the welds.

5. Ascertain whether the assumed size of the cover plates satisfies the AISC Specification

Using the appropriate AISC *Manual* section, we find 7.56/0.375 = 20.2 < 32, which is acceptable; ½(10 − 7.56)/0.375 = 3.25 < 16, which is acceptable.

6. Test the adequacy of the trial section

Calculate the section modulus of the trial section. Referring to Fig. 65a, we see I = 984 + 2(3.75)(9.31)^2 = 1634 in^4 (68,012.1 cm^4); S = I/c = 1634/9.5 = 172.0 in^3 (2819.08 cm^3). The reinforced section is therefore satisfactory.

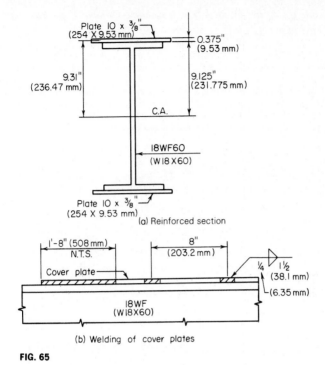

Plate 10 x ⅜"
(254 X 9.53 mm)

0.375"
(9.53 mm)

9.31"
(236.47 mm)

9.125"
(231.775 mm)

C.A.

18WF60
(W18X60)

Plate 10 x ⅜"
(254 X 9.53 mm)

(a) Reinforced section

1'-8" (508 mm)
N.T.S.

8"
(203.2 mm)

Cover plate

¼   1½
(38.1 mm)

(6.35 mm)

18WF
(W18X60)

(b) Welding of cover plates

FIG. 65

7. Locate the points at which the cover plates are not needed

To locate the points at which the cover plates may theoretically be dispensed with, calculate the moment capacity of the wide-flange shape alone. Thus, $M = fS = 24(107.8)/12 = 215.6$ ft·kips (292.3 kN·m).

8. Locate the points at which the computed moment occurs

These points are F and G (Fig. 64). Thus, $M_F = 35.2y_1 - 8(y_1 - 4) - \frac{1}{2}(1.2y_1^2) = 215.6$; $y_1 = 8.25$ ft (2.515 m); $M_G = 30.8y_2 - \frac{1}{2}(1.2y_2^2) = 215.6$; $y_2 = 8.36$ ft (2.548 m).

Alternatively, locate F by considering the area under the shear diagram between E and F. Thus $M_F = 340.3 - \frac{1}{2}(1.2y_3^2) = 215.6$; $y_3 = 14.42$ ft (4.395 m); $y_1 = 22.67 - 14.42 = 8.25$ ft (2.515 m).

For symmetry, center the cover plates about midspan, placing the theoretical cutoff points at 8 ft 3 in (2.51 m) from each support.

9. Calculate the axial force in the cover plate

Calculate the axial force P lb (N) in the cover plate at its end by computing the mean bending stress. Determine the length of fillet weld required to transmit this force to the wide-flange shape. Thus $f_{mean} = My/I = 215,600(12)(9.31)/1634 = 14,740$ lb/in² (101,632.3 kPa). Then $P = Af_{mean} = 3.75(14,740) = 55,280$ lb (245,885.4 N). Use a ¼-in (6.35-mm) fillet weld, which satisfies the requirements of the *Specification*. The capacity of the weld = $4(600) = 2400$ lb/lin in (420,304.3 N/m). Then the length L required for this weld is $L = 55,280/2400 = 23.0$ in (584.20 mm).

10. Extend the cover plates

In accordance with the *Specification*, extend the cover plates 20 in (508.0 mm) beyond the theoretical cutoff point at each end, and supply a continuous ¼-in fillet weld along both edges in this extension. This requirement yields 40 in (1016.0 mm) of weld as compared with the 23 in (584.2 mm) needed to develop the plate.

11. *Calculate the horizontal shear flow at the inner surface of the cover plate*

Choose F or G, whichever is larger. Design the intermittent fillet weld to resist this shear flow. Thus $V_F = 35.2 - 8 - 1.2(8.25) = 17.3$ kips (76.95 kN); $V_G = -30.8 + 1.2(8.36) = -20.8$ kips (-92.51 kN). Then $q = VQ/I = 20,800(3.75)(9.31)/1634 = 444$ lb/lin in (77,756.3 N/m).

The *Specification* calls for a minimum weld length of 1.5 in (38.10 mm). Let s denote the center-to-center spacing as governed by shear. Then $s = 2(1.5)(2400)/444 = 16.2$ in (411.48 mm). However, the *Specification* imposes additional restrictions on the weld spacing. To preclude the possibility of error in fabrication, provide an identical spacing at the top and bottom. Thus, $s_{max} = 21(0.375) = 7.9$ in (200.66 mm). Therefore, use a ¼-in (6.35-mm) fillet weld, 1.5 in (38.10 mm) long, 8 in (203.2 mm) on centers, as shown in Fig. 65b.

DESIGN OF A CONTINUOUS BEAM

The beam in Fig. 66a is continuous from A to D and is laterally supported at 5-ft (1.5-m) intervals. Design the member.

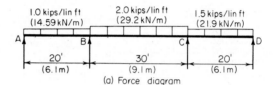

(a) Force diagram

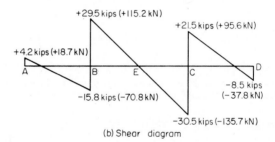

(b) Shear diagram

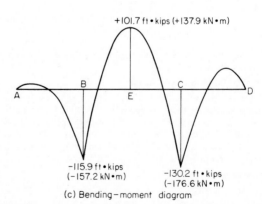

(c) Bending–moment diagram

FIG. 66

Calculation Procedure:

1. Find the bending moments at the interior supports; calculate the reactions and construct shear and bending-moment diagrams

The maximum moments are $+101.7$ ft·kips (137.9 kN·m) and -130.2 ft·kips (176.55 kN·m).

2. Calculate the modified maximum moments

Calculate these moments in the manner prescribed in the AISC *Specification*. The clause covering this calculation is based on the postelastic behavior of a continuous beam. (Refer to a later calculation procedure for an analysis of this behavior.)

Modified maximum moments: $+101.7 + 0.1(0.5)(115.9 + 130.2) = +114.0$ ft·kips (154.58 kN·m); $0.9(-130.2) = -117.2$ ft·kips (-158.92 kN·m); design moment $= 117.2$ ft·kips (158.92 kN·m).

3. Select the beam size

Thus, $S = M/f = 117.2(12)/24 = 58.6$ in^3 (960.45 cm^3). Use 16WF40 (W16 × 40) with $S = 64.4$ in^3 (1055.52 cm^3); $L_c = 7.6$ ft (2.32 m).

SHEARING STRESS IN A BEAM—EXACT METHOD

Calculate the maximum shearing stress in an 18WF55 (W18 × 55) beam at a section where the vertical shear is 70 kips (311.4 kN).

Calculation Procedure:

1. Record the relevant properties of the member

The shearing stress is a maximum at the centroidal axis and is given by $v = VQ/(It)$. The static moment of the area above this axis is found by applying the properties of the ST9WF27.5 (WT9 × 27.5), which are presented in the AISC *Manual*. Note that the T section considered is one-half the wide-flange section being used. See Fig. 67.

The properties of these sections are $I_{WF} = 890$ in^4 ($37,044.6$ cm^4); $A_T = 8.10$ in^2 (52.261 cm^2); $t_w = 0.39$ in (9.906 mm); $y_m = 9.06 - 2.16 = 6.90$ in (175.26 mm).

2. Calculate the shearing stress at the centroidal axis

Substituting gives $Q = 8.10(6.90) = 55.9$ in^3 (916.20 cm^3); then $v = 70,000(55.9)/[890(0.39)] = 11,270$ lb/in^2 ($77,706.7$ kPa).

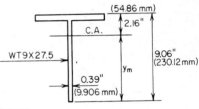

FIG. 67

SHEARING STRESS IN A BEAM—APPROXIMATE METHOD

Solve the previous calculation procedure, using the approximate method of determining the shearing stress in a beam.

Calculation Procedure:

1. Assume that the vertical shear is resisted solely by the web

Consider the web as extending the full depth of the section and the shearing stress as uniform across the web. Compare the results obtained by the exact and the approximate methods.

2. Compute the shear stress

Take the depth of the web as 18.12 in (460.248 mm), $v = 70,000/[18.12(0.39)] = 9910$ lb/in^2 ($68,329.45$ kPa). Thus, the ratio of the computed stresses is $11,270/9910 = 1.14$.

Since the error inherent in the approximate method is not unduly large, this method is applied in assessing the shear capacity of a beam. The allowable shear V for each rolled section is recorded in the allowable-uniform-load tables of the AISC *Manual*.

The design of a rolled section is governed by the shearing stress only in those instances where the ratio of maximum shear to maximum moment is extraordinarily large. This condition exists in a heavily loaded short-span beam and a beam that carries a large concentrated load near its support.

MOMENT CAPACITY OF A WELDED PLATE GIRDER

A welded plate girder is composed of a 66 × ⅜ in (1676.4 × 9.53 mm) web plate and two 20 × ¾ in (508.0 × 19.05 mm) flange plates. The unbraced length of the compression flange is 18 ft (5.5 m). If $C_b = 1$, what bending moment can this member resist?

Calculation Procedure:

1. Compute the properties of the section

The tables in the AISC *Manual* are helpful in calculating the moment of inertia. Thus $A_f = 15$ in² (96.8 cm²); $A_w = 24.75$ in² (159.687 cm²); $I = 42,400$ in⁴ (176.481 dm⁴); $S = 1256$ in³ (20,585.8 cm³).

For the T section comprising the flange and one-sixth the web, $A = 15 + 4.13 = 19.13$ in² (123.427 cm²); then $I = (1/12)(0.75)(20)^3 = 500$ in⁴ (2081.1 dm⁴); $r = (500/19.13)^{0.5} = 5.11$ in (129.794 mm); $L'/r = 18(12)/5.11 = 42.3$.

2. Ascertain if the member satisfies the AISC Specification

Let h denote the clear distance between flanges, in (cm). Then: flange, ½(20)/0.75 = 13.3 < 16— this is acceptable; web, $h/t_w = 66/0.375 = 176 < 320$—this is acceptable.

3. Compute the allowable bending stress

Use $f_1 = 22,000 - 0.679(L'/r)^2/C_b$, or $f_1 = 22,000 - 0.679(42.3)^2 = 20,800$ lb/in² (143,416.0 kPa); $f_2 = 12,000,000/(L'd/A_f) = 12,000,000(15)/[18(12)(67.5)] = 12,300$ lb/in² (84,808.5 kPa). Therefore, use 20,800 lb/in² (143,416.0 kPa) because it is the larger of the two stresses.

4. Reduce the allowable bending stress in accordance with the AISC Specification

Using the equation given in the *Manual* yields $f_3 = 20,800\{1 - 0.005(24.75/15)[176 - 24,000/(20,800)^{0.5}]\} = 20,600$ lb/in² (142,037.0 kPa).

5. Determine the allowable bending moment

Use $M = f_3 S = 20.6(1256)/12 = 2156$ ft·kips (2923.5 kN·m).

ANALYSIS OF A RIVETED PLATE GIRDER

A plate girder is composed of one web plate 48 × ⅜ in (1219.2 × 9.53 mm); four flange angles 6 × 4 × ¾ in (152.4 × 101.6 × 19.05 mm); two cover plates 14 × ½ in (355.6 × 12.7 mm). The flange angles are set 48.5 in (1231.90 mm) back to back with their 6-in (152.4-mm) legs outstanding; they are connected to the web plate by ⅞-in (22.2-mm) rivets. If the member has continuous lateral support, what bending moment may be applied? What spacing of flange-to-web rivets is required in a panel where the vertical shear is 180 kips (800.6 kN)?

Calculation Procedure:

1. Obtain the properties of the angles from the AISC Manual

Record the angle dimensions as shown in Fig. 68.

2. Check the cover plates for compliance with the AISC Specification

The cover plates are found to comply with the pertinent sections of the *Specification*.

3. Compute the gross flange area and rivet-hole area

Ascertain whether the *Specification* requires a reduction in the flange area. Thus gross flange area = 2(6.94) + 7.0 = 20.88 in² (134.718 cm²); area of rivet holes = 2(½)(1) + 4(¾)(1) = 4.00 in² (25.808 cm²); allowable area of holes = 0.15(20.88) = 3.13. The excess area = hole area −

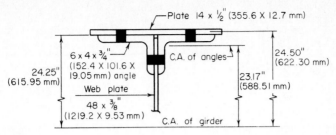

FIG. 68

allowable area $= 4.00 - 3.13 = 0.87$ in^2 (5.613 cm^2). Consider that this excess area is removed from the outstanding legs of the angles, at both the top and the bottom.

4. Compute the moment of inertia of the net section

	in^4	dm^4
One web plate, I_0	3,456	14.384
Four flange angles, I_0	35	0.1456
$Ay^2 = 4(6.94)(23.17)^2$	14,900	62.0184
Two cover plates:		
$Ay^2 = 2(7.0)(24.50)^2$	8,400	34.9634
I of gross section	26,791	111.5123
Deduct $2(0.87)(23.88)^2$ for excess area	991	4.12485
I of net section	25,800	107.387

5. Establish the allowable bending stress

Use the *Specification*. Thus $h/t_w = (48.5 - 8)/0.375 < 24,000/(22,000)^{0.5}$; $\therefore$ use 22,000 lb/in^2 (151,690.0 kPa). Also, $M = fI/c = 22(25,800)/[24.75(12)] = 1911$ ft·kips (2591.3 kN·m).

6. Calculate the horizontal shear flow to be resisted

Here Q of flange $= 13.88(23.17) + 7.0(24.50) - 0.87(23.88) = 472$ in^3 (7736.1 cm^3); $q = VQ/I = 180,000(472)/25,800 = 3290$ lb/lin in (576,167.2 N/m).

From a previous calculation procedure, $R_{ds} = 18,040$ lb (80,241.9 N); $R_b = 42,440(0.375) = 15,900$ lb (70,723.2 N); $s = 15,900/3290 = 4.8$ in (121.92 mm), where $s =$ allowable rivet spacing, in (mm). Therefore, use a 4¾-in (120.65-mm) rivet pitch. This satisfies the requirements of the *Specification*.

Note: To determine the allowable rivet spacing, divide the horizontal shear flow into the rivet capacity.

DESIGN OF A WELDED PLATE GIRDER

A plate girder of welded construction is to support the loads shown in Fig. 69a. The distributed load will be applied to the top flange, thereby offering continuous lateral support. At its ends, the girder will bear on masonry buttresses. The total depth of the girder is restricted to approximately 70 in (1778.0 mm). Select the cross section, establish the spacing of the transverse stiffeners, and design both the intermediate stiffeners and the bearing stiffeners at the supports.

Calculation Procedure:

1. Construct the shear and bending-moment diagrams

These diagrams are shown in Fig. 69.

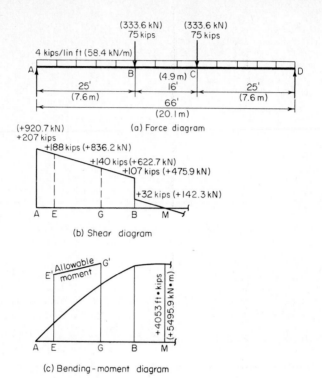

(a) Force diagram

(b) Shear diagram

(c) Bending-moment diagram

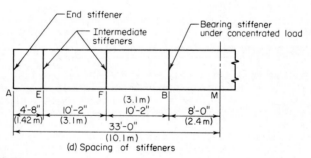

(d) Spacing of stiffeners

FIG. 69

2. Choose the web-plate dimensions

Since the total depth is limited to about 70 in (1778.0 mm), use a 68-in (1727.2-mm) deep web plate. Determine the plate thickness, using the *Specification* limits, which are a slenderness ratio h/t_w of 320. However, if an allowable bending stress of 22,000 lb/in^2 (151,690.0 kPa) is to be maintained, the *Specification* imposes an upper limit of $24,000/(22,000)^{0.5} = 162$. Then $t_w = h/162 = 68/162 = 0.42$ in (10.668 mm); use a $\frac{7}{16}$-in (11.112-mm) plate. Hence, the area of the web $A_w = 29.75$ in^2 (191.947 cm^2).

3. Select the flange plates

Apply the approximation $A_f = Mc/(2fy^2) - A_w/6$, where y = distance from the neutral axis to the centroidal axis of the flange, in (mm).

Assume 1-in (25.4-mm) flange plates. Then $A_f = 4053(12)(35)/[2(22)(34.5)^2] - 29.75/6 = 27.54$ in^2 (177.688 cm^2). Try $22 \times 1\frac{1}{4}$ in (558.8 × 31.75 mm) plates with $A_f = 27.5$ in^2 (177.43 cm^2). The width-thickness ratio of projection $= 11/1.25 = 8.8 < 16$. This is acceptable.

Thus, the trial section will be one web plate 68 × $\frac{7}{16}$ in (1727 × 11.11 mm); two flange plates $22 \times 1\frac{1}{4}$ in (558.8 × 31.75 mm).

4. Test the adequacy of the trial section

For this test, compute the maximum flexural and shearing stresses. Thus, $I = (1/12)(0.438)(68)^3 + 2(27.5)(34.63)^2 = 77,440$ in^3 (1,269,241.6 cm^3); $f = Mc/I = 4053(12)(35.25)/77,440 = 22.1$ kips/in^2 (152.38 MPa). This is acceptable. Also, $v = 207/29.75 = 6.96 < 14.5$ kips/in^2 (99.98 MPa). This is acceptable. Hence, the trial section is satisfactory.

5. Determine the distance of the stiffeners from the girder ends

Refer to Fig. 69d for the spacing of the intermediate stiffeners. Establish the length of the end panel AE. The *Specification* stipulates that the smaller dimension of the end panel shall not exceed $11,000(0.438)/(6960)^{0.5} = 57.8 < 68$ in (1727.2 mm). Therefore, provide stiffeners at 56 in (1422.4 mm) from the ends.

6. Ascertain whether additional intermediate stiffeners are required

See whether stiffeners are required in the interval EB by applying the *Specification* criteria.

Stiffeners are not required when $h/t_w < 260$ and the shearing stress within the panel is below the value given by either of two equations in the *Specification*, whichever equation applies. Thus $EB = 396 - (56 + 96) = 244$ in (6197.6 mm); $h/t_w = 68/0.438 = 155 < 260$; this is acceptable. Also, $a/h = 244/68 = 3.59$.

In lieu of solving either of the equations given in the *Specification*, enter the table of a/h, h/t_w values given in the AISC *Manual* to obtain the allowable shear stress. Thus, with $a/h > 3$ and $h/t_w = 155$, $v_{\text{allow}} = 3.45$ kips/in^2 (23.787 MPa) from the table.

At E, $V = 207 - 4.67(4) = 188$ kips (836.2 kN); $v = 188/29.75 = 6.32$ kips/in^2 (43.576 MPa) > 3.45 kips/in^2 (23.787 MPa); therefore, intermediate stiffeners are required in EB.

7. Provide stiffeners, and investigate the suitability of their tentative spacing

Provide stiffeners at F, the center of EB. See whether this spacing satisfies the *Specification*. Thus $[260/(h/t_w)]^2 = (260/155)^2 = 2.81$; $a/h = 122/68 = 1.79 < 2.81$. This is acceptable.

Entering the table referred to in step 6 with $a/h = 1.79$ and $h/t_w = 155$ shows $v_{\text{allow}} = 7.85 > 6.32$. This is acceptable.

Before we conclude that the stiffener spacing is satisfactory, it is necessary to investigate the combined shearing and bending stress and the bearing stress in interval EB.

8. Analyze the combination of shearing and bending stress

This analysis should be made throughout EB in the light of the *Specification* requirements. The net effect is to reduce the allowable bending moment whenever $V > 0.6V_{\text{allow}}$. Thus, $V_{\text{allow}} = 7.85(29.75) = 234$ kips (1040.8 kN); and $0.6(234) = 140$ kips (622.7 kN).

In Fig. 69b, locate the boundary section G where $V = 140$ kips (622.7 kN). The allowable moment must be reduced to the left of G. Thus, $AG = (207 - 140)/4 = 16.75$ ft (5.105 m); $M_G = 2906$ ft·kips (3940.5 kN·m); $M_E = 922$ ft·kips (1250.2 kN·m). At G, $M_{\text{allow}} = 4053$ ft·kips (5495.8 kN·m). At E, $f_{\text{allow}} = [0.825 - 0.375(188/234)](36) = 18.9$ kips/in^2 (130.31 MPa); $M_{\text{allow}} = 18.9(77,440)/[35.25(12)] = 3460$ ft·kips (4691.8 kN·m).

In Fig. 69c, plot points E' and G' to represent the allowable moments and connect these points with a straight line. In all instances, $M < M_{\text{allow}}$.

9. Use an alternative procedure, if desired

As an alternative procedure in step 8, establish the interval within which $M > 0.75M_{\text{allow}}$ and reduce the allowable shear in accordance with the equation given in the *Specification*.

10. Compare the bearing stress under the uniform load with the allowable stress

The allowable stress given in the *Specification* is $f_{b,\text{allow}} = [5.5 + 4/(a/h)^2]10,000/(h/t_w)^2$ kips/in^2 (MPa), or, for this girder, $f_{b,\text{allow}} = (5.5 + 4/1.79^2)10,000/155^2 = 2.81$ kips/in^2 (19.374 MPa). Then $f_b = 4/[12(0.438)] = 0.76$ kips/in^2 (5.240 MPa). This is acceptable. The stiffener spacing in interval EB is therefore satisfactory in all respects.

11. *Investigate the need for transverse stiffeners in the center interval*

Considering the interval BC, V = 32 kips (142.3 kN); v = 1.08 kips/in^2 (7.447 MPa); a/h = $192/68$ = 2.82 $\simeq [260/(h/t_w)]^2$.

The *Manual* table used in step 6 shows that v_{allow} > 1.08 kips/in^2 (7.447 MPa); $f_{b,\text{allow}}$ = (5.5 + 4/2.82^2)10,000/155^2 = 2.49 kips/in^2 (17.169 MPa) > 0.76 kips/in^2 (5.240 MPa). This is acceptable. Since all requirements are satisfied, stiffeners are not needed in interval BC.

12. *Design the intermediate stiffeners in accordance with the* Specification

For the interval EB, the preceding calculations yield these values: v = 6.32 kips/in^2 (43.576 MPa); v_{allow} = 7.85 kips/in^2 (54.125 MPa). Enter the table mentioned in step 6 with a/h = 1.79 and h/t_w = 155 to obtain the percentage of web area, shown in italics in the table. Thus, A_{st} required = 0.0745(29.75)(6.32/7.85) = 1.78 in^2 (11.485 cm^2). Try two 4 $\times$ ¼ in (101.6 $\times$ 6.35 mm) plates; A_{st} = 2.0 in^2 (12.90 cm^2); width-thickness ratio = 4/0.25 = 16. This is acceptable. Also, $(h/50)^4$ = $(68/50)^4$ = 3.42 in^4 (142.351 cm^4); I = (1/12)(0.25)(8.44)^3 = 12.52 in^4 (521.121 cm^4) > 3.42 in^4 (142.351 cm^4). This is acceptable.

The stiffeners must be in intimate contact with the compression flange, but they may terminate 1¾ in (44.45 mm) from the tension flange. The connection of the stiffeners to the web must transmit the vertical shear specified in the *Specification*.

13. *Design the bearing stiffeners at the supports*

Use the directions given in the *Specification*. The stiffeners are considered to act in conjunction with the tributary portion of the web to form a column section, as shown in Fig. 70. Thus, area

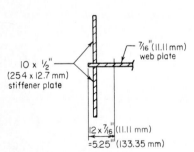

FIG. 70   Effective column section.

of web = 5.25(0.438) = 2.30 in^2 (14.839 cm^2). Assume an allowable stress of 20 kips/in^2 (137.9 MPa). Then, plate area required = 207/20 − 2.30 = 8.05 in^2 (51.938 cm^2).

Try two plates 10 $\times$ ½ in (254.0 $\times$ 12.7 mm), and compute the column capacity of the section. Thus, A = 2(10)(0.5) + 2.30 = 12.30 in^2 (79.359 cm^2); I = (1/12)(0.5)(20.44)^3 = 356 in^4 (1.4818 dm^4); r = (356/12.30)$^{0.5}$ = 5.38 in (136.652 mm); L/r = 0.75(68)/5.38 = 9.5.

Enter the table of slenderness ratio and allowable stress in the *Manual* with the slenderness ratio of 9.5, and obtain an allowable stress of 21.2 kips/in^2 (146.17 MPa). Then f = 207/12.30 = 16.8 kips/in^2 (115.84 MPa) < 21.2 kips/in^2 (146.17 MPa). This is acceptable.

Compute the bearing stress in the stiffeners. In computing the bearing area, assume that each stiffener will be clipped 1 in (25.4 mm) to clear the flange-to-web welding. Then f = 207/[2(9)(0.5)] = 23 kips/in^2 (158.6 MPa). The *Specification* provides an allowable stress of 33 kips/in^2 (227.5 MPa).

The 10 $\times$ ½ in (254.0 $\times$ 12.7 mm) stiffeners at the supports are therefore satisfactory with respect to both column action and bearing.

Steel Columns and Tension Members

The general remarks appearing at the opening of the previous part apply to this part as well.

A column is a compression member having a length that is very large in relation to its lateral dimensions. The *effective* length of a column is the distance between adjacent points of contraflexure in the buckled column or in the imaginary extension of the buckled column, as shown in Fig. 71. The column length is denoted by L, and the effective length by KL. Recommended design values of K are given in the AISC *Manual*.

The capacity of a column is a function of its effective length and the properties of its cross section. It therefore becomes necessary to formulate certain principles pertaining to the properties of an area.

Consider that the moment of inertia I of an area is evaluated with respect to a group of concurrent axes. There is a distinct value of I associated with each axis, as given by earlier equations in this section. The *major* axis is the one for which I is maximum; the *minor* axis is the one for which I is minimum. The major and minor axes are referred to collectively as the *principal* axes.

With reference to the equation given earlier, namely, $I_{x''} = I_{x'} \cos^2 \theta + I_{y'} \sin^2 \theta - P_{x'y'} \sin 2\theta$, the orientation of the principal axes relative to the given x' and y' axes is found by differentiating $I_{x''}$ with respect to θ, equating this derivative to zero, and solving for θ to obtain $\tan 2\theta = 2P_{x'y'}/(I_{y'} - I_{x'})$, Fig. 15.

The following statements are corollaries of this equation:

1. The principal axes through a given point are mutually perpendicular, since the two values of θ that satisfy this equation differ by 90°.

2. The product of inertia of an area with respect to its principal axes is zero.

3. Conversely, if the product of inertia of an area with respect to two mutually perpendicular axes is zero, these are principal axes.

4. An axis of symmetry is a principal axis, for the product of inertia of the area with respect to this axis and one perpendicular thereto is zero.

Let A_1 and A_2 denote two areas, both of which have a radius of gyration r with respect to a given axis. The radius of gyration of their composite area is found in this manner: $I_c = I_1 + I_2 = A_1 r^2 + A_2 r^2 = (A_1 + A_2) r^2$. But $A_1 + A_2 = A_c$. Substituting gives $I_c = A_c r^2$; therefore, $r_c = r$.

This result illustrates the following principle: If the radii of gyration of several areas with respect to a given axis are all equal, the radius of gyration of their composite area equals that of the individual areas.

The equation $I_x = \Sigma I_0 + \Sigma A k^2$, when applied to a single area, becomes $I_x = I_0 + A k^2$. Then $A r_x^2 = A r_0^2 + A k^2$, or $r_x = (r_0^2 + k^2)^{0.5}$. If the radius of gyration with respect to a centroidal axis is known, the radius of gyration with respect to an axis parallel thereto may be readily evaluated by applying this relationship.

The Euler equation for the strength of a slender column reveals that the member tends to buckle about the minor centroidal axis of its cross section. Consequently, all column design equations, both those for slender members and those for intermediate-length members, relate the capacity of the column to its minimum radius of gyration. The first step in the investigation of a column, therefore, consists in identifying the minor centroidal axis and evaluating the corresponding radius of gyration.

CAPACITY OF A BUILT-UP COLUMN

A compression member consists of two 15-in (381.0-mm) 40-lb (177.92-N) channels laced together and spaced 10 in (254.0 mm) back to back with flanges outstanding, as shown in Fig. 72. What axial load may this member carry if its effective length is 22 ft (6.7 m)?

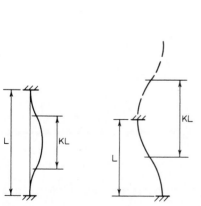

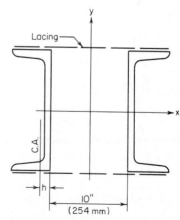

FIG. 71   Effective column lengths.

FIG. 72   Built-up column.

Calculation Procedure:

1. Record the properties of the individual channel

Since x and y are axes of symmetry, they are the principal centroidal axes. However, it is not readily apparent which of these is the minor axis, and so it is necessary to calculate both r_x and r_y. The symbol r, without a subscript, is used to denote the *minimum* radius of gyration, in inches (centimeters).

Using the AISC *Manual*, we see that the channel properties are $A = 11.70$ in^2 (75.488 cm^2); $h = 0.78$ in (19.812 mm); $r_1 = 5.44$ in (138.176 mm); $r_2 = 0.89$ in (22.606 mm).

2. Evaluate the minimum radius of gyration of the built-up section; determine the slenderness ratio

Thus, $r_x = 5.44$ in (138.176 mm); $r_y = (r_2^2 + 5.78^2)^{0.5} > 5.78$ in (146.812 mm); therefore, $r = 5.44$ in (138.176 mm); $KL/r = 22(12)/5.44 = 48.5$.

3. Determine the allowable stress in the column

Enter the *Manual* slenderness-ratio allowable-stress table with a slenderness ratio of 48.5 to obtain the allowable stress $f = 18.48$ kips/in^2 (127.420 MPa). Then, the column capacity $= P = Af = 2(11.70)(18.48) = 432$ kips (1921.5 kN).

CAPACITY OF A DOUBLE-ANGLE STAR STRUT

A star strut is composed of two 5 × 5 × ⅜ in (127.0 × 127.0 × 9.53 mm) angles intermittently connected by ⅜-in (9.53-mm) batten plates in both directions. Determine the capacity of the member for an effective length of 12 ft (3.7 m).

Calculation Procedure:

1. Identify the minor axis

Refer to Fig. 73a. Since p and q are axes of symmetry, they are the principal axes; p is manifestly the minor axis because the area lies closer to p than q.

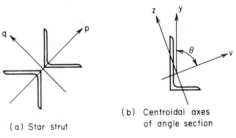

(b) Centroidal axes
of angle section

(a) Star strut

FIG. 73

2. Determine r_v^2

Refer to Fig. 73b, where v is the major and z the minor axis of the angle section. Apply $I_{x'} = I_{x'} \cos^2 \theta + I_{y'} \sin^2 \theta - P_{x'y'} \sin 2\theta$, and set $P_{vz} = 0$ to obtain $r_y^2 = r_v^2 \cos^2 \theta + r_z^2 \sin^2 \theta$; therefore, $r_v^2 = r_y^2 \sec^2 \theta - r_z^2 \tan^2 \theta$. For an equal-leg angle, $\theta = 45°$, and this equation reduces to $r_v^2 = 2r_y^2 - r_z^2$.

3. Record the member area and computer r_v

From the *Manual*, $A = 3.61$ in^2 (23.291 cm^2); $r_y = 1.56$ in (39.624 mm); $r_z = 0.99$ in (25.146 mm); $r_v = (2 \times 1.56^2 - 0.99^2)^{0.5} = 1.97$ in (50.038 mm).

4. Determine the minimum radius of gyration of the built-up section; compute the strut capacity

Thus, $r = r_p = 1.97$ in (50.038 mm); $KL/r = 12(12)/1.97 = 73$. From the *Manual*, $f = 16.12$ kips/in^2 (766.361 MPa). Then $P = Af = 2(3.61)(16.12) = 116$ kips (515.97 kN).

SECTION SELECTION FOR A COLUMN WITH TWO EFFECTIVE LENGTHS

A 30-ft (9.2-m) long column is to carry a 200-kip (889.6-kN) load. The column will be braced about both principal axes at top and bottom and braced about its minor axis at midheight. Architectural details restrict the member to a nominal depth of 8 in (203.2 mm). Select a section of A242 steel by consulting the allowable-load tables in the AISC *Manual* and then verify the design.

Calculation Procedure:

1. Select a column section

Refer to Fig. 74. The effective length with respect to the minor axis may be taken as 15 ft (4.6 m). Then $K_xL = 30$ ft (9.2 m) and $K_yL = 15$ ft (4.6 m).

The allowable column loads recorded in the *Manual* tables are calculated on the premise that the column tends to buckle about the minor axis. In the present instance, however, this premise is not necessarily valid. It is expedient for design purposes to conceive of a uniform-strength column, i.e., one for which K_x and K_y bear the same ratio as r_x and r_y, thereby endowing the column with an identical slenderness ratio with respect to the two principal axes.

Select a column section on the basis of the K_yL value; record the value of r_x/r_y of this section. Using linear interpolation in the *Manual* Table shows that an 8WF40 (W8 × 40) column has a capacity of 200 kips (889.6 kN) when $K_yL = 15.3$ ft (4.66 m); at the bottom of the table it is found that $r_x/r_y = 1.73$.

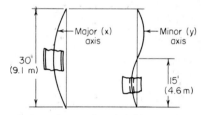

FIG. 74 Effective column lengths.

2. Compute the value of K_xL associated with a uniform-strength column, and compare this with the actual value

Thus, $K_xL = 1.73(15.3) = 26.5$ ft (8.1 m) < 30 ft (9.2 m). The section is therefore inadequate.

3. Try a specific column section of larger size

Trying 8WF48 (W8 × 48), the capacity = 200 kips (889.6 kN) when $K_yL = 17.7$ ft (5.39 m). For uniform strength, $K_xL = 1.74(17.7) = 30.8 > 30$ ft (9.39 m > 9.2 m). The 8WF48 therefore appears to be satisfactory.

4. Verify the design

To verify the design, record the properties of this section and compute the slenderness ratios. For this grade of steel and thickness of member, the yield-point stress is 50 kips/in² (344.8 MPa), as given in the *Manual*. Thus, $A = 14.11$ in² (91.038 cm²); $r_x = 3.61$ in (91.694 mm); $r_y = 2.08$ in (52.832 mm). Then $K_xL/r_x = 30(12)/3.61 = 100$; $K_yL/r_y = 15(12)/2.08 = 87$.

5. Determine the allowable stress and member capacity

From the *Manual*, $f = 14.71$ kips/in² (101.425 MPa) with a slenderness ratio of 100. Then $P = 14.11(14.71) = 208$ kips (925.2 kN). Therefore, use 8WF48 because the capacity of the column exceeds the intended load.

STRESS IN COLUMN WITH PARTIAL RESTRAINT AGAINST ROTATION

The beams shown in Fig. 75a are rigidly connected to a 14WF95 (W14 × 95) column of 28-ft (8.5-m) height that is pinned at its foundation. The column is held at its upper end by cross bracing lying in a plane normal to the web. Compute the allowable axial stress in the column in the absence of bending stress.

Calculation Procedure:

1. Draw schematic diagrams to indicate the restraint conditions

Show these conditions in Fig. 75b. The cross bracing prevents sidesway at the top solely with respect to the minor axis, and the rigid beam-to-column connections afford partial fixity with respect to the major axis.

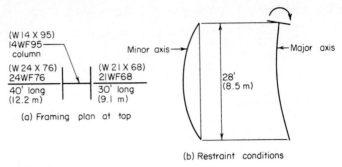

(a) Framing plan at top

(b) Restraint conditions

FIG. 75

2. Record the I_x values of the column and beams

	I_x	
Section	in⁴	cm⁴
14WF95	1064	44,287
24WF76	2096	87,242
21WF68	1478	61,519

3. Calculate the rigidity of the column relative to that of the restraining members at top and bottom

Thus, $I_c/L_c = 1064/28 = 38$. At the top, $\Sigma(I_g/L_g) = 2096/40 + 1478/30 = 101.7$. At the top, the rigidity $G_t = 38/101.7 = 0.37$.

In accordance with the instructions in the *Manual*, set the rigidity at the bottom $G_b = 10$.

4. Determine the value of K_x

Using the *Manual* alignment chart, determine that $K_x = 1.77$.

5. Compute the slenderness ratio with respect to both principal axes, and find the allowable stress

Thus, $K_x L/r_x = 1.77(28)(12)/6.17 = 96.4$; $K_y L/r_y = 28(12)/3.71 = 90.6$.

Using the larger value of the slenderness ratio, find from the *Manual* the allowable axial stress in the absence of bending $= f = 13.43$ kips/in² (92.600 MPa).

LACING OF BUILT-UP COLUMN

Design the lacing bars and end tie plates of the member in Fig. 76. The lacing bars will be connected to the channel flanges with ½-in (12.7-mm) rivets.

Calculation Procedure:

1. Establish the dimensions of the lacing system to conform to the AISC Specification

The function of the lacing bars and tie plates is to preserve the integrity of the column and to prevent local failure.

Refer to Fig. 76. The standard gage in 15-in (381.0-mm) channel = 2 in (50.8 mm), from the AISC *Manual*. Then $h = 14 < 15$ in (381.0 mm); therefore, use single lacing.

Try $\theta = 60°$; then, $v = 2(14) \cot 60° = 16.16$ in (410.5 mm). Set $v = 16$ in (406.4 mm); therefore, $d = 16.1$ in (408.94 mm). For the built-up section, $KL/r = 48.5$; for the single channel, $KL/r = 16/0.89 < 48.5$. This is acceptable. The spacing of the bars is therefore satisfactory.

2. Design the lacing bars

The lacing system must be capable of transmitting an assumed transverse shear equal to 2 percent of the axial load; this shear is carried by two bars, one on each side. A lacing bar is classified as a secondary member. To compute the transverse shear, assume that the column will be loaded to its capacity of 432 kips (1921.5 N).

Then force per bar = $\frac{1}{2}(0.02)(432)(16.1/14) = 5.0$ kips (22.24 N). Also, $L/r \leq 140$; therefore, $r = 16.1/140 = 0.115$ in (2.9210 mm).

For a rectangular section of thickness t, $r = 0.289t$. Then $t = 0.115/0.289 = 0.40$ in (10.160 mm). Set $t = \frac{7}{16}$ in (11.11 mm); $r = 0.127$ in (3.226 mm); $L/r = 16.1/0.127 = 127$; $f = 9.59$ kips/in^2 (66.123 MPa); $A = 5.0/9.59 = 0.52$ in^2 (3.355 cm^2). From the *Manual*, the minimum width required for $\frac{1}{2}$-in (12.7 mm) rivets = $1\frac{1}{2}$ in (38.1 mm). Therefore, use a flat bar $1\frac{1}{2} \times \frac{7}{16}$ in (38.1 $\times$ 11.11 mm); $A = 0.66$ in^2 (4.258 cm^2).

3. Design the end tie plates in accordance with the Specification

The minimum length = 14 in (355.6 mm); $t = 14/50 = 0.28$. Therefore, use plates 14 $\times$ $\frac{5}{16}$ in (355.6 $\times$ 7.94 mm). The rivet pitch is limited to six diameters, or 3 in (76.2 mm).

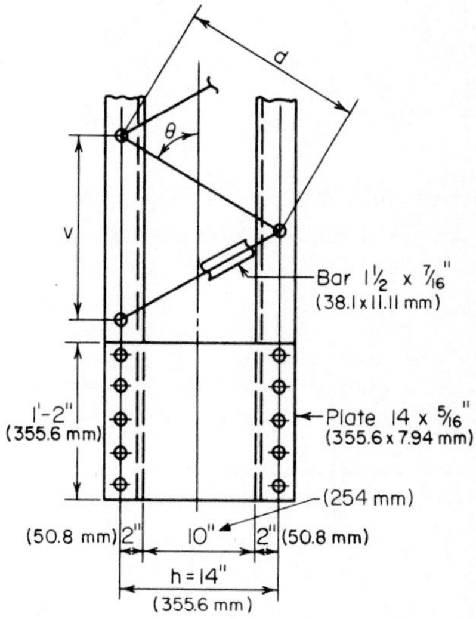

FIG. 76 Lacing and tie plates.

SELECTION OF A COLUMN WITH A LOAD AT AN INTERMEDIATE LEVEL

A column of 30-ft (9.2-m) length carries a load of 130 kips (578.2 kN) applied at the top and a load of 56 kips (249.1 kN) applied to the web at midheight. Select an 8-in (203.2-mm) column of A242 steel, using $K_xL = 30$ ft (9.2 m) and $K_yL = 15$ ft (4.6 m).

Calculation Procedure:

1. Compute the effective length of the column with respect to the major axis

The following procedure affords a rational method of designing a column subjected to a load applied at the top and another load applied approximately at the center. Let m = load at intermediate level, kips per total load, kips (kilonewtons). Replace the factor K with a factor K' defined by $K' = K(1 - m/2)^{0.5}$. Thus, for this column, $m = 56/186 = 0.30$. And $K_x'L = 30(1 - 0.15)^{0.5} = 27.6$ ft (8.41 m).

2. Select a trial section on the basis of the K_yL value

From the AISC *Manual* for an 8WF40 (W8 $\times$ 40), capacity = 186 kips (827.3 kN) when $K_yL = 16.2$ ft (4.94 m) and $r_x/r_y = 1.73$.

3. Determine whether the selected section is acceptable

Compute the value of K_xL associated with a uniform-strength column, and compare this with the actual effective length. Thus, $K_xL = 1.73(16.2) = 28.0 > 27.6$ ft (8.41 m). Therefore, the 8WF40 is acceptable.

DESIGN OF AN AXIAL MEMBER FOR FATIGUE

A web member in a welded truss will sustain precipitous fluctuations of stress caused by moving loads. The structure will carry three load systems having the following characteristics:

System	Force induced in member, kips (kN)		No. of times applied
	Maximum compression	Maximum tension	
A	46 (204.6)	18 (80.1)	60,000
B	40 (177.9)	9 (40.0)	1,000,000
C	32 (142.3)	8 (35.6)	2,500,000

The effective length of the member is 11 ft (3.4 m). Design a double-angle member.

Calculation Procedure:

1. Calculate for each system the design load, and indicate the yield-point stress on which the allowable stress is based

The design of members subjected to a repeated variation of stress is regulated by the AISC *Specification*. For each system, calculate the design load and indicate the yield-point stress on which the allowable stress is based. Where the allowable stress is less than that normally permitted, increase the design load proportionally to compensate for this reduction. Let + denote tension and − denote compression. Then

System	Design load, kips (kN)	Yield-point stress, kips/in^2 (MPa)
A	$-46 - \tfrac{5}{9}(18) = -58 \ (-257.9)$	36 (248.2)
B	$-40 - \tfrac{5}{9}(9) = -46 \ (-204.6)$	33 (227.5)
C	$1.5(-32 - \tfrac{3}{4} \times 8) = -57 \ (-253.5)$	33 (227.5)

2. Select a member for system A and determine if it is adequate for system C

From the AISC *Manual*, try two angles 4 × 3½ × ⅜ in (101.6 × 88.90 × 9.53 mm), with long legs back to back; the capacity is 65 kips (289.1 kN). Then $A = 5.34$ in^2 (34.453 cm^2); $r = r_x = 1.25$ in (31.750 mm); $KL/r = 11(12)/1.25 = 105.6$.

From the *Manual*, for a yield-point stress of 33 kips/in^2 (227.5 MPa), $f = 11.76$ kips/in^2 (81.085 MPa). Then the capacity $P = 5.34(11.76) = 62.8$ kips (279.3 kN) > 57 kips (253.5 kN). This is acceptable. Therefore, use two angles 4 × 3½ × ⅜ in (101.6 × 88.90 × 9.53 mm), long legs back to back.

INVESTIGATION OF A BEAM COLUMN

A 12WF53 (W12 × 53) column with an effective length of 20 ft (6.1 m) is to carry an axial load of 160 kips (711.7 kN) and the end moments indicated in Fig. 77. The member will be secured against sidesway in both directions. Is the section adequate?

Calculation Procedure:

1. Record the properties of the section

The simultaneous set of values of axial stress and bending stress must satisfy the inequalities set forth in the AISC *Specification*.

The properties of the section are $A = 15.59$ in^2 (100.586 cm^2); $S_x = 70.7$ in^3 (1158.77 cm^3); $r_x = 5.23$ in (132.842 mm); $r_y = 2.48$ in (62.992 mm). Also, from the *Manual*, $L_c = 10.8$ ft (3.29 m); $L_u = 21.7$ ft (6.61 m).

2. Determine the stresses listed below

The stresses that must be determined are the axial stress f_a; the bending stress f_b; the axial stress F_a, which would be permitted in the absence of bending; and the bending stress F_b, which would be permitted in the absence of axial load. Thus, $f_a = 160/15.59 = 10.26$ kips/in^2 (70.742 MPa); $f_b = 31.5(12)/70.7 = 5.35$ kips/in^2 (36.888 MPa); $KL/r = 240/2.48 = 96.8$; therefore, $F_a = 13.38$ kips/in^2 (92.255 MPa); $L_u < KL < L_c$; therefore, $F_b = 22$ kips/in^2 (151.7 MPa). (Although this consideration is irrelevant in the present instance, note that the *Specification* establishes two maximum d/t ratios for a compact section. One applies to a beam, the other to a beam column.)

3. Calculate the moment coefficient C_m

Since the algebraic sign of the bending moment remains unchanged, M_1/M_2 is positive. Thus, $C_m = 0.6 + 0.4(15.2/31.5) = 0.793$.

4. Apply the appropriate criteria to test the adequacy of the section

Thus, $f_a/F_a = 10.26/13.38 = 0.767 > 0.15$. The following requirements therefore apply: $f_a/F_a + [C_m/(1 - fa/F'_e)](f_b/F_b) \leq 1$; $f_a/(0.6f_y) + f_b/F_b \leq 1$, where $F'_e = 149,000/(KL/r)^2$ kips/in^2 and KL and r are evaluated with respect to the plane of bending.

Evaluating gives $F'_e = 149,000(5.23)^2/240^2 = 70.76$ kips/in^2 (487.890 MPa); $f_a/F'_e = 10.26/70.76 = 0.145$. Substituting in the first requirements equation yields $0.767 + (0.793/0.855)(5.35/22) = 0.993$. This is acceptable. Substituting in the second requirements equation, we find $10.26/22 + 5.35/22 = 0.709$. This section is therefore satisfactory.

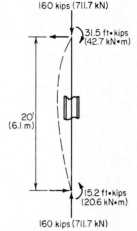

160 kips (711.7 kN)

31.5 ft·kips (42.7 kN•m)

20' (6.1 m)

15.2 ft·kips (20.6 kN•m)

160 kips (711.7 kN)

FIG. 77 Beam column.

APPLICATION OF BEAM-COLUMN FACTORS

For the previous calculation procedure, investigate the adequacy of the 12WF53 section by applying the values of the beam-column factors B and a given in the AISC *Manual.*

Calculation Procedure:

1. Record the basic values of the previous calculation procedure

The beam-column factors were devised in an effort to reduce the labor entailed in analyzing a given member as a beam column when $f_a/F_a > 0.15$. They are defined by $B = A/S$ per inch (decimeter); $a = 0.149 \times 10^6 I$ in^4 (6201.9I dm^4).

Let P denote the applied axial load and P_{allow} the axial load that would be permitted in the absence of bending. The equations given in the previous procedure may be transformed to $P + BMC_m(F_a/F_b)a/[a - P(KL)^2] \leq P_{\text{allow}}$, and $PF_a/(0.6f_y) + BMF_a/F_b \leq P_{\text{allow}}$, where KL, B, and a are evaluated with respect to the plane of bending.

The basic values of the previous procedure are $P = 160$ kips (711.7 kN); $M = 31.5$ ft·kips (42.71 kN·m); $F_b = 22$ kips/in^2 (151.7 MPa); $C_m = 0.793$.

2. Obtain the properties of the section

From the *Manual* for a 12WF53, $A = 15.59$ in^2 (100.587 cm^2); $B_x = 0.221$ per inch (8.70 per meter); $a_x = 63.5 \times 10^6$ in^4 (264.31 $\times 10^3$ dm^4). Then when $KL = 20$ ft (6.1 m), $P_{\text{allow}} = 209$ kips (929.6 kN).

3. Substitue in the first transformed equation

Thus, $F_a = P_{\text{allow}}/A = 209/15.59 = 13.41$ kips/in^2 (92.461 MPa), $P(KL)^2 = 160(240)^2 = 9.22 \times 10^6$ kip·in^2 (2.648 $\times 10^4$ kN·m^2), and $a_x/[a_x - P(KL)^2] = 63.5/(63.5 - 9.22) = 1.17$; then $160 + 0.221(31.5)(12)(0.793)(13.41/22)(1.17) = 207 < 209$ kips (929.6 kN). This is acceptable.

4. Substitute in the second transformed equation

Thus, $160(13.41/22) + 0.221(31.5)(12)(13.41/22) = 148 < 209$ kips (929.6 kN). This is acceptable. The 12WF53 section is therefore satisfactory.

NET SECTION OF A TENSION MEMBER

The $7 \times \frac{1}{4}$ in (177.8×6.35 mm) plate in Fig. 78 carries a tensile force of 18,000 lb (80,064.0 N) and is connected to its support with three $\frac{3}{4}$-in (19.05-mm) rivets in the manner shown. Compute the maximum tensile stress in the member.

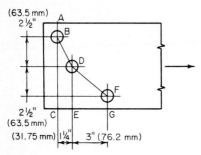

FIG. 78

Calculation Procedure:

1. Compute the net width of the member at each section of potential rupture

The AISC *Specification* prescribes the manner of calculating the net section of a tension member. The effective diameter of the holes is considered to be $\frac{1}{8}$ in (3.18 mm) greater than that of the rivets.

After computing the net width of each section, select the minimum value as the effective width. The *Specification* imposes an upper limit of 85 percent of the gross width.

Refer to Fig. 78: From B to D, $s = 1.25$ in (31.750 mm), $g = 2.5$ in (63.50 mm); from D to F, $s = 3$ in (76.2 mm), $g = 2.5$ in (63.50 mm); $w_{AC} = 7 - 0.875 = 6.12$ in (155.45 mm); $w_{ABDE} = 7 - 2(0.875) + 1.25^2/[4(2.5)] = 5.41$ in (137.414 mm); $w_{ABDFG} = 7 - 3(0.875) + 1.25^2/(4 \times 2.5) + 3^2/(4 \times 2.5) = 5.43$ in (137.922 mm); $w_{max} = 0.85(7) = 5.95$ in (151.13 mm). Selecting the lowest value gives $w_{eff} = 5.41$ in (137.414 mm).

2. Compute the tensile stress on the effective net section

Thus, $f = 18,000/[5.41(0.25)] = 13,300$ lb/in^2 (91,703.5 kPa).

DESIGN OF A DOUBLE-ANGLE TENSION MEMBER

The bottom chord of a roof truss sustains a tensile force of 141 kps (627.2 kN). The member will be spliced with $\frac{3}{4}$-in (19.05-mm) rivets as shown in Fig. 79a. Design a double-angle member and specify the minimum rivet pitch.

Calculation Procedure:

1. Show one angle in its developed form

Cut the outstanding leg, and position it to be coplanar with the other one, as in Fig. 79b. The gross width of the angle w_g is the width of the equivalent plate thus formed; it equals the sum of the legs of the angle less the thickness.

2. Determine the gross width in terms of the thickness

Assume tentatively that 2.5 rivet holes will be deducted to arrive at the net width. Express w_g in terms of the thickness t of each angle. Then net area required $= 141/22 = 6.40$ in^2 (41.292 cm^2); also, $2t(w_g - 2.5 \times 0.875) = 6.40$; $w_g = 3.20/t + 2.19$.

3. Assign trial thickness values, and determine the gross width

Construct a tabulation of the computed values. Then select the most economical size of member. Thus

t, in (mm)	w_g, in (mm)	$w_g + t$, in (mm)	Available size, in (mm)	Area, in^2 (cm^2)
$\frac{1}{2}$ (12.7)	8.59 (218.186)	9.09 (230.886)	$6 \times 3\frac{1}{2} \times \frac{1}{2}$ (152.4 $\times$ 88.9 $\times$ 12.7)	4.50 (29.034)
$\frac{7}{16}$ (11.11)	9.50 (241.300)	9.94 (252.476)	$6 \times 4 \times \frac{7}{16}$ (152.4 $\times$ 101.6 $\times$ 11.11)	4.18 (26.969)
$\frac{3}{8}$ (9.53)	10.72 (272.228)	11.10 (281.940)	None	

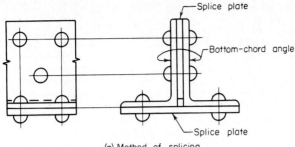

(a) Method of splicing

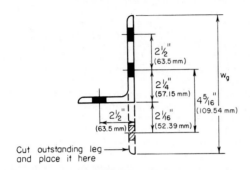

(b) Development of angle for net section

FIG. 79

The most economical member is the one with the least area. Therefore, use two angles 6 × 4 × ⁷⁄₁₆ in (152.4 × 101.6 × 11.11 mm).

4. Record the standard gages

Refer to the *Manual* for the standard gages, and record the values shown in Fig. 79*b*.

5. Establish the rivet pitch

Find the minimum value of s to establish the rivet pitch. Thus, net width required = ½[6.40/(7/16)] = 7.31 in (185.674 mm); gross width = 6 + 4 − 0.44 = 9.56 in (242.824 mm). Then 9.56 − 3(0.875) + s^2/(4 × 2.5) + s^2/(4 × 4.31) = 7.31; s = 1.55 in (39.370 mm).

For convenience, use the standard pitch of 3 in (76.2 mm). This results in a net width of 7.29 in (185.166 mm); the deficiency is negligible.

Plastic Design of Steel Structures

Consider that a structure is subjected to a gradually increasing load until it collapses. When the yield-point stress first appears, the structure is said to be in a state of *initial yielding*. The load that exists when failure impends is termed the *ultimate load*.

In elastic design, a structure has been loaded to capacity when it attains initial yielding, on the theory that plastic deformation would annul the utility of the structure. In plastic design, on the other hand, it is recognized that a structure may be loaded beyond initial yielding if:

1. The tendency of the fiber at the yield-point stress toward plastic deformation is resisted by the adjacent fibers.
2. Those parts of the structure that remain in the elastic-stress range are capable of supporting this incremental load.

The ultimate load is reached when these conditions cease to exist and thus the structure collapses.

Thus, elastic design is concerned with an allowable *stress*, which equals the yield-point stress divided by an appropriate factor of safety. In contrast, plastic design is concerned with an allowable *load*, which equals the ultimate load divided by an appropriate factor called the *load factor*. In reality, however, the distinction between elastic and plastic design has become rather blurred because specifications that ostensibly pertain to elastic design make covert concessions to plastic behavior. Several of these are underscored in the calculation procedures that follow.

In the plastic analysis of flexural members, the following simplifying assumptions are made:

1. As the applied load is gradually increased, a state is eventually reached at which all fibers at the section of maximum moment are stressed to the yield-point stress, in either tension or compression. The section is then said to be in a state of *plastification*.

2. While plastification is proceeding at one section, the adjacent sections retain their linear-stress distribution.

Although the foregoing assumptions are fallacious, they introduce no appreciable error.

When plastification is achieved at a given section, no additional bending stress may be induced in any of its fibers, and the section is thus rendered impotent to resist any incremental bending moment. As loading continues, the beam behaves as if it had been constructed with a hinge at the given section. Consequently, the beam is said to have developed a *plastic hinge* (in contradistinction to a true hinge) at the plastified section.

The *yield moment* M_y of a beam section is the bending moment associated with initial yielding. The plastic moment M_p is the bending moment associated with plastification.

The *plastic modulus* Z of a beam section, which is analogous to the section modulus used in elastic design, is defined by $Z = M_p/f_y$, where f_y denotes the yield-point stress. The *shape factor* SF is the ratio of M_p to M_y, being so named because its value depends on the shape of the section. Then SF $= M_p/M_y = f_yZ/(f_yS) = Z/S$.

In the following calculation procedures, it is understood that the members are made of A36 steel.

ALLOWABLE LOAD ON BAR SUPPORTED BY RODS

A load is applied to a rigid bar that is symmetrically supported by three steel rods as shown in Fig. 80. The cross-sectional areas of the rods are: rods A and C, 1.2 in^2 (7.74 cm^2); rod B, 1.0 in^2 (6.45 cm^2). Determine the maximum load that may be applied, (*a*) using elastic design with an allowable stress of 22,000 lb/in^2 (151,690.0 kPa); (*b*) using plastic design with a load factor of 1.85.

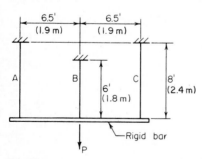

FIG. 80

Calculation Procedure:

1. Express the relationships among the tensile stresses in the rods

The symmetric disposition causes the bar to deflect vertically without rotating, thereby elongating the three rods by the same amount. As the first method of solving this problem, assume that the load is gradually increased from zero to its allowable value.

Expressing the relationships among the tensile stresses, we have $\Delta L = s_AL_a/E = s_BL_B/E = s_CL_C/E$; therefore, $s_A = s_C$, and $s_A = s_BL_B/L_A = 0.75s_B$ for this arrangement of rods. Since s_B is the maximum stress, the allowable stress first appears in rod B.

2. Evaluate the stresses at the instant the load attains its allowable value

Calculate the load carried by each rod, and sum these loads to find P_{allow}. Thus $s_B = 22,000$ lb/in^2 (151,690.0 kPa); $s_A = 0.75(22,000) = 16,500$ lb/in^2 (113,767.5 kPa); $P_A = P_C = 16,500(1.2) = 19,800$ lb (88,070.4 N); $P_B = 22,000(1.0) = 22,000$ lb (97,856.0 N); $P_{allow} = 2(19,800) + 22,000 = 61,600$ lb (273,996.8 N).

Next, consider that the load is gradually increased from zero to its ultimate value. When rod B attains its yield-point stress, its tendency to deform plastically is inhibited by rods A and C because the rigidity of the bar constrains the three rods to elongate uniformly. The structure therefore remains stable as the load is increased beyond the elastic range until rods A and C also attain their yield-point stress.

3. Find the ultimate load

To find the ultimate load P_u, equate the stress in each rod to f_y, calculate the load carried by each rod, and sum these loads to find the ultimate load P_u. Thus, $P_A = P_C = 36,000(1.2) = 43,200$ lb (192,153.6 N); $P_B = 36,000(1.0) = 36,000$ lb (160,128.0 N); $P_u = 2(43,200) + 36,000 = 122,400$ lb (544,435.2 N).

4. Apply the load factor to establish the allowable load

Thus, $P_{allow} = P_u/LF = 122,400/1.85 = 66,200$ lb (294,457.6 N).

DETERMINATION OF SECTION SHAPE FACTORS

Without applying the equations and numerical values of the plastic modulus given in the AISC *Manual*, determine the shape factor associated with a rectangle, a circle, and a 16WF40 (W16 × 40). Explain why the circle has the highest and the wide-flange section the lowest factor of the three.

Calculation Procedure:

1. Calculate M_y for each section

Use the equation $M_y = Sf_y$ for each section. Thus, for a rectangle, $M_y = bd^2f_y/6$. For a circle, using the properties of a circle as given in the *Manual*, we find $M_y = \pi d^3 f_y/32$. For a 16WF40, $A = 11.77$ in^2 (75.940 cm^2), $S = 64.4$ in^3 (1055.52 cm^3), and $M_y = 64.4 f_y$.

2. Compute the resultant forces associated with plastification

In Fig. 81, the resultant forces are C and T. Once these forces are known, their action lines and M_p should be computed.

Thus, for a rectangle, $C = bd f_y/2$, $a = d/2$, and $M_p = aC = bd^2 f_y/4$. For a circle, $C = \pi d^2 f_y/8$, $a = 4d/(3\pi)$, and $M_p = aC = d^3 f_y/6$. For a 16WF40, $C = \frac{1}{2}(11.77 f_y) = 5.885 f_y$.

To locate the action lines, refer to the *Manual* and note the position of the centroidal axis of the ST8WF20 (WT8 × 20) section, i.e., a section half the size of that being considered. Thus, $a = 2(8.00 - 1.82) = 12.36$ in (313.944 mm); $M_p = aC = 12.36(5.885 f_y) = 72.7 f_y$.

3 Divide M_p by M_y to obtain the shape factor

For a rectangle, SF $= (bd^2/4)/(bd^2/6) = 1.50$. For a circle, SF $= (d^3/6)/(\pi d^3/32) = 1.70$. For a 16WF40, SF $= 72.7/64.4 = 1.13$.

4. Explain the relative values of the shape factor

To explain the relative values of the shape factor, express the resisting moment contributed by a given fiber at plastification and at initial yielding,

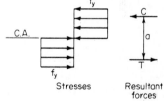

FIG. 81 Conditions at section of plastification.

and compare the results. Let dA denote the area of the given fiber and y its distance from the neutral axis. At plastification, $dM_p = f_y y dA$. At initial yielding, $f = f_y y/c$; $dM_y = f_y y^2 dA/c$; $dM_p/dM_y = c/y$.

By comparing a circle and a hypothetical wide-flange section having the same area and depth, the circle is found to have a larger shape factor because of its relatively low values of y.

As this analysis demonstrates, the process of plastification mitigates the detriment that accrues from placing any area near the neutral axis, since the stress at plastification is independent of the position of the fiber. Consequently, a section that is relatively inefficient with respect to flexure has a relatively high shape factor. The AISC *Specification* for elastic design implicitly recognizes the value of the shape factor by assigning an allowable bending stress of $0.75 f_y$ to rectangular bearing plates and $0.90 f_y$ to pins.

DETERMINATION OF ULTIMATE LOAD BY THE STATIC METHOD

The 18WF45 (W18 × 45) beam in Fig. 82a is simply supported at A and fixed at C. Disregarding the beam weight, calculate the ultimate load that may be applied at B (a) by analyzing the behavior of the beam during its two phases; (b) by analyzing the bending moments that exist at impending collapse. (The first part of the solution illustrates the postelastic behavior of the member.)

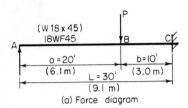

(a) Force diagram

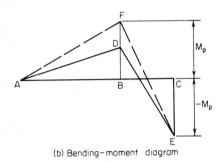

(b) Bending–moment diagram

FIG. 82

Calculation Procedure:

1. Calculate the ultimate-moment capacity of the member

Part a: As the load is gradually increased from zero to its ultimate value, the beam passes through two phases. During phase 1, the *elastic phase*, the member is restrained against rotation at C. This phase terminates when a plastic hinge forms at that end. During phase 2—the *postelastic*, or *plastic, phase*—the member functions as a simply supported beam. This phase terminates when a plastic hinge forms at B, since the member then becomes unstable.

Using data from the AISC *Manual*, we have $Z = 89.6$ in^3 (1468.54 cm^3). Then $M_p = f_y Z = 36(89.6)/12 = 268.8$ ft·kips (364.49 kN·m).

2. Calculate the moment BD

Let P_1 denote the applied load at completion of phase 1. In Fig. 82b, construct the bending-moment diagram ADEC corresponding to this load. Evaluate P_1 by applying the equations for case 14 in the AISC *Manual*. Calculate the moment BD. Thus, $CE = -ab(a + L)P_1/(2L^2) = -20(10)(50)P_1/[2(900)] = -268.8$; $P_1 = 48.38$ kips (215.194 kN); $BD = ab^2(a + 2L)P_1/(2L^3) = 20(100)(80)(48.38)/[2(27,000)] = 143.3$ ft·kips (194.31 kN·m).

3. Determine the incremental load at completion of phase 2

Let P_2 denote the incremental applied load at completion of phase 2, i.e., the actual load on the beam minus P_1. In Fig. 82b, construct the bending-moment diagram AFEC that exists when phase 2 terminates. Evaluate P_2 by considering the beam as simply supported. Thus, $BF = 268.8$ ft·kips (364.49 kN·m); $DF = 268.8 - 143.3 = 125.5$ ft·kips (170.18 kN·m); but $DF = abP_2/L = 20(10)P_2/30 = 125.5$; $P_2 = 18.82$ kips (83.711 kN).

4. Sum the results to obtain the ultimate load

Thus, $P_u = 48.38 + 18.82 = 67.20$ kips (298.906 kN).

5. Construct the force and bending-moment diagrams for the ultimate load

Part b: The following considerations are crucial: The bending-moment diagram always has vertices at B and C, and formation of two plastic hinges will cause failure of the beam. Therefore, the plastic moment occurs at B and C at impending failure. *The sequence in which the plastic hinges are formed at these sections is immaterial.*

These diagrams are shown in Fig. 83. Express M_p in terms of P_u, and evaluate P_u. Thus, $BF = 20R_A = 268.8$; therefore, $R_A = 13.44$ kips (59.781 kN). Also, $CE = 30R_A - 10P_u = 30 \times 13.44 - 10P_u = -268.8$; $P_u = 67.20$ kips (298.906 kN).

Here is an alternative method: $BF = (abP_u/L) - aM_p/L = M_p$, or $20(10)P_u/30 = 50M_p/30$; $P_u = 67.20$ kips (298.906 kN).

This solution method used in part b is termed the *static*, or *equilibrium*, method. As this solution demonstrates, it is unnecessary to trace the stress history of the member as it passes through its successive phases, as was done in part a; the analysis can be confined to the conditions that exist

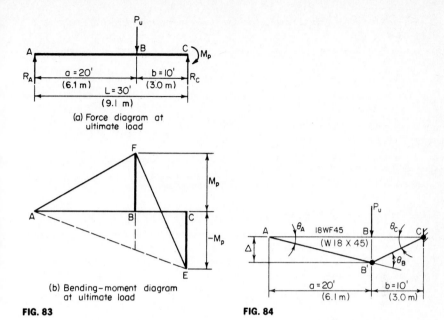

(a) Force diagram at ultimate load

(b) Bending-moment diagram at ultimate load

FIG. 83

FIG. 84

at impending failure. This procedure also illustrates the following important characteristics of plastic design:

1. Plastic design is far simpler than elastic design.

2. Plastic design yields results that are much more reliable than those secured through elastic design. For example, assume that the support at C does not completely inhibit rotation at that end. This departure from design conditions will invalidate the elastic analysis but will in no way affect the plastic analysis.

DETERMINING THE ULTIMATE LOAD BY THE MECHANISM METHOD

Use the mechanism method to solve the problem given in the previous calculation procedure.

Calculation Procedure:

1. Indicate, in hyperbolic manner, the virtual displacement of the member from its initial to a subsequent position

To the two phases of beam behavior previously considered, it is possible to add a third. Consider that when the ultimate load is reached, the member is subjected to an incremental deflection. This will result in collapse, but the behavior of the member can be analyzed during an infinitesimally small deflection from its stable position. This is termed a *virtual* deflection, or displacement.

Since the member is incapable of supporting any load beyond that existing at completion of phase 2, this virtual deflection is not characterized by any change in bending stress. Rotation therefore occurs solely at the real and plastic hinges. Thus, during phase 3, the member behaves as a mechanism (i.e., a constrained chain of pin-connected rigid bodies, or links).

In Fig. 84, indicate, in hyperbolic manner, the virtual displacement of the member from its initial position ABC to a subsequent position AB'C. Use dots to represent plastic hinges. (The initial position may be represented by a straight line for simplicity because the analysis is concerned solely with the deformation that occurs *during* phase 3.)

2. Express the linear displacement under the load and the angular displacement at every plastic hinge

Use a convenient unit to express these displacements. Thus, $\Delta = a\theta_A = b\theta_C$; therefore, $\theta_C = a\theta_A/b = 2\theta_A$; $\theta_B = \theta_A + \theta_C = 3\theta_A$.

3. Evaluate the external and internal work associated with the virtual displacement

The work performed by a constant force equals the product of the force and its displacement parallel to its action line. Also, the work performed by a constant moment equals the product of the moment and its angular displacement. Work is a positive quantity when the displacement occurs in the direction of the force or moment. Thus, the external work $W_E = P_u\Delta = P_u a\theta_A = 20P_u\theta_A$. And the internal work $W_I = M_p(\theta_B + \theta_C) = 5M_p\theta_A$.

4. Equate the external and internal work to evaluate the ultimate load

Thus, $20P_u\theta_A = 5M_p\theta_A$; $P_u = (5/20)(268.8) = 67.20$ kips (298.906 kN).

The solution method used here is also termed the *virtual-work*, or *kinematic*, method.

ANALYSIS OF A FIXED-END BEAM UNDER CONCENTRATED LOAD

If the beam in the two previous calculation procedures is fixed at A as well as at C, what is the ultimate load that may be applied at B?

Calculation Procedure:

1. Determine when failure impends

When hinges form at A, B, and C, failure impends. Repeat steps 3 and 4 of the previous calculation procedure, modifying the calculations to reflect the revised conditions. Thus $W_E = 20P_u\theta_A$; $W_I = M_p(\theta_A + \theta_B + \theta_C) = 6M_p\theta_A$; $20P_u\theta_A = 6M_p\theta_A$; $P_u = (6/20)(268.8) = 80.64$ kips (358.687 kN).

2. Analyze the phases through which the member passes

This member passes through three phases until the ultimate load is reached. Initially, it behaves as a beam fixed at both ends, then as a beam fixed at the left end only, and finally as a simply supported beam. However, as already discussed, these considerations are extraneous in plastic design.

ANALYSIS OF A TWO-SPAN BEAM WITH CONCENTRATED LOADS

The continuous 18WF45 (W18 × 45) beam in Fig. 85 carries two equal concentrated loads having the locations indicated. Disregarding the weight of the beam, compute the ultimate value of these loads, using both the static and the mechanism method.

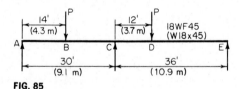

FIG. 85

Calculation Procedure:

1. Construct the force and bending-moment diagrams

The continuous beam becomes unstable when a plastic hinge forms at C and at another section. The bending-moment diagram has vertices at B and D, but it is not readily apparent at which of these sections the second hinge will form. The answer is found by assuming a plastic hinge at B and at D, in turn, computing the corresponding value of P_u, and selecting the lesser value as the correct result. Part a will use the static method; part b, the mechanism method.

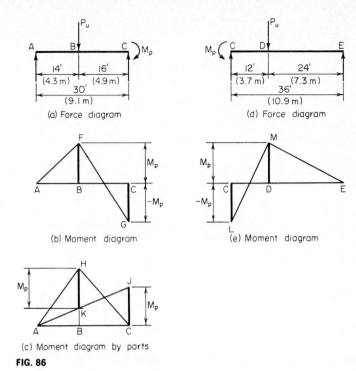

(a) Force diagram

(b) Moment diagram

(c) Moment diagram by parts

(d) Force diagram

(e) Moment diagram

FIG. 86

Assume, for part a, a plastic hinge at B and C. In Fig. 86, construct the force diagram and bending-moment diagram for span AC. The moment diagram may be drawn in the manner shown in Fig. 86b or c, whichever is preferred. In Fig. 86c, ACH represents the moments that would exist in the absence of restraint at C, and ACJ represents, in absolute value, the moments induced by this restraint. Compute the load P_u associated with the assumed hinge location. From previous calculation procedures, $M_p = 268.8$ ft·kips (364.49 kN·m); then $M_B = 14 \times 16P_u/30 - 14M_p/30 = M_p$; $P_u = 44(268.8)/224 = 52.8$ kips (234.85 kN).

2. Assume another hinge location and compute the ultimate load associated with this location

Now assume a plastic hinge at C and D. In Fig. 86, construct the force diagram and bending-moment diagram for CE. Computing the load P_u associated with this assumed location, we find $M_D = 12 \times 24P_u/36 - 24M_p/36 = M_p$; $P_u = 60(268.8)/288 = 56.0$ kips (249.09 kN).

3. Select the lesser value of the ultimate load

The correct result is the lesser of these alternative values, or $P_u = 52.8$ kips (234.85 kN). At this load, plastic hinges exist at B and C but not at D.

4. For the mechanism method, assume a plastic-hinge location

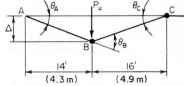

FIG. 87

It will be assumed that plastic hinges are located at B and C (Fig. 87). Evaluate P_u. Thus, $\theta_C = 14\theta_A/16$; $\theta_B = 30\theta_A/16$; $\Delta = 14\theta_A$; $W_E = P_u\Delta = 14P_u\theta_A$; $W_I = M_p(\theta_B + \theta_C) = 2.75M_p\theta_A$; $14P_u\theta_A = 2.75M_p\theta_A$; $P_u = 52.8$ kips (234.85 kN).

5. *Assume a plastic hinge at another location*

Select C and D for the new location. Repeat the above procedure. The result will be identical with that in step 2.

SELECTION OF SIZES FOR A CONTINUOUS BEAM

Using a load factor of 1.70, design the member to carry the working loads (with beam weight included) shown in Fig. 88a. The maximum length that can be transported is 60 ft (18.3 m).

Calculation Procedure:

1. *Determine the ultimate loads to be supported*

Since the member must be spliced, it will be economical to adopt the following design:

a. Use the particular beam size required for each portion, considering that the two portions will fail simultaneously at ultimate load. Therefore, three plastic hinges will exist at failure—one at the interior support and one in the interior of each span.

b. Extend one beam beyond the interior support, splicing the member at the point of contraflexure in the adjacent span. Since the maximum simple-span moment is greater for AB than for BC, it is logical to assume that for economy the left beam rather than the right one should overhang the support.

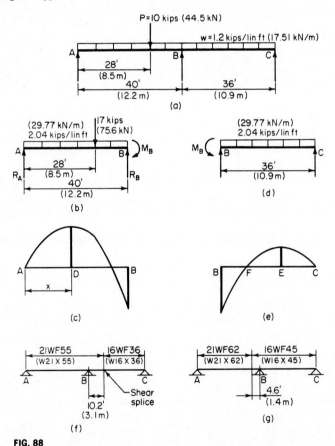

FIG. 88

Multiply the working loads by the load factor to obtain the ulitmate loads to be supported. Thus, $w = 1.2$ kips/lin ft (17.51 kN/m); $w_u = 1.70(1.2) = 2.04$ kips/lin ft (29.77 kN/m); $P = 10$ kips (44.5 kN); $P_u = 1.70(10) = 17$ kips (75.6 kN).

2. Construct the ultimate-load and corresponding bending-moment diagram for each span

Set the maximum positive moment M_D in span AB and the negative moment at B equal to each other in absolute value.

3. Evaluate the maximum positive moment in the left span

Thus, $R_A = 45.9 - M_B/40$; $x = R_A/2.04$; $M_D = \frac{1}{2}R_Ax = R_A^2/4.08 = M_B$. Substitue the value of R_A and solve. Thus, $M_D = 342$ ft·kips (463.8 kN·m).

An indirect but less cumbersome method consists of assigning a series of trial values to M_B and calculating the corresponding value of M_D, continuing the process until the required equality is obtained.

4. Select a section to resist the plastic moment

Thus, $Z = M_p/f_y = 342(12)/36 = 114$ in³ (1868.5 cm³). Referring to the AISC *Manual*, use a 21WF55 (W21 × 55) with $Z = 125.4$ in³ (2055.31 cm³).

5. Evaluate the maximum positive moment in the right span

Equate M_B to the true plastic-moment capacity of the 21WF55. Evaluate the maximum positive moment M_E in span BC, and locate the point of contraflexure. Thus, $M_B = -36(125.4)/12 = -376.2$ ft·kips (-510.13 kN·m); $M_E = 169.1$ ft·kips (229.30 kN·m); $BF = 10.2$ ft (3.11 m).

6. Select a section to resist the plastic moment

The moment to be resisted is M_E. Thus, $Z = 169.1(12)/36 = 56.4$ in³ (924.40 cm³). Use 16WF36 (W16 × 36) with $Z = 63.9$ in³ (1047.32 cm³).

The design is summarized in Fig. 88f. By inserting a hinge at F, the continuity of the member is destroyed and its behavior is thereby modified under gradually increasing load. However, the ultimate-load conditions, which constitute the only valid design criteria, are not affected.

7. Alternatively, design the member with the right-hand beam overhanging the support

Compare the two designs for economy. The latter design is summarized in Fig. 88g. The total beam weight associated with each scheme is as shown in the following table.

Design 1	Design 2
55(50.2) = 2,761 lb (12,280.9 N)	62(35.4) = 2,195 lb (9,763.4 N)
36(25.8) = 929 lb (4,132.2 N)	45(40.6) = 1,827 lb (8,126.5 N)
Total 3,690 lb (16,413.1 N)	4,022 lb (17,889.9 N)

For completeness, the column sizes associated with the two schemes should also be compared.

MECHANISM-METHOD ANALYSIS OF A RECTANGULAR PORTAL FRAME

Calculate the plastic moment and the reactions at the supports at ultimate load of the prismatic frame in Fig. 89a. Use a load factor of 1.85, and apply the mechanism method.

Calculation Procedure:

1. Compute the ultimate loads to be resisted

There are three potential modes of failure to consider:

a. Failure of the beam BD through the formation of plastic hinges at B, C, and D (Fig. 89b)

b. Failure by sidesway through the formation of plastic hinges at B and D (Fig. 89c)

c. A composite of the foregoing modes of failure, characterized by the formation of plastic hinges at C and D

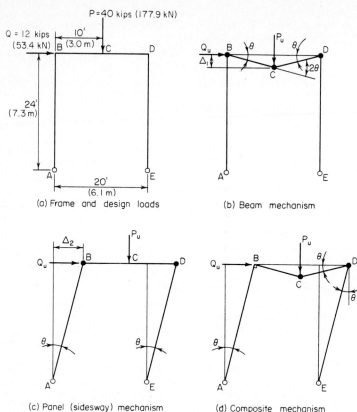

(a) Frame and design loads

(b) Beam mechanism

(c) Panel (sidesway) mechanism

(d) Composite mechanism

FIG. 89

Since the true mode of failure is not readily discernible, it is necessary to analyze each of the foregoing. The true mode of failure is the one that yields the highest value of M_p.

Although the work quantities are positive, it is advantageous to supply each angular displacement with an algebraic sign. A rotation is considered positive if the angle on the interior side of the frame increases. The algebraic sum of the angular displacements must equal zero.

Computing the ultimate loads to be resisted yields $P_u = 1.85(40) = 74$ kips (329.2 kN); $Q_u = 1.85(12) = 22.2$ kips (98.75 kN).

2. Assume the mode of failure in Fig. 89b and compute M_p

Thus, $\Delta_1 = 10\theta$; $W_E = 74(10\theta) = 740\theta$. Then indicate in a tabulation, such as that shown here, where the plastic moment occurs. Include all significant sections for completeness.

Section	Angular displacement	Moment	W_I
A			
B	$-\theta$	M_p	$M_p\theta$
C	$+2\theta$	M_p	$2M_p\theta$
D	$-\theta$	M_p	$M_p\theta$
E	. .	. .	. .
Total			$4M_p\theta$

Then $4M_p\theta = 740\theta$; $M_p = 185$ ft·kips (250.9 kN·m).

3. Repeat the foregoing procedure for failure by sidesway

Thus, $\Delta_2 = 24\theta$; $W_E = 22.2(24\theta) = 532.8\theta$.

Section	Angular displacement	Moment	W_I
A	$-\theta$		
B	$+\theta$	M_p	$M_p\theta$
C			
D	$-\theta$	M_p	$M_p\theta$
E	$+\theta$		
Total			$2M_p\theta$

Then $2M_p\theta = 532.8\theta$; $M_p = 266.4$ ft·kips (361.24 kN·m).

4. Assume the composite mode of failure and compute M_p

Since this results from superposition of the two preceding modes, the angular displacements and the external work may be obtained by adding the algebraic values previously found. Thus, $W_E = 740\theta + 532.8\theta = 1272.8\theta$. Then the tabulation is as shown:

Section	Angular displacement	Moment	W_I
A	$-\theta$		
B			
C	$+2\theta$	M_p	$2M_p\theta$
D	-2θ	M_p	$2M_p\theta$
E	$+\theta$		
Total			$4M_p\theta$

Then $4M_p\theta = 1272.8\theta$; $M_p = 318.2$ ft·kips (431.48 kN·m).

5. Select the highest value of M_p as the correct result

Thus, $M_p = 318.2$ ft·kips (431.48 kN·m). The structure fails through the formation of plastic hinges at C and D. That a hinge should appear at D rather than at B is plausible when it is considered that the bending moments induced by the two loads are of like sign at D but of opposite sign at B.

6. Compute the reactions at the supports

Draw a free-body diagram of the frame at ultimate load (Fig. 90). Compute the reactions at the supports by applying the computed values of M_C and M_D. Thus, $\Sigma M_E = 20V_A + 22.2(24) - 74(10) = 0$; $V_A = 10.36$ kips (46.081 kN); $V_E = 74 - 10.36 = 63.64$ kips (283.071 kN); $M_C = 10V_A + 24H_A = 103.6 + 24H_A = 318.2$; $H_A = 8.94$ kips (39.765 kN); $H_E = 22.2 - 8.94 = 13.26$ kips (58.980 kN); $M_D = -24H_E = -24(13.26) = -318.2$ ft·kips (−431.48 kN·m). Thus, the results are verified.

ANALYSIS OF A RECTANGULAR PORTAL FRAME BY THE STATIC METHOD

Compute the plastic moment of the frame in Fig. 89a by using the static method.

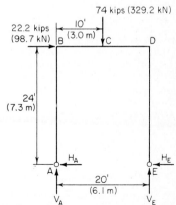

FIG. 90

Calculation Procedure:

1. Determine the relative values of the bending moments

Consider a bending moment as positive if the fibers on the interior side of the neutral plane are in tension. Consequently, as the mechanisms in Fig. 89 reveal, the algebraic sign of the plastic moment at a given section agrees with that of its angular displacement during collapse.

Determine the relative values of the bending moments at B, C, and D. Refer to Fig. 90. As previously found by statics, $V_A = 10.36$ kips (46.081 kN), $M_B = 24H_A$, $M_C = 24H_A + 10V_A$; therefore, $M_C = M_B + 103.6$, Eq. a. Also, $M_D = 24H_A + 20V_A - 74(10)$; $M_D = M_B - 532.8$, Eq. b; or $M_D = M_C - 636.4$, Eq. c.

2. Assume the mode the failure in Fig. 89b

This requires that $M_B = M_D = -M_p$. This relationship is incompatible with Eq. b, and the assumed mode of failure is therefore incorrect.

3. Assume the mode of failure in Fig. 89c

This requires that $M_B = M_p$, and $M_C < M_p$; therefore, $M_C < M_B$. This relationship is incompatible with Eq. a, and the assumed mode of failure is therefore incorrect.

By a process of elimination, it has been ascertained that the frame will fail in the manner shown in Fig. 89d.

4. Compute the value of M_p for the composite mode of failure

Thus, $M_C = M_p$, and $M_D = -M_p$. Substitute these values in Eq. c. Or, $-M_p = M_p - 636.4$; $M_p = 318.2$ ft·kips (431.48 kN·m).

THEOREM OF COMPOSITE MECHANISMS

By analyzing the calculations in the calculation procedure before the last one, establish a criterion to determine when a composite mechanism is significant (i.e., under what conditions it may yield an M_p value greater than that associated with the basic mechanisms).

Calculation Procedure:

1. Express the external and internal work associated with a given mechanism

Thus, $W_E = e\theta$, and $W_I = iM_p\theta$, where the coefficients e and i are obtained by applying the mechanism method. Then $M_p = e/i$.

2. Determine the significance of mechanism sign

Let the subscripts 1 and 2 refer to the basic mechanisms and the subscript 3 to their composite mechanism. Then $M_{p1} = e_1/i_1$; $M_{p2} = e_2/i_2$.

When the basic mechanisms are superposed, the values of W_E are additive. If the two mechanisms do not produce rotations of opposite sign at any section, the values of W_I are also additive, and $M_{p3} = e_3/i_3 = (e_1 + e_2)/(i_1 + i_2)$. This value is intermediate between M_{p1} and M_{p2}, and the composite mechanism therefore lacks significance. But if the basic mechanisms produce rotations of opposite sign at any section whatsoever, M_{p3} *may* exceed both M_{p1} and M_{p2}.

In summary, a composite mechanism is significant only if the two basic mechanisms of which it is composed produce rotations of opposite sign at any section. This theorem, which establishes a necessary but not sufficient condition, simplifies the analysis of a complex frame by enabling the engineer to discard the nonsignificant composite mechanisms at the outset.

ANALYSIS OF AN UNSYMMETRIC RECTANGULAR PORTAL FRAME

The frame in Fig. 91a sustains the ultimate loads shown. Compute the plastic moment and ultimate-load reactions.

Calculation Procedure:

1. Determine the solution method to use

Apply the mechanism method. In Fig. 91b, indicate the basic mechanisms.

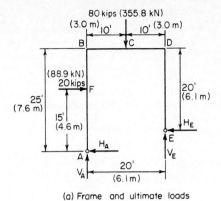

(a) Frame and ultimate loads

Mechanism I

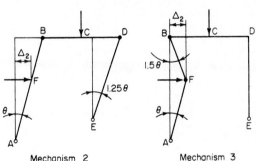

Mechanism 2   Mechanism 3

(b) Basic mechanisms

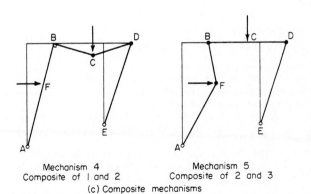

Mechanism 4   Mechanism 5
Composite of I and 2   Composite of 2 and 3

(c) Composite mechanisms

FIG. 91

103

2. Identify the significant composite mechanisms

Apply the theorem of the previous calculation procedure. Using this theorem, identify the significant composite mechanisms. For mechanisms 1 and 2, the rotations at B are of opposite sign; their composite therefore warrants investigation.

For mechanisms 1 and 3, there are no rotations of opposite sign; their composite therefore fails the test. For mechanisms 2 and 3, the rotations at B are of opposite sign; their composite therefore warrants investigation.

3. Evaluate the external work associated with each mechanism

Mechanism	W_E
1	$80\Delta_1 = 80(10\theta) = 800\theta$
2	$20\Delta_2 = 20(15\theta) = 300\theta$
3	300θ
4	1100θ
5	600θ

4. List the sections at which plastic hinges form; record the angular displacement associated with each mechanism

Use a list such as the following:

Mechanism	Section			
	B	C	D	F
1	$-\theta$	$+2\theta$	$-\theta$	
2	$+\theta$	. .	-1.25θ	
3	-1.5θ	. .	. . .	$+2.5\theta$
4	. .	$+2\theta$	-2.25θ	
5	-0.5θ	. .	-1.25θ	$+2.5\theta$

5. Evaluate the internal work associated with each mechanism

Equate the external and internal work to find M_p. Thus, $M_{p1} = 800/4 = 200$; $M_{p2} = 300/2.25 = 133.3$; $M_{p3} = 300/4 = 75$; $M_{p4} = 1100/4.25 = 258.8$; $M_{p5} = 600/4.25 = 141.2$. Equate the external and internal work to find M_p.

6. Select the highest value as the correct result

Thus, $M_p = 258.8$ ft·kips (350.93 kN·m). The frame fails through the formation of plastic hinges at C and D.

7. Determine the reactions at ultimate load

To verify the foregoing solution, ascertain that the bending moment does not exceed M_p in absolute value anywhere in the frame. Refer to Fig. 91a.

Thus, $M_D = -20H_E = -258.8$; therefore, $H_E = 12.94$ kips (57.557 kN); $M_C = M_D + 10V_E = 258.8$; therefore, $V_E = 51.76$ kips (230.23 kN); then $H_A = 7.06$ kips (31.403 kN); $V_A = 28.24$ kips (125.612 kN).

Check the moments. Thus $\Sigma M_E = 20V_A + 5H_A + 20(10) - 80(10) = 0$; this is correct. Also, $M_F = 15H_A = 105.9$ ft·kips (143.60 kN·m) $< M_p$. This is correct. Last, $M_B = 25H_A - 20(10) = -23.5$ ft·kips (-31.87 kN·m) $> -M_p$. This is correct.

ANALYSIS OF GABLE FRAME BY STATIC METHOD

The prismatic frame in Fig. 92a carries the ultimate loads shown. Determine the plastic moment by applying the static method.

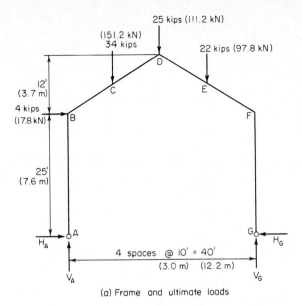

25 kips (111.2 kN)

(151.2 kN)
34 kips

22 kips (97.8 kN)

12'
(3.7 m)

4 kips
(17.8 kN)

25'
(7.6 m)

H_A

4 spaces @ 10' = 40'
(3.0 m)   (12.2 m)

V_A V_G H_G

(a) Frame and ultimate loads

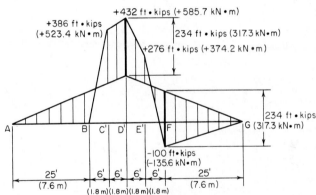

+432 ft • kips (+585.7 kN • m)

+386 ft • kips
(+523.4 kN • m)

234 ft • kips (317.3 kN•m)

+276 ft • kips (+374.2 kN • m)

234 ft • kips
G (317.3 kN• m)

−100 ft•kips
(−135.6 kN•m)

25'
(7.6 m)

6' 6' 6' 6'
(1.8 m)(1.8 m)(1.8 m)(1.8 m)

25'
(7.6 m)

A   B   C' D' E'   F

(b) Projected bending – moment diagram

FIG. 92

Calculation Procedure:

1. *Compute the vertical shear V_A and the bending moment at every significant section, assuming $H_A = 0$*

Thus, $V_A = 41$ kips (182.4 kN). Then $M_B = 0$; $M_C = 386$; $M_D = 432$; $M_E = 276$; $M_F = -100$.

Note that failure of the frame will result from the formation of two plastic hinges. It is helpful, therefore, to construct a "projected" bending-moment diagram as an aid in locating these hinges. The computed bending moments are used in plotting the projected bending-moment diagram.

2. *Construct a projected bending-moment diagram*

To construct this diagram, consider the rafter BD to be projected onto the plane of column AB and the rafter FD to be projected onto the plane of column GF. Juxtapose the two halves, as shown in Fig. 92*b*. Plot the values calculated in step 1 to obtain the bending-moment diagram corresponding to the assumed condition of $H_A = 0$.

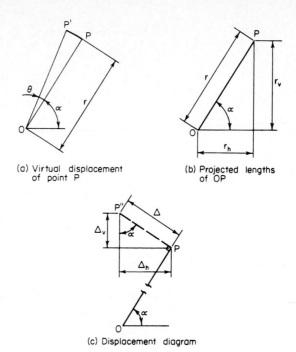

(a) Virtual displacement of point P

(b) Projected lengths of OP

(c) Displacement diagram

FIG. 93

The bending moments caused solely by a specific value of H_A are represented by an isosceles triangle with its vertex at D'. The true bending moments are obtained by superposition. It is evident by inspection of the diagram that plastic hinges form at D and F and that H_A is directed to the right.

3. Evaluate the plastic moment

Apply the true moments at D and F. Thus, $M_D = M_p$ and $M_F = -M_p$; therefore, $432 - 37H_A = -(-100 - 25H_A)$; $H_A = 5.35$ kips (23.797 kN) and $M_p = 234$ ft·kips (317 kN·m).

THEOREM OF VIRTUAL DISPLACEMENTS

In Fig. 93a, point P is displaced along a virtual (infinitesimally small) circular arc PP' centered at O and having a central angle θ. Derive expressions for the horizontal and vertical displacement of P in terms of the given data. (These expressions are applied later in analyzing a gable frame by the mechanism method.)

Calculation Procedure:

1. Construct the displacement diagram

In Fig. 93b, let r_h = length of horizontal projection of OP; r_v = length of vertical projection of OP; Δ_h = horizontal displacement of P; Δ_v = vertical displacement of P.

In Fig. 93c, construct the displacement diagram. Since PP' is infinitesimally small, replace this circular arc with the straight line PP'' that is tangent to the arc at P and therefore normal to radius OP.

2. Evaluate Δ_h and Δ_v, considering only absolute values

Since θ is infinitesimally small, set $PP'' = r\theta$; $\Delta_h = PP'' \sin \alpha = r\theta \sin \alpha$; $\Delta_v = PP'' \cos \alpha = r\theta \cos \alpha$. But $r \sin \alpha = r_v$ and $r \cos \alpha = r_h$; therefore, $\Delta_h = r_v\theta$ and $\Delta_v = r_h\theta$.

These results may be combined and expressed verbally thus: If a point is displaced along a virtual circular arc, its displacement as projected on the u axis equals the displacement angle times the length of the radius as projected on an axis normal to u.

GABLE-FRAME ANALYSIS BY USING THE MECHANISM METHOD

For the frame in Fig. 92a, assume that plastic hinges form at D and F. Calculate the plastic moment associated with this assumed mode of failure by applying the mechanism method.

Calculation Procedure:

1. *Indicate the frame configuration following a virtual displacement*

During collapse, the frame consists of three rigid bodies: ABD, DF, and GF. To evaluate the external and internal work performed during a virtual displacement, it is necessary to locate the instantaneous center of rotation of each body.

In Fig. 94, indicate by dash lines the configuration of the frame following a virtual displacement. In Fig. 94, D is displaced to D' and F to F'. Draw a straight line through A and D intersecting the prolongation of GF at H.

Since A is the center of rotation of ABD, DD' is normal to AD and HD; since G is the center of rotation of GF, FF' is normal to GF and HF. Therefore, H is the instantaneous center of rotation of DF.

2. *Record the pertinent dimensions and rotations*

Record the dimensions a, b, and c in Fig. 94, and express θ_2 and θ_3 in terms of θ_1. Thus, $\theta_2/\theta_1 = HD/AD$; $\therefore \theta_2 = \theta_1$. Also, $\theta_3/\theta_1 = HF/GF = 49/25$; $\therefore \theta_3 = 1.96\theta_1$.

3. *Determine the angular displacement, and evaluate the internal work*

Determine the angular displacement (in absolute value) at D and F, and evaluate the internal work in terms of θ_1. Thus, $\theta_D = \theta_1 + \theta_2 = 2\theta_1$; $\theta_F = \theta_1 + \theta_3 = 2.96\theta_1$. Then $W_I = M_p (\theta_D + \theta_F) = 4.96 M_p \theta_1$.

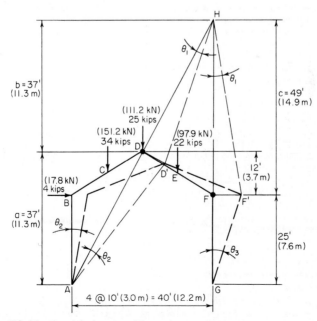

FIG. 94 Virtual displacement of frame.

4. Apply the theorem of virtual displacements to determine the displacement of each applied load

Determine the displacement of each applied load in the direction of the load. Multiply the displacement by the load to obtain the external work. Record the results as shown:

	Load		Displacement in direction of load		External work	
Section	kips	kN	ft	m	ft·kips	kN·m
B	4	17.8	$\Delta_h = 25\theta_2 = 25\ \theta_1$	$7.6\ \theta_1$	$100\ \theta_1$	$135.6\ \theta_1$
C	34	151.2	$\Delta_v = 10\theta_2 = 10\ \theta_1$	$3.0\ \theta_1$	$340\ \theta_1$	$461.0\ \theta_1$
D	25	111.2	$\Delta_v = 20\ \theta_1$	$6.1\ \theta_1$	$500\ \theta_1$	$678.0\ \theta_1$
E	22	97.9	$\Delta_v = 10\ \theta_1$	$3.0\ \theta_1$	$220\ \theta_1$	$298.3\ \theta_1$
Total					$1160\ \theta_1$	$1572.9\ \theta_1$

5. Equate the external and internal work to find M_p

Thus, $4.96M_p\theta_1 = 1160\theta_1$; $M_p = 234$ ft·kips (317.3 kN·m).

Other modes of failure may be assumed and the corresponding value of M_p computed in the same manner. The failure mechanism analyzed in this procedure (plastic hinges at D and F) yields the highest value of M_p and is therefore the true mechanism.

REDUCTION IN PLASTIC-MOMENT CAPACITY CAUSED BY AXIAL FORCE

A 10WF45 (W10 × 45) beam-column is subjected to an axial force of 84 kips (373.6 kN) at ultimate load. (a) Applying the exact method, calculate the plastic moment this section can develop with respect to the major axis. (b) Construct the interaction diagram for this section, and then calculate the plastic moment by assuming a linear interaction relationship that approximates the true relationship.

Calculation Procedure:

1. Record the relevant properties of the member

Let P = applied axial force, kips (kN); P_y = axial force that would induce plastification if acting alone, kips (kN) = Af_y; M'_p = plastic-moment capacity of the section in combination with P, ft·kips (kN·m).

A typical stress diagram for a beam-column at plastification is shown in Fig. 95a. To simplify the calculations, resolve this diagram into the two parts shown at the right. This procedure is tantamount to assuming that the axial load is resisted by a central core and the moment by the outer segments of the section, although in reality they are jointly resisted by the integral action of the entire section.

From the AISC Manual, for a 10WF45: $A = 13.24$ in^2 (85.424 cm^2); $d = 10.12$ in (257.048 mm); $t_f = 0.618$ in (15.6972 mm); $t_w = 0.350$ in (8.890 mm); $d_w = 10.12 - 2(0.618) = 8.884$ in (225.6536 mm); $Z = 55.0$ in^3 (901.45 cm^3).

2. Assume that the central core that resists the 84-kip (373.6-kN) load is encompassed within the web; determine the core depth

Calling the depth of the core g, refer to Fig. 95d. Then $g = 84/[0.35(36)] = 6.67 < 8.884$ in (225.6536 mm).

3. Compute the plastic modulus of the core, the plastic modulus of the remaining section, and the value of M'_p

Using data from the Manual for the plastic modulus of a rectangle, we find $Z_c = \frac{1}{4}t_w g^2 = \frac{1}{4}(0.35)(6.67)^2 = 3.9$ in^3 (63.92 cm^3); $Z_r = 55.0 - 3.9 = 51.1$ in^3 (837.53 cm^3); $M'_p = 51.1(36)/$

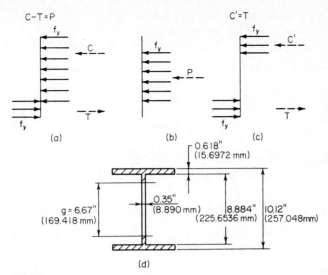

FIG. 95

$12 = 153.3$ ft·kips (207.87 kN·m). This constitutes the solution of part a. The solution of part b is given in steps 4 through 6.

4. *Assign a series of values to the parameter g, and compute the corresponding sets of values of P and M'_p*

Apply the results to plot the interaction diagram in Fig. 96. This comprises the parabolic curves CB and BA, where the points A, B, and C correspond to the conditions $g = 0$, $g = d_w$, and $g = d$, respectively.

The interaction diagram is readily analyzed by applying the following relationships: $dP/dg = f_y t$; $dM'_p/dg = -\frac{1}{2}f_y tg$; $\therefore dP/dM'_p = -2/g$. This result discloses that the change in slope along CB is very small, and the curvature of this arc is negligible.

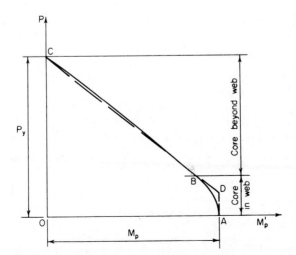

FIG. 96 Interaction diagram for axial force and moment.

5. Replace the true interaction diagram with a linear one

Draw a vertical line $AD = 0.15P_y$, and then draw the straight line CD (Fig. 96). Establish the equation of CD. Thus, slope of $CD = -0.85P_y/M_p$; $\therefore P = P_y - 0.85P_yM'_p/M_p$, or $M'_p = 1.18(1 - P/P_y)M_p$.

The provisions of one section of the AISC *Specification* are based on the linear interaction diagram.

6. Ascertain whether the data are represented by a point on AD or CD; calculate M'_p accordingly

Thus, $P_y = Af_y = 13.24(36) = 476.6$ kips (2119.92 kN); $P/P_y = 84/476.6 = 0.176$; therefore, apply the last equation given in step 5. Thus, $M_p = 55.0(36)/12 = 165$ ft·kips (223.7 kN·m); $M'_p = 1.18(1 - 0.176)(165) = 160.4$ ft·kips (217.50 kN·m). This result differs from that in part a by 4.6 percent.

Timber Engineering

In designing timber members, the following references are often used: *Wood Handbook*, Forest Products Laboratory, U.S. Department of Agriculture, and *National Design Specification for Stress-Grade Lumber and Its Fastenings*, National Forest Products Association. The members are assumed to be continuously dry and subject to normal loading conditions.

For most species of lumber, the true or *dressed* dimensions are less than the nominal dimensions by the following amounts: ⅜ in (9.53 mm) for dimensions less than 6 in (152.4 mm); ½ in (12.7 mm) for dimensions of 6 in (152.4 mm) or more. The average weight of timber is 40 lb/ft³ (6.28 kN/m³). The width and depth of the transverse section are denoted by b and d, respectively.

BENDING STRESS AND DEFLECTION OF WOOD JOISTS

A floor is supported by 3×8 in (76.2 × 203.2 mm) wood joists spaced 16 in (406.4 mm) on centers with an effective span of 10 ft (3.0 m). The total floor load transmitted to the joists is 107 lb/in² (5.123 kN/m²). Compute the maximum bending stress and initial deflection, using $E = 1,760,000$ lb/in² (12,135 kPa).

Calculation Procedure:

1. Calculate the beam properties or extract them from a table

Thus, $A = 2⅝(7½) = 19.7$ in² (127.10 cm²); beam weight = $(A/144)$ (lumber density, lb/ft³) = $(19.7/144)(40) = 5$ lb/lin ft (73.0 N/m); $I = (1/12)(2⅝)(7½)^3 = 92.3$ in⁴ (3841.81 cm⁴); $S = 92.3/3.75 = 24.6$ in³ (403.19 cm³).

2. Compute the unit load carried by the joists

Thus, the unit load $w = 107(1.33) + 5 = 148$ lb/lin ft (2159.9 N/m), where the factor 1.33 is the width, ft, of the floor load carried by each joist and 5 = the beam weight, lb/lin ft.

3. Compute the maximum bending stress in the joist

Thus, the bending moment in the joist is $M = (1/8)wL^2 12$, where $M =$ bending moment, in·lb (N·m); $L =$ joist length, ft (m). Substituting gives $M = (1/8)(148)(10)^2(12) = 22,200$ in·lb (2508.2 N·m). Then for the stress in the beam, $f = M/S$, where $f =$ stress, lb/in² (kPa), and $S =$ beam section modulus, in³ (cm³); or $f = 22,200/24.6 = 902$ lb/in² (6219.3 kPa).

4. Compute the initial deflection at midspan

Using the AISC *Manual* deflection equation, we see that the deflection Δ in (mm) = $(5/384)wL^4/(EI)$, where $I =$ section moment of inertia, in⁴ (cm⁴) and other symbols are as before. Substituting yields $\Delta = 5(148)(10)^4(1728)/[384(1,760,000)(92.3)] = 0.205$ in (5.2070 mm). In this relation, the factor 1728 converts cubic feet to cubic inches.

SHEARING STRESS CAUSED BY STATIONARY CONCENTRATED LOAD

A 3×10 in (76.2×254.0 mm) beam on a span of 12 ft (3.7 m) carries a concentrated load of 2730 lb (12,143.0 N) located 2 ft (0.6 m) from the support. If the allowable shearing stress is 120 lb/in^2 (827.4 kPa), determine whether this load is excessive. Neglect the beam weight.

Calculation Procedure:

1. Calculate the reaction at the adjacent support

In a rectangular section, the shearing stress varies parabolically with the depth and has the maximum value of $v = 1.5V/A$, where V = shear, lb (N).

The *Wood Handbook* notes that checks are sometimes present near the neutral axis of timber beams. The vitiating effect of these checks is recognized in establishing the allowable shearing stresses. However, these checks also have a beneficial effect, for they modify the shear distribution and thereby reduce the maximum stress. The amount of this reduction depends on the position of the load. The maximum shearing stress to be applied in design is given by $v = 10(a/d)^2 v'/\{9[2 + (a/d)^2]\}$, where v = true maximum shearing stress, lb/in^2 (kPa); v' = nominal maximum stress computed from $1.5V/A$; a = distance from load to adjacent support, in (mm).

Computing the reaction R at the adjacent support gives $R = V_{max} = 2730(12 - 2)/12 = 2275$ lb (10,119.2 N). Then $v' = 1.5V/A = 1.5(2275)/24.9 = 137$ lb/in^2 (944.6 kPa).

2. Find the design stress

Using the equation given in step 1, we get $(a/d)^2 = (24/9.5)^2 = 6.38$; $v = 10(6.38)(137)/[9(8.38)] = 116$ lb/in^2 (799.8 kPa) < 120 lb/in^2 (827.4 kPa). The load is therefore not excessive.

SHEARING STRESS CAUSED BY MOVING CONCENTRATED LOAD

A 4×12 in (101.6×304.8 mm) beam on a span of 10 ft (3.0 m) carries a total uniform load of 150 lb/lin ft (2189.1 N/m) and a moving concentrated load. If the allowable shearing stress is 130 lb/in^2 (896.4 kPa), what is the allowable value of the moving load as governed by shear?

Calculation Procedure:

1. Calculate the reaction at the support

The transient load induces the absolute maximum shearing stress when it lies at a certain critical distance from the support rather than directly above it. This condition results from the fact that as the load recedes from the support, the reaction decreases but the shear-redistribution effect becomes less pronounced. The approximate method of analysis recommended in the *Wood Handbook* affords an expedient means of finding the moving-load capacity.

Place the moving load P at a distance of $3d$ or $\frac{1}{4}L$ from the support, whichever is less. Calculate the reaction at the support, disregarding the load within a distance of d therefrom.

Thus, $3d = 2.9$ ft (0.884 m) and $\frac{1}{4}L = 2.5$ ft (0.762 m); then $R = V_{max} = 150(5 - 0.96) + \frac{3}{4}P = 610 + \frac{3}{4}P$.

2. Calculate the allowable shear

Thus, $V_{allow} = \frac{2}{3}vA = \frac{2}{3}(130)(41.7) = 3610$ lb (16,057.3 N). Then $610 + \frac{3}{4}P = 3610$; $P = 4000$ lb (17,792.0 N).

STRENGTH OF DEEP WOODEN BEAMS

If the allowable bending stress in a shallow beam is 1500 lb/in^2 (10,342.5 kPa), what is the allowable bending moment in a 12×20 in (304.8×508.0 mm) beam?

Calculation Procedure:

1. Calculate the depth factor F

An increase in depth of a rectangular beam is accompanied by a decrease in the modulus of rupture. For beams more than 16 in (406.4 mm) deep, it is necessary to allow for this reduction in strength by introducing a *depth factor F*.

Thus, $F = 0.81(d^2 + 143)/(d^2 + 88)$, where d = dressed depth of beam, in. Substituting gives $F = 0.81(19.5^2 + 143)/(19.5^2 + 88) = 0.905$.

2. Apply the result of step 1 to obtain the moment capacity

Use the relation $M = FfS$, where the symbols are as given earlier. Thus, $M = 0.905 \times (1.5)(728.8)/12 = 82.4$ ft·kips (111.73 kN·m).

DESIGN OF A WOOD-PLYWOOD BEAM

A girder having a 36-ft (11.0-m) span is to carry a uniform load of 550 lb/lin ft (8026.6 N/m), which includes its estimated weight. Design a box-type member of glued construction, using the allowable stresses given in the table. The modulus of elasticity of both materials is 1,760,000 lb/in² (12,135.2 MPa), and the ratio of deflection to span cannot exceed 1/360. Architectural details limit the member depth to 40 in (101.6 cm).

	Lumber	Plywood
Tension, lb/in² (kPa)	1,500 (10,342.5)	2,000 (13,790.0)
Compression parallel to grain, lb/in² (kPa)	1,350 (9,308.3)	1,460 (10,066.7)
Compression normal to grain, lb/in² (kPa)	390 (2,689.1)	405 (2,792.5)
Shear parallel to plane of plies, lb/in² (kPa)	. . . .	72° (496.4)
Shear normal to plane of plies, lb/in² (kPa)	. . . .	192 (1,323.8) ‹

°Use 36 lb/in² (248.2 kPa) at contact surface of flange and web to allow for stress concentration.

Calculation Procedure:

1. Compute the maximum shear and bending moment

Thus, $V = \frac{1}{2}(550)(36) = 9900$ lb (44,035.2 N); $M = \frac{1}{8}(wL^2)12 = \frac{1}{8}(550)(36)^2 12 = 1,070,000$ in·lb (120,888.6 N·m). To preclude the possibility of field error, make the tension and compression flanges alike.

2. Calculate the beam depth for a balanced condition

Assume that the member precisely satisfies the requirements for flexure and deflection, and calculate the depth associated with this balanced condition. To allow for the deflection caused by shear, which is substantial when a thin web is used, increase the deflection as computed in the conventional manner by one-half. Thus, $M = fI/c = 2fI/d = 2700I/d$, Eq. a. $\Delta = (7.5/48)L^2M/(EI) = L/360$, Eq. b.

Substitute in Eq. b the value of M given by Eq. a; solve for d to obtain $d = 37.3$ in (947.42 mm). Use the permissible depth of 40 in (1016.0 mm). As a result of this increase in depth, a section that satisfies the requirement for flexure will satisfy the requirement for deflection as well.

3. Design the flanges

Approximate the required area of the compression flange; design the flanges. For this purpose, assume that the flanges will be 5½ in (139.7 mm) deep. The lever arm of the resultant forces in the flanges will be 34.8 in (883.92 mm), and the average fiber stress will be 1165 lb/in² (8032.7 kPa). Then $A = 1,070,000/[1165(34.8)] = 26.4$ in² (170.33 cm²). Use three 2 × 6 in (50.8 × 152.4 mm) sections with glued vertical laminations for both the tension and compression flange. Then $A = 3(8.93) = 26.79$ in² (170.268 cm²); $I_o = 3(22.5) = 67.5$ in⁴ (2809.56 cm⁴).

4. Design the webs

Use the approximation $t_w = 1.25V/dv_n = 1.25(9900)/[40(192)] = 1.61$ in (40.894 mm). Try two ⅞-in (22.2-mm) thick plywood webs. A catalog of plywood properties reveals that the ⅞-in (22.2-mm) member consists of seven plies and that the parallel plies have an aggregate thickness of 0.5 in (12.7 mm). Draw the trial section as shown in Fig. 97.

5. Check the bending stress in the member

For simplicity, disregard the webs in evaluating the moment of inertia. Thus, the moment of inertia of the flanges $I_f = 2(67.5 + 26.79 \times 17.25^2) = 16,080$ in⁴ (669,299.448 cm⁴); then the stress $f = Mc/I = 1,070,000(20)/16,080 = 1330 < 1350$ lb/in² (9308.25 kPa). This is acceptable.

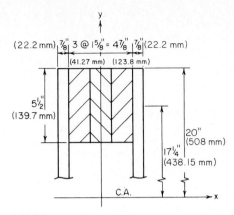

FIG. 97

6. Check the shearing stress at the contact surface of the flange and web

Use the relation $Q_f = Ad = 26.79(17.25) = 462$ in^3 (7572.2 cm^3). The q per surface $= VQ_f/(2I_f) = 9900(462)/[2(16,080)] = 142$ lb/lin in (24.8 kN/m). Assume that the shearing stress is uniform across the surface, and apply 36 lb/in^2 (248.2 kPa), as noted earlier, as the allowable stress. Then, $v = 142/5.5 = 26$ lb/in^2 (179.3 kPa) < 36 lb/in^2 (248.2 kPa). This is acceptable.

7. Check the shearing stress in the webs

For this purpose, include the webs in evaluating the moment of inertia but apply solely the area of the parallel plies. At the neutral axis $Q = Q_f + Q_w = 462 + 2(0.5)(20)(10) = 662$ in^3 (10,850.2 cm^3); $I = I_f + I_w = 16,080 + 2(1/12)(0.5)(40)^3 = 21,410$ in^4 (89.115 dm^4). Then $v = VQ/(It) = 9900(662)/[21,410(2)(0.875)] = 175$ lb/in^2 (1206.6 kPa) < 192 lb/in^2 (1323.8 kPa). This is acceptable.

8. Check the deflection, applying the moment of inertia of only the flanges

Thus, $\Delta = (7.5/384)wL^4/(EI_f) = 7.5(550)(36)^4(1728)/[384(1,760,000)(16,080)] = 1.10$ in (27.94 mm); $\Delta/L = 1.10/[36(12)] < 1/360$. This is acceptable, and the trial section is therefore satisfactory in all respects.

9. Establish the allowable spacing of the bridging

To do this, compare the moments of inertia with respect to the principal axes. Thus, $I_y = 2(1/12)(5.5)(4.875)^3 + 2(0.5)(40)(2.875)^2 = 433$ in^4 (18,022.8 cm^4); then $I_x/I_y = 16,080/433 = 37.1$.

For this ratio, the *Wood Handbook* specifies that "the beam should be restrained by bridging or other bracing at intervals of not more than 8 ft (2.4 m)."

DETERMINING THE CAPACITY OF A SOLID COLUMN

An 8 × 10 in (203.2 × 254 mm) column has an unbraced length of 10 ft 6 in (3.20 m). The allowable compressive stress is 1500 lb/in^2 (10,342.5 kPa), and $E = 1,760,000$ lb/in^2 (12,135.2 MPa). Calculate the allowable load on this column (*a*) by applying the recommendations of the *Wood Handbook*; (*b*) by applying the provisions of the *National Design Specification*.

Calculation Procedure:

1. Record the properties of the member; evaluate K; classify the column

Let L = unbraced length of column, in (mm); d = smaller side of rectangular section, in (mm); f_c = allowable compressive stress parallel to the grain in short column of the same species, lb/in^2 (kPa); f = allowable compressive stress parallel to grain in column under investigation, lb/in^2 (kPa).

The *Wood Handbook* divides columns into three categories: short, intermediate, and long. Let K denote a parameter defined by the equation $K = 0.64(E/f_c)^{0.5}$.

The range of the slenderness ratio and the allowable stress for each category of column are as follows: *short column,* $L/d \leq 11$ and $f = f_c$; *intermediate column,* $11 < L/d \leq K$ and $f = f_c[1 - \frac{1}{3}(L/d/K)^4]$; *long column,* $L/d > K$ and $f = 0.274E/(L/d)^2$.

For this column, the area $A = 71.3$ in^2 (460.03 cm^2), using the dressed dimensions. Then $L/d = 126/7.5 = 16.8$. Also, $K = 0.64(1,760,000/1500)^{0.5} = 21.9$. Therefore, this is an intermediate column because L/d lies between K and 11.

2. Compute the capacity of the member

Use the relation capacity, lb (N) $= P = Af = 71.3(1500)[1 - \frac{1}{3}(16.8/21.9)^4] = 94,600$ lb (420,780.8 N). This constitutes the solution to part a, using data from the *Wood Handbook*. For part b, data from the *National Design Specification* are used.

3. Compute the capacity of the column

Determine the stress from $f = 0.30E/(L/d)^2 = 0.30(1,760,000)/16.8^2 = 1870$ lb/in^2 (12,893.6 kPa). Setting $f = 1500$ lb/in^2 (10,342.5 kPa) gives $P = Af = 71.3(1500) = 107,000$ lb (475,936 N). Note that the smaller stress value is used when the column capacity is computed.

DESIGN OF A SOLID WOODEN COLUMN

A 12-ft (3.7-m) long wooden column supports a load of 98 kips (435.9 kN). Design a solid section in the manner recommended in the *Wood Handbook*, using $f_c = 1400$ lb/in^2 (9653 kPa) and $E = 1,760,000$ lb/in^2 (12,135.2 MPa).

Calculation Procedure:

1. Assume that $d = 7.5$ in (190.5 mm), and classify the column

Thus, $L/d = 144/7.5 = 19.2$ and $K = 0.64(1,760,000/1400)^{0.5} = 22.7$. This is an intermediate column if the assumed dimension is correct.

2. Compute the required area and select a section

For an intermediate column, the stress $f = 1400[1 - \frac{1}{3}(19.2/22.7)^4] = 1160$ lb/in^2 (7998.2 kPa). Then $A = P/f = 98,000/1160 = 84.5$ in^2 (545.19 cm^2).

Study of the required area shows that an 8 × 12 in (203.2 × 304.8 mm) column having an area of 86.3 in^2 (556.81 cm^2) should be used.

INVESTIGATION OF A SPACED COLUMN

The wooden column in Fig. 98 is composed of three 3 × 8 in (76.2 × 203.2 mm) sections. Determine the capacity of the member if $f_c = 1400$ lb/in^2 (9653 kPa) and $E = 1,760,000$ lb/in^2 (12,135.2 MPa).

Calculation Procedure:

1. Record the properties of the elemental section

In analyzing a spaced column, it is necessary to assess both the aggregate strength of the elements and the strength of the built-up section. The end spacer blocks exert a restraining effect on the elements and thereby enhance their capacity. This effect is taken into account by multiplying the modulus of elasticity by a *fixity factor F*.

The area of the column $A = 19.7$ in^2 (127.10 cm^2) when the dressed sizes are used. Also, $L/d = 114/2.625 = 43.4$; $F = 2.5$; $K = 0.64(2.5 \times 1,760,000/1400)^{0.5} = 35.9$. Therefore, this is a long column.

2. Calculate the aggregate strength of the elements

Thus, $f = 0.274E/(L/d)^2$ for a long column, or $f = 0.274(2.5)(1,760,000)/(43.4)^2 = 640$ lb/in^2 (4412.8 kPa). $P = 3(19.7)(640) = 37,800$ lb (168,134.4 N).

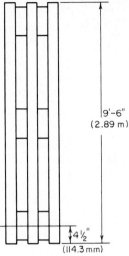

FIG. 98 Spaced column.

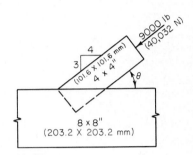

FIG. 99

3. Repeat the foregoing steps for the built-up member

Thus, $L/d = 114/7.5 \times 15.2$; $K = 22.7$; therefore, this is an intermediate column. Then $f = 1400[1 - \frac{1}{3}(15.2/22.7)^4] = 1306$ lb/in^2 (9004.9 kPa) > 640 lb/in^2 (4412.8 kPa).

The column capacity is therefore limited by the elements and $P = 37{,}800$ lb (168,134.4 N).

COMPRESSION ON AN OBLIQUE PLANE

Determine whether the joint in Fig. 99 is satisfactory with respect to bearing if the allowable compressive stresses are 1400 and 400 lb/in^2 (9653 and 2758 kPa) parallel and normal to the grain, respectively.

Calculation Procedure:

1. Compute the compressive stress

Thus, $f = P/A = 9000/3.625^2 = 685$ lb/in^2 (4723.1 kPa).

2. Compute the allowable compression stress in the main member

Apply Hankinson's equation: $N = PQ/(P \sin^2 \theta + Q \cos^2 \theta)$, where P = allowable compressive stress parallel to grain, lb/in^2 (kPa); Q = allowable compressive stress normal to grain; lb/in^2 (kPa); N = allowable compressive stress inclined to the grain, lb/in^2 (kPa); θ = angle between action line of N and direction of grain. Thus, $\sin^2 \theta = 0.36$, $\cos^2 \theta = (4/5)^2 = 0.64$; then $N = 1400(400)/(1400 \times 0.36 + 400 \times 0.64) = 737$ lb/in^2 (5081.6 kPa) > 685 lb/in^2 (4723.1 kPa). Therefore, the joint is satisfactory.

3. Alternatively, solve Hankinson's equation by using the nomogram in the Wood Handbook

DESIGN OF A NOTCHED JOINT

In Fig. 100, $M1$ is a 4×4, $F = 5500$ lb (24,464 N), and $\phi = 30°$. The allowable compressive stresses are $P = 1200$ lb/in^2 (8274 kPa) and $Q = 390$ lb/in^2 (2689.1 kPa). The projection of $M1$ into $M2$ is restricted to a vertical distance of 2.5 in (63.5 mm). Design a suitable notch.

Calculation Procedure:

1. Record the values of the trigonometric functions of ϕ and $\phi/2$

The most feasible type of notch is the one shown in Fig. 100, in which AC and BC bisect the angles between the intersecting edges. The allowable bearing pressures on these faces are therefore identical for the two members.

With $\phi = 30°$, $\sin 30° = 0.500$; $\sin 15° = 0.259$; $\cos 15° = 0.966$; $\tan 15° = 0.268$.

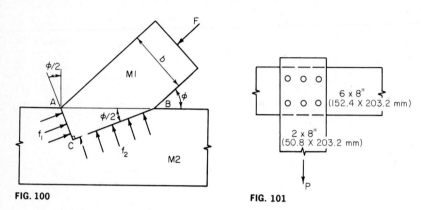

FIG. 100   FIG. 101

2. Find the lengths AC and BC

Express these two lengths as functions of AB. Or, $AB = b/\sin \phi$; $AC = [b \sin (\phi/2)]/\sin \phi$; $BC = [b \cos (\phi/2)]/\sin \phi$; $AC = 3.625(0.259/0.500) = 1.9$ in (48.26 mm); $BC = 3.625(0.966/0.500) = 7.0$ in (177.8 mm). The projection into $M2$ is therefore not excessive.

3. Evaluate the stresses f_1 and f_2

Resolve F into components parallel to AC and BC. Thus, $f_1 = (F \sin \phi)/(A \tan \phi/2)$; $f_2 = (F \sin \phi)[\tan (\phi/2)]/A$, where A = crossectional area of $M1$. Substituting gives $f_1 = 783$ lb/in^2 (5399 kPa); $f_2 = 56$ lb/in^2 (386.1 kPa).

4. Calculate the allowable stresses

Compute the allowable stresses N_1 and N_2 on AC and BC, respectively, and compare these with the actual stresses. Thus, by using Hankinson's equation from the previous calculation procedure, $N_1 = 1200(390)/(1200 \times 0.259^2 + 390 \times 0.966^2) = 1053$ lb/in^2 (7260.4 kPa). This is acceptable because it is greater than the actual stress. Also, $N_2 = 1200(390)/(1200 \times 0.966^2 + 390 \times 0.259^2) = 408$ lb/in^2 (2813.2 kPa). This is also acceptable, and the joint is therefore satisfactory.

ALLOWABLE LATERAL LOAD ON NAILS

In Fig. 101, the Western hemlock members are connected with six 50d common nails. Calculate the lateral load P that may be applied to this connection.

Calculation Procedure:

1. Determine the member group

The capacity of this connection is calculated in conformity with Part VIII of the *National Design Specification*. Refer to the *Specification* to ascertain the classification of the species. Western hemlock is in group III.

2. Determine the properties of the nail

Refer to the *Specification* to determine the properties of the nail. Calculate the penetration-diameter ratio, and compare this value with that stipulated in the *Specification*. Thus, length = 5.5 in

(139.7 mm); diameter = 0.244 in (6.1976 mm); penetration/diameter ratio = $(5.5 - 1.63)/0.244$ = 15.9 > 13. This is acceptable.

3. Find the capacity of the connection

Using the *Specification*, find the capacity of the nail. Then the capacity of the connection = P = 6(165) = 990 lb (4403.5 N).

CAPACITY OF LAG SCREWS

In Fig. 102, the cottonwood members are connected wtih three ⅝-in (15.88-mm) lag screws 8 in (203.2 mm) long. Determine the load P that may be applied to this connection.

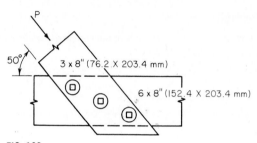

50°

3 x 8" (76.2 X 203.4 mm)

6 x 8" (152.4 X 203.4 mm)

FIG. 102

Calculation Procedure:

1. Determine the member group

The *National Design Specification* shows that cottonwood is classified in group IV.

2. Find the allowable screw loads

The *National Design Specification* gives the following values for each screw: allowable load parallel to grain = 550 lb (2446.4 N); allowable load normal to grain = 330 lb (1467.8 N).

3. Compute the allowable load on the connection

Use the Scholten nomogram, or $N = PQ/(P \sin^2 \theta + Q \cos^2 \theta)$, with $\theta = 50°$, and solve as given earlier. Either solution gives $P = 3(395) = 1185$ lb (5270.9 N).

DESIGN OF A BOLTED SPLICE

A 6 × 12 in (152.4 × 304.8 mm) southern pine member carrying a tensile force of 56 kips (249.1 kN) parallel to the grain is to be spliced with steel side plates. Design the splice.

Calculation Procedure:

1. Determine the number of bolts, and bolt size, required

Find the bolt capacity from the *National Design Specification*. The *Specification* allows a 25 percent increase in capacity of the parallel-to-grain loading when steel plates are used as side members.

Determine the number of bolts from $n = P/\text{capacity per bolt}$, lb (N), where P = load, lb (N). By assuming ⅞-in (22.2-mm) diameter bolts, $n = 56,000/[3940(1.25)] = 11.4$; use 12 bolts. The value 1.25 in the denominator is the increase in bolt load mentioned above.

As a trial, use three rows of four bolts each, as shown in Fig. 103.

2. Determine whether the joint complies with the Specification

Assume ¹⁵⁄₁₆-in (23.8-mm) diameter bolt holes. The gross area of the dressed lumber is 63.25 in² (408.089 cm²). The net area = gross area − area of the bolt holes = $63.25 - 3(0.94)(5.5)$ =

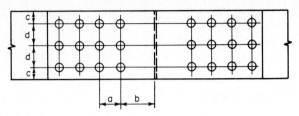

FIG. 103

47.74 in^2 (308.018 cm^2). The bearing area under the bolts = number of bolts [bolt diameter, in (mm)] [width, in (mm)] = $12(0.875)(5.5)$ = 57.75 in^2 (372.603 cm^2). The ratio of the net to bearing area is $47.74/57.75$ = $0.83 > 0.80$. This is acceptable, according to the *Specification*. The joint is therefore satisfactory, and the assumptions are usable in the design.

3. Establish the longitudinal bolt spacing

Using the *Specification*, we find a = $4(\frac{7}{8})$ = 3.5 in (88.90 mm); b_{min} = $7(\frac{7}{8})$ = 6⅛ in (155.58 mm).

4. Establish the transverse bolt spacing

Using the *Specification* gives L/D = $5.5(\frac{7}{8})$ = $6.3 > 6$. Make c = 2 in (50.8 mm) and d = 3¾ in (95.25 mm).

INVESTIGATION OF A TIMBER-CONNECTOR JOINT

The members in Fig. 104a have the following sizes: A, 4 × 8 in (101.6 × 203.2 mm); B, 3 × 8 in (76.2 × 203.2 mm). They are connected by six 4-in (101.6-mm) split-ring connectors, in the manner shown. The lumber is dense structural redwood. Investigate the adequacy of this joint, and establish the spacing of the connectors.

Calculation Procedure:

1. Determine the allowable stress

The *National Design Specification* shows that the allowable stress is 1700 lb/in² (11,721.5 kPa).

2. Find the lumber group

The *Specification* shows this species is classified in group C.

3. Compute the capacity of the connectors

The *Specification* shows that the capacity of a connector in parallel-to-grain loading for group C lumber is 4380 lb (19,482.2 N). With six connectors, the total capacity is $6(4380)$ = 26,280 lb (116,890 N). This is acceptable.

The *Specification* requires a minimum edge distance of 2¾ in (69.85 mm). The edge distance in the present instance is 3¾ in (95.25 mm).

4. Calculate the net area of member A

Apply the dimensions of the groove, which are recorded in the *Specification*. Referring to Fig. 104b, gross area = 27.19 in^2 (175.430 cm^2). The projected area of the groove and bolt hole = $4.5(1.00) + 0.813(2.625)$ = 6.63 in^2 (42.777 cm^2). The net area = $27.19 - 6.63$ = 20.56 in^2 (132.7 cm^2).

5. Calculate the stress at the net section; compare with the allowable stress

The stress f = load/net area = $26,000/20.56$ = 1260 lb/in² (8688 kPa). From the *Specification*, the allowable stress is f_{allow} = $(\frac{7}{8})(1700)$ = 1488 lb/in² (10,260 kPa). Also from the *Specification*, f_{allow} = 1650 lb/in² (11,377 kPa). The joint is therefore satisfactory in all respects.

6. Establish the connector spacing

Using the *Specification*, apply the recorded values without reduction because the connectors are stressed almost to capacity. Thus, a = 7 in (177.8 mm) and b = 9 in (228.6 mm).

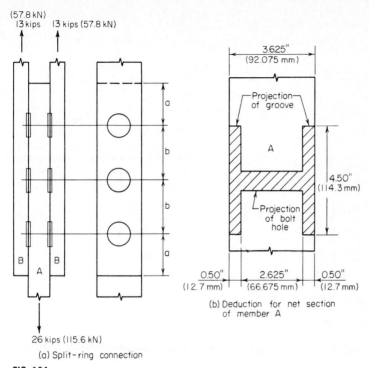

FIG. 104

(a) Split-ring connection

(b) Deduction for net section of member A

Reinforced Concrete

The design of reinforced-concrete members in this handook is executed in accordance with the specification titled *Building Code Requirements for Reinforced Concrete* of the American Concrete Institute (ACI). The ACI *Reinforced Concrete Design Handbook* contains many useful tables that expedite design work. The designer should become thoroughly familiar with this handbook and use the tables it contains whenever possible.

The spacing of steel reinforcing bars in a concrete member is subject to the restrictions imposed by the ACI *Code*. With reference to the beam and slab shown in Fig. 105, the reinforcing steel is assumed, for simplicity, to be concentrated at its centroidal axis, and the effective depth of the flexural member is taken as the distance from the extreme compression fiber to this axis. (The term *depth* hereafter refers to the *effective* rather than the overall depth of the beam.) For design purposes, it is usually assumed that the distance from the exterior surface to the center of the first row of steel bars is 2½ in (63.5 mm) in a beam with web stirrups, 2 in (50.8 mm) in a beam without stirrups, and 1 in (25.4 mm) in a slab. Where two rows of steel bars are provided, it is usually assumed that the distance from the exterior surface to the centroidal axis of the reinforcement is 3½ in (88.9 mm). The ACI *Handbook* gives the minimum beam widths needed to accommodate various combinations of bars in one row.

In a well-proportioned beam, the width-depth ratio lies between 0.5 and 0.75. The width and overall depth are usually an even number of inches.

The basic notational system pertaining to reinforced concrete beams is as follows: f'_c = ultimate compressive strength of concrete, lb/in² (kPa); f_c = maximum compressive stress in concrete, lb/in² (kPa); f_s = tensile stress in steel, lb/in² (kPa); f_y = yield-point stress in steel, lb/in² (kPa); ϵ_c = strain of extreme compression fiber; ϵ_s = strain of steel; b = beam width, in (mm); d = beam depth, in (mm); A_s = area of tension reinforcement, in² (cm²); p = tension-reinforcement ratio, $A_s/(bd)$; q = tension-reinforcement index, pf_y/f'_c; n = ratio of modulus of elasticity of steel to that of concrete, E_s/E_c; C = resultant compressive force on transverse section, lb (N); T = resultant tensile force on transverse section, lb (N).

Where the subscript b is appended to a symbol, it signifies that the given quantity is evaluated at balanced-design conditions.

Design of Flexural Members by Ultimate-Strength Method

In the ultimate-strength design of a reinforced-concrete structure, as in the plastic design of a steel structure, the capacity of the structure is found by determining the load that will cause failure and dividing this result by the prescribed load factor. The load at impending failure is termed the *ultimate load,* and the maximum bending moment associated with this load is called the *ultimate moment.*

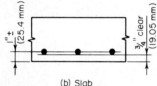

(a) Beam with stirrups

(b) Slab

FIG. 105 Spacing of reinforcing bars.

Since the tensile strength of concrete is relatively small, it is generally disregarded entirely in analyzing a beam. Consequently, the effective beam section is considered to comprise the reinforcing steel and the concrete on the compression side of the neutral axis, the concrete between these component areas serving merely as the ligature of the member.

The following notational system is applied in ultimate-strength design: a = depth of compression block, in (mm); c = distance from extreme compression fiber to neutral axis, in (mm); ϕ = capacity-reduction factor.

Where the subscript u is appended to a symbol, it signifies that the given quantity is evaluated at ultimate load.

For simplicity (Fig. 106), designers assume that when the ultimate moment is attained at a given section, there is a uniform stress in the concrete extending across a depth a, and that $f_c = 0.85f'_c$, and $a = k_1c$, where k_1 has the value stipulated in the ACI *Code.*

A reinforced-concrete beam has three potential modes of failure: crushing of the concrete, which is assumed to occur when ϵ_c reaches the value of 0.003; yielding of the steel, which begins when f_s reaches the value f_y; and the simultaneous crushing of the concrete and yielding of the steel. A beam that tends to fail by the third mode is said to be in *balanced design.* If the value of p exceeds that corresponding to balanced design (i.e., if there is an excess of reinforcement), the beam tends to fail by crushing of the concrete. But if the value of p is less than that corresponding to balanced design, the beam tends to fail by yielding of the steel.

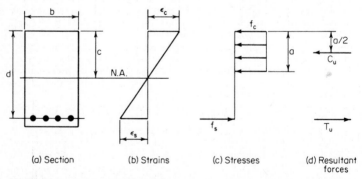

(a) Section   (b) Strains   (c) Stresses   (d) Resultant forces

FIG. 106 Conditions at ultimate moment.

Failure of the beam by the first mode would occur precipitously and without warning, whereas failure by the second mode would occur gradually, offering visible evidence of progressive failure. Therefore, to ensure that yielding of the steel would occur prior to failure of the concrete, the ACI *Code* imposes an upper limit of $0.75p_b$ on p.

To allow for material imperfections, defects in workmanship, etc., the *Code* introduces the capacity-reduction factor ϕ. A section of the *Code* sets $\phi = 0.90$ with respect to flexure and $\phi = 0.85$ with respect to diagonal tension, bond, and anchorage.

The basic equations for the ultimate-strength design of a rectangular beam reinforced solely in tension are

$$C_u = 0.85abf_c' \qquad T_u = A_s f_y \tag{1}$$

$$q = \frac{[A_s/(bd)]f_y}{f_c'} \tag{2}$$

$$a = 1.18qd \qquad c = \frac{1.18qd}{k_1} \tag{3}$$

$$M_u = \phi A_s f_y \left(d - \frac{a}{2} \right) \tag{4}$$

$$M_u = \phi A_s f_y d(1 - 0.59q) \tag{5}$$

$$M_u = \phi b d^2 f_c' q(1 - 0.59q) \tag{6}$$

$$A_s = \frac{bdf_c - [(bdf_c)^2 - 2bf_c M_u/\phi]^{0.5}}{f_y} \tag{7}$$

$$p_b = \frac{0.85k_1 f_c'}{f_y} \frac{87,000}{87,000 + f_y} \tag{8}$$

$$q_b = 0.85k_1 \left(\frac{87,000}{87,000 + f_y} \right) \tag{9}$$

In accordance with the *Code*,

$$q_{max} = 0.75q_b = 0.6375k_1 \left(\frac{87,000}{87,000 + f_y} \right) \tag{10}$$

Figure 107 shows the relationship between M_u and A_s for a beam of given size. As A_s increases, the internal forces C_u and T_u increase proportionately, but M_u increases by a smaller proportion because the action line of C_u is depressed. The M_u-A_s diagram is parabolic, but its curvature is small. By comparing the coordinates of two points P_a and P_b, the following result is obtained, in which the subscripts correspond to that of the given point:

$$\frac{M_{ua}}{A_{sa}} > \frac{M_{ub}}{A_{sb}} \qquad \text{where } A_{sa} < A_{sb} \tag{11}$$

CAPACITY OF A RECTANGULAR BEAM

A rectangular beam having a width of 12 in (304.8 mm) and an effective depth of 19.5 in (495.3 mm) is reinforced with steel bars having an area of 5.37 in² (34.647 cm²). The beam is made of 2500-lb/in² (17,237.5-kPa) concrete, and the steel has a yield-point stress of 40,000 lb/in² (275,800 kPa). Compute the ultimate moment this beam may resist (*a*) without referring to any design tables and without apply-

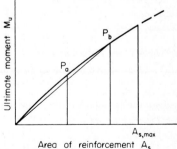

FIG. 107

ing the basic equations of ultimate-strength design except those that are readily apparent; (b) by applying the basic equations.

Calculation Procedure:

1. Compute the area of reinforcement for balanced design

Use the relation $\epsilon_s = f_y/E_s = 40,000/29,000,000 = 0.00138$. For balanced design, $c/d = \epsilon_c/(\epsilon_c + \epsilon_s) = 0.003/(0.003 + 0.00138) = 0.685$. Solving for c by using the relation for c/d, we find $c = 13.36$ in (339.344 mm). Also, $a = k_1 c = 0.85(13.36) = 11.36$ in (288.544 mm). Then $T_u = C_u = ab(0.85)f'_c = 11.36(12)(0.85)(2500) = 290,000$ lb (1,289,920 N); $A_s = T_u/f_y = 290,000/40,000 = 7.25$ in^2 (46,777 cm^2); and $0.75A_s = 5.44$ in^2 (35.097 cm^2). In the present instance, $A_s = 5.37$ in^2 (34.647 cm^2). This is acceptable.

2. Compute the ultimate-moment capacity of this member

Thus $T_u = A_s f_y = 5.37(40,000) = 215,000$ lb (956,320 N); $C_u = ab(0.85)f'_c = 25,500a = 215,000$ lb (956,320 N); $a = 8.43$ in (214.122 mm); $M_u = \phi T_u(d - a/2) = 0.90(215,000)(19.5 - 8.43/2) = 2,960,000$ in·lb (334,421 N·m). These two steps comprise the solution to part a. The next two steps comprise the solution of part b.

3. Apply Eq. 10; ascertain whether the member satisfies the Code

Thus, $q_{max} = 0.6375k_1(87,000)/(87,000 + f_y) = 0.6375(0.85)(87/127) = 0.371$; $q = [A_s/(bd)]f_y/f'_c = [5.37/(12 \times 19.5)]40/2.5 = 0.367$. This is acceptable.

4. Compute the ultimate-moment capacity

Applying Eq. 5 yields $M_u = \phi A_s f_y d(1 - 0.59q) = 0.90(5.37)(40,000)(19.5)(1 - 0.59 \times 0.367) = 2,960,000$ in·lb (334,421 N·m). This agrees exactly with the result computed in step 2.

DESIGN OF A RECTANGULAR BEAM

A beam on a simple span of 20 ft (6.1 m) is to carry a uniformly distributed live load of 1670 lb/lin ft (24,372 N/m) and a dead load of 470 lb/lin ft (6859 N/m), which includes the estimated weight of the beam. Architectural details restrict the beam width to 12 in (304.8 mm) and require that the depth be made as small as possible. Design the section, using $f'_c = 3000$ lb/in^2 (20,685 kPa) and $f_y = 40,000$ lb/in^2 (275,800 kPa).

Calculation Procedure:

1. Compute the ultimate load for which the member is to be designed

The beam depth is minimized by providing the maximum amount of reinforcement permitted by the Code. From the previous calculation procedure, $q_{max} = 0.371$.

Use the load factors given in the Code: $w_{DL} = 470$ lb/lin ft (6859 N/m); $w_{LL} = 1670$ lb/lin ft (24,372 N/m); $L = 20$ ft (6.1 m). Then $w_u = 1.5(470) + 1.8(1670) = 3710$ lb/lin ft (54,143 N/m); $M_u = \frac{1}{8}(3710)(20)^2 12 = 2,230,000$ in·lb (251,945.4 N·m).

2. Establish the beam size

Solve Eq. 6 for d. Thus, $d^2 = M_u/[\phi b f'_c q(1 - 0.59q)] = 2,230,000/[0.90(12)(3000) \times (0.371)(0.781)]$; $d = 15.4$ in (391.16 mm).

Set $d = 15.5$ in (393.70 mm). Then the corresponding reduction in the value of q is negligible.

3. Select the reinforcing bars

Using Eq. 2, we find $A_s = qbd f'_c/f_y = 0.371(12)(15.5)(3/40) = 5.18$ in^2 (33.421 cm^2). Use four no. 9 and two no. 7 bars, for which $A_s = 5.20$ in^2 (33.550 cm^2). This group of bars cannot be accommodated in the 12-in (304.8-mm) width and must therefore be placed in two rows. The overall beam depth will therefore be 19 in (482.6 mm).

4. Summarize the design

Thus, the beam size is 12×19 in (304.8×482.6 mm); reinforcement, four no. 9 and two no. 7 bars.

DESIGN OF THE REINFORCEMENT IN A RECTANGULAR BEAM OF GIVEN SIZE

A rectangular beam 9 in (228.6 mm) wide with a 13.5-in (342.9-mm) effective depth is to sustain an ultimate moment of 95 ft·kips (128.8 kN·m). Compute the area of reinforcement, using f'_c = 3000 lb/in² (20,685 kPa) and f_y = 40,000 lb/in² (275,800 kPa).

Calculation Procedure:

1. Investigate the adequacy of the beam size

From previous calculation procedures, q_{max} = 0.371. By Eq. 6, $M_{u,max}$ = 0.90 × (9)(13.5)²(3)(0.371)(0.781) = 1280 in·kips (144.6 kN·m); M_u = 95(12) = 1140 in·kips (128.8 kN·m). This is acceptable.

2. Apply Eq. 7 to evalute A_s

Thus, f_c = 0.85(3) = 2.55 kips/in² (17.582 MPa); bdf_c = 9(13.5)(2.55) = 309.8 kips (1377.99 kN); A_s = [309.8 − (309.8² − 58,140)⁰·⁵]/40 = 2.88 in² (18.582 cm²).

CAPACITY OF A T BEAM

Determine the ultimate moment that may be resisted by the T beam in Fig. 108a if f'_c = 3000 lb/in² (20,685 kPa) and f_y = 40,000 lb/in² (275,800 kPa).

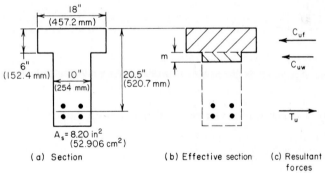

(a) Section   (b) Effective section   (c) Resultant forces

FIG. 108

Calculation Procedure:

1. Compute T_u and the resultant force that may be developed in the flange

Thus, T_u = 8.20(40,000) = 328,000 lb (1,458,944 N); f_c = 0.85(3000) = 2550 lb/in² (17,582.3 kPa); C_{uf} = 18(6)(2550) = 275,400 lb (1,224,979 N). Since $C_{uf} < T_u$, the deficiency must be supplied by the web.

2. Compute the resultant force developed in the web and the depth of the stress block in the web

Thus, C_{uw} = 328,000 − 275,400 = 52,600 lb (233,964.8 N); m = depth of the stress block = 52,600/[2550(10)] = 2.06 in (52.324 mm).

3. Evaluate the ultimate-moment capacity

Thus, M_u = 0.90[275,400(20.5 − 3) + 52,600(20.5 − 6 − 1.03)] = 4,975,000 in·lb (562,075.5 N·m).

4. Determine if the reinforcement complies with the Code

Let b' = width of web, in (mm); A_{s1} = area of reinforcement needed to resist the compressive force in the overhanging portion of the flange, in² (cm²); A_{s2} = area of reinforcement needed to resist the compressive force in the remainder of the section, in^2 (cm^2). Then $p_2 = A_{s2}/(b'd)$; A_{s1} = 2550(6)(18 − 10)/40,000 = 3.06 in^2 (19.743 cm^2); A_{s2} = 8.20 − 3.06 = 5.14 in^2 (33.163 cm^2). Then p_2 = 5.14/[10(20.5)] = 0.025.

A section of the ACI *Code* subjects the reinforcement ratio p_2 to the same restriction as that in a rectangular beam. By Eq. 8, $p_{2,\text{max}}$ = 0.75p_b = 0.75(0.85)(0.85)(3/40)(87/127) = 0.0278 > 0.025. This is acceptable.

CAPACITY OF A T BEAM OF GIVEN SIZE

The T beam in Fig. 109 is made of 3000-lb/in^2 (20,685-kPa) concrete, and f_y = 40,000 lb/in^2 (275,800 kPa). Determine the ultimate-moment capacity of this member if it is reinforced in tension only.

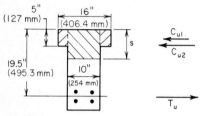

FIG. 109

Calculation Procedure:

1. Compute C_{u1}, $C_{u2,\text{max}}$, and s_{max}

Let the subscript 1 refer to the overhanging portion of the flange and the subscript 2 refer to the remainder of the compression zone. Then f_c = 0.85(3000) = 2550 lb/in^2 (17,582.3 kPa); C_{u1} = 2550(5)(16 − 10) = 76,500 lb (340,272 N). From the previous calculation procedure, $p_{2,\text{max}}$ = 0.0278. Then $A_{s2,\text{max}}$ = 0.0278(10)(19.5) = 5.42 in^2 (34.970 cm^2); $C_{u2,\text{max}}$ = 5.42(40,000) = 216,800 lb (964,326.4 N); s_{max} = 216,800/[10(2550)] = 8.50 in (215.9 mm).

2. Compute the ultimate-moment capacity

Thus, $M_{u,\text{max}}$ = 0.90[76,500(19.5 − 5/2) + 216,800(19.5 − 8.50/2)] = 4,145,000 in·lb (468,300 N·m).

DESIGN OF REINFORCEMENT IN A T BEAM OF GIVEN SIZE

The T beam in Fig. 109 is to resist an ultimate moment of 3,960,000 in·lb (447,400.8 N·m). Determine the required area of reinforcement, using f_c' = 3000 lb/in^2 (20,685 kPa) and f_y = 40,000 lb/in^2 (275,800 kPa).

Calculation Procedure:

1. Obtain a moment not subject to reduction

From the previous calculation procedure, the ultimate-moment capacity of this member is 4,145,000 in·lb (468,300 N·m). To facilitate the design, divide the given ultimate moment M_u by the capacity-reduction factor to obtain a moment M_u' that is not subject to reduction. Thus M_u' = 3,960,000/0.9 = 4,400,000 in·lb (497,112 N·m).

2. Compute the value of s associated with the given moment

From step 2 in the previous calculation procedure, M_{u1}' = 1,300,000 in·lb (146,874 N·m). Then M_{u2}' = 4,400,000 − 1,300,000 = 3,100,000 in·lb (350,238 N·m). But M_{u2}' = 2550(10s)(19.5 − s/2), so s = 7.79 in (197.866 mm).

3. Compute the area of reinforcement

Thus, $C_{u2} = M_{u2}'/(d − \tfrac{1}{2}s)$ = 3,100,000/(19.5 − 3.90) = 198,700 lb (883,817.6 N). From step 1 of the previous calculation procedure, C_{u1} = 76,500 lb (340,272 N); T_u = 76,500 + 198,700 = 275,200 lb (1,224,089.6 N); A_s = 275,200/40,000 = 6.88 in^2 (174.752 mm).

4. Verify the solution

To verify the solution, compute the ultimate-moment capacity of the member. Use the notational system given in earlier calculation procedures. Thus, C_{uf} = 16(5)(2550) = 204,000 lb (907,392

N); $C_{uw} = 275,200 - 204,000 = 71,200$ lb (316,697.6 N); $m = 71,200/[2550(10)] = 2.79$ in (70.866 mm); $M_u = 0.90[204,000(19.5 - 2.5) + 71,200(19.5 - 5 - 1.40)] = 3,960,000$ in·lb (447,400.8 N·m). Thus, the result is verified because the computed moment equals the given moment.

REINFORCEMENT AREA FOR A DOUBLY REINFORCED RECTANGULAR BEAM

A beam that is to resist an ultimate moment of 690 ft·kips (935.6 kN·m) is restricted to a 14-in (355.6-mm) width and 24-in (609.6-mm) total depth. Using $f'_c = 5000$ lb/in² and $f_y = 50,000$ lb/in² (344,750 kPa), determine the area of reinforcement.

Calculation Procedure:

1. Compute the values of q_b, q_{max}, and p_{max} for a singly reinforced beam

As the following calculations will show, it is necessary to reinforce the beam both in tension and in compression. In Fig. 110, let A_s = area of tension reinforcement, in² (cm²); A'_s = area of

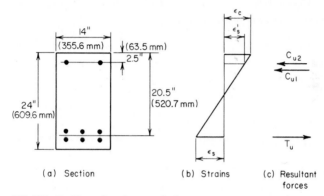

(a) Section   (b) Strains   (c) Resultant forces

FIG. 110   Doubly reinforced rectangular beam.

compression reinforcement, in² (cm²); d' = distance from compression face of concrete to centroid of compression reinforcement, in (mm); f_s = stress in tension steel, lb/in² (kPa); f'_s = stress in compression steel, lb/in² (kPa); ϵ'_s = strain in compression steel; $p = A_s/(bd)$; $p' = A'_s/(bd)$; $q = pf_y/f'_c$; M_u = ultimate moment to be resisted by member, in·lb (N·m); M_{u1} = ultimate-moment capacity of member if reinforced solely in tension; M_{u2} = increase in ultimate-moment capacity resulting from use of compression reinforcement; C_{u1} = resultant force in concrete, lb (N); C_{u2} = resultant force in compression steel, lb (N).

If $f'_s = f_y$, the tension reinforcement may be resolved into two parts having areas of $A_s - A'_s$ and A'_s. The first part, acting in combination with the concrete, develops the moment M_{u1}. The second part, acting in combination with the compression reinforcement, develops the moment M_{u2}.

To ensure that failure will result from yielding of the tension steel rather than crushing of the concrete, the ACI *Code* limits $p - p'$ to a maximum value of $0.75p_b$, where p_b has the same significance as for a singly reinforced beam. Thus the *Code*, in effect, permits setting $f'_s = f_y$ if inception of yielding in the compression steel will precede or coincide with failure of the concrete at balanced-design ultimate moment. This, however, introduces an inconsistency, for the limit imposed on $p - p'$ precludes balanced design.

By Eq. 9, $q_b = 0.85(0.80)(87/137) = 0.432$; $q_{max} = 0.75(0.432) = 0.324$; $p_{max} = 0.324(5/50) = 0.0324$.

2. Compute M_{u1}, M_{u2}, and C_{u2}

Thus, $M_u = 690,000(12) = 8,280,000$ in·lb (935,474.4 N·m). Since two rows of tension bars are probably required, $d = 24 - 3.5 = 20.5$ in (520.7 mm). By Eq. 6, $M_{u1} = 0.90(14)(20.5)^2(5000)$

$\times (0.324)(0.809) = 6,940,000$ in·lb ($784,081.2$ N·m); $M_{u2} = 8,280,000 - 6,940,000 = 1,340,000$ in·lb ($151,393.2$ N·m); $C_{u2} = M_{u2}/(d - d') = 1,340,000/(20.5 - 2.5) = 74,400$ lb ($330,931.2$ N).

3. Compute the value of ϵ'_s under the balanced-design ultimate moment

Compare this value with the strain at incipient yielding. By Eq. 3, $c_b = 1.18q_b d/k_1 = 1.18(0.432)(20.5)/0.80 = 13.1$ in (332.74 mm); $\epsilon'_s/\epsilon_c = (13.1 - 2.5)/13.1 = 0.809$; $\epsilon'_s = 0.809(0.003) = 0.00243$; $\epsilon_y = 50/29,000 = 0.0017 < \epsilon'_s$. The compression reinforcement will therefore yield before the concrete fails, and $f'_s = f_y$ may be used.

4. Alternatively, test the compression steel for yielding

Apply

$$p - p' \geq \frac{0.85k_1 f'_c d'(87,000)}{f_y d(87,000 - f_y)} \tag{12}$$

If this relation obtains, the compression steel will yield. The value of the right-hand member is $0.85(0.80)(5/50)(2.5/20.5)(87/37) = 0.0195$. From the preceding calculations, $p - p' = 0.0324 > 0.0195$. This is acceptable.

5. Determine the areas of reinforcement

By Eq. 2, $A_s = A'_s = q_{max}bdf'_c/f_y = 0.324(14)(20.5)(5/50) = 9.30$ in² (60.00 cm²); $A'_s = C_{u2}/(\phi f_y) = 74,400/[0.90(50,000)] = 1.65$ in² (10.646 cm²); $A_s = 9.30 + 1.65 = 10.95$ in² (70.649 cm²).

6. Verify the solution

Apply the following equations for the ultimate-moment capacity:

$$a = \frac{(A_s - A'_s)f_y}{0.85f'_c b} \tag{13}$$

So $a = 9.30(50,000)/[0.85(5000)(14)] = 7.82$ in (198.628 mm). Also,

$$M_u = \phi f_y \left[(A_s - A'_s)\left(d - \frac{a}{2}\right) + A'_s(d - d') \right] \tag{14}$$

So $M_u = 0.90(50,000)(9.30 \times 16.59 + 1.65 \times 18) = 8,280,000$ in·lb ($935,474.4$ N·m), as before. Therefore, the solution has been verified.

DESIGN OF WEB REINFORCEMENT

A 15-in (381-mm) wide 22.5-in (571.5-mm) effective-depth beam carries a uniform ultimate load of 10.2 kips/lin ft (148.86 kN/m). The beam is simply supported, and the clear distance between supports is 18 ft (5.5 m). Using $f'_c = 3000$ lb/in² (20,685 kPa) and $f_y = 40,000$ lb/in² (275,800 kPa), design web reinforcement in the form of vertical U stirrups for this beam.

Calculation Procedure:

1. Construct the shearing-stress diagram for half-span

The ACI Code provides two alternative methods for computing the allowable shearing stress on an unreinforced web. The more precise method recognizes the contribution of both the shearing stress and flexural stress on a cross section in producing diagonal tension. The less precise and more conservative method restricts the shearing stress to a stipulated value that is independent of the flexural stress.

For simplicity, the latter method is adopted here. A section of the Code sets $\phi = 0.85$ with respect to the design of web reinforcement. Let v_u = nominal ultimate shearing stress, lb/in² (kPa); v_c = shearing stress resisted by concrete, lb/in² (kPa); v'_u = shearing stress resisted by the web reinforcement, lb/in² (kPa); A_v = total cross-sectional area of stirrup, in² (cm²); V_u = ultimate vertical shear at section, lb (N); s = center-to-center spacing of stirrups, in (mm).

The shearing-stress diagram for half-span is shown in Fig. 111. Establish the region AF within

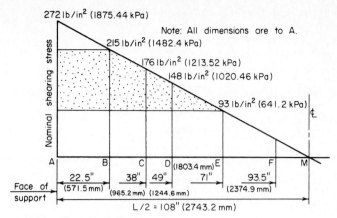

FIG. 111 Shearing-stress diagram.

which web reinforcement is required. The *Code* sets the allowable shearing stress in the concrete at

$$v_c = 2\phi(f_c')^{0.5} \tag{15}$$

The equation for nominal ultimate shearing stress is

$$v_u = \frac{V_u}{bd} \tag{16}$$

Then, $v_c = 2(0.85)(3000)^{0.5} = 93$ lb/in^2 (641.2 kPa).

At the face of the support, $V_u = 9(10,200) = 91,800$ lb (408,326.4 N); $v_u = 91,800/[15(22.5)]$ = 272 lb/in^2 (1875.44 kPa). The slope of the shearing-stress diagram = $-272/108 = -2.52$ lb/ (in^2·in) (-0.684 kPa/mm). At distance d from the face of the support, $v_u = 272 - 22.5(2.52)$ = 215 lb/in^2 (1482.4 kPa); $v_u' = 215 - 93 = 122$ lb/in^2 (841.2 kPa).

Let E denote the section at which $v_u = v_c$. Then, $AE = (272 - 93)/2.52 = 71$ in (1803.4 mm). A section of the *Code* requires that web reinforcement be continued for a distance d beyond the section where $v_u = v_c$; $AF = 71 + 22.5 = 93.5$ in (2374.9 mm).

2. Check the beam size for Code compliance

Thus, $v_{u,\max} = 10\phi(f_c')^{0.5} = 466 > 215$ lb/in^2 (1482.4 kPa). This is acceptable.

3. Select the stirrup size

Equate the spacing near the support to the minimum practical value, which is generally considered to be 4 in (101.6 mm). The equation for stirrup spacing is

$$s = \frac{\phi A_v f_y}{v_u' b} \tag{17}$$

Then $A_v = s v_u' b/(\phi f_y) = 4(122)(15)/[0.85(40,000)] = 0.215$ in^2 (1.3871 cm^2). Since each stirrup is bent into the form of a U, the total cross-sectional area is twice that of a straight bar. Use no. 3 stirrups for which $A_v = 2(0.11) = 0.22$ in^2 (1.419 cm^2).

4. Establish the maximum allowable stirrup spacing

Apply the criteria of the *Code*, or $s_{\max} = d/4$ if $v_u > 6\phi(f_c')^{0.5}$. The right-hand member of this inequality has the value 279 lb/in^2 (1923.70 kPa), and this limit therefore does not apply. Then $s_{\max} = d/2 = 11.25$ in (285.75 mm), or $s_{\max} = A_v/(0.0015b) = 0.22/[0.0015(15)] = 9.8$ in (248.92 mm). The latter limit applies, and the stirrup spacing will therefore be restricted to 9 in (228.6 mm).

5. Locate the beam sections at which the required stirrup spacing is 6 in (152.4 mm) and 9 in (228.6 mm)

Use Eq. 17. Then $\phi A_v f_y/b = 0.85(0.22)(40,000)/15 = 499$ lb/in (87.38 kN/m). At C: $v_u' = 499/6 = 83$ lb/in^2 (572.3 kPa); $v_u = 83 + 93 = 176$ lb/in^2 (1213.52 kPa); $AC = (272 - 176)/2.52 = 38$ in (965.2 mm). At D: $v_u' = 499/9 = 55$ lb/in^2 (379.2 kPa); $v_u = 55 + 93 = 148$ lb/in^2 (1020.46 kPa); $AD = (272 - 148)/2.52 = 49$ in (1244.6 mm).

6. Devise a stirrup spacing conforming to the computed results

The following spacing, which requires 17 stirrups for each half of the span, is satisfactory and conforms with the foregoing results:

Quantity	Spacing, in (mm)	Total, in (mm)	Distance from last stirrup to face of support, in (mm)
1	2 (50.8)	2 (50.8)	2 (50.8)
9	4 (101.6)	36 (914.4)	38 (965.2)
2	6 (152.4)	12 (304.8)	50 (1270)
5	9 (228.6)	45 (1143)	95 (2413)

DETERMINATION OF BOND STRESS

A beam of 4000-lb/in^2 (27,580-kPa) concrete has an effective depth of 15 in (381 mm) and is reinforced with four no. 7 bars. Determine the ultimate bond stress at a section where the ultimate shear is 72 kips (320.3 kN). Compare this with the allowable stress.

Calculation Procedure:

1. Determine the ultimate shear flow h_u

The adhesion of the concrete and steel must be sufficiently strong to resist the horizontal shear flow. Let u_u = ultimate bond stress, lb/in^2 (kPa); V_u = ultimate vertical shear, lb (N); Σo = sum of perimeters of reinforcing bars, in (mm). Then the ultimate shear flow at any plane between the neutral axis and the reinforcing steel is $h_u = V_u/(d - a/2)$.

In conformity with the notational system of the working-stress method, the distance $d - a/2$ is designated as jd. Dividing the shear flow by the area of contact in a unit length and introducing the capacity-reduction factor yield

$$u_u = \frac{V_u}{\phi \Sigma o j d} \tag{18}$$

A section of the ACI *Code* sets $\phi = 0.85$ with respect to bond, and j is usually assigned the approximate value of 0.875 when this equation is used.

2. Calculate the bond stress

Thus, $\Sigma o = 11.0$ in (279.4 mm), from the ACI *Handbook*. Then $u_u = 72,000/[0.85(11.0)(0.875) \times (15)] = 587$ lb/in^2 (4047.4 kPa).

The allowable stress is given in the *Code* as

$$u_{u,\text{allow}} = \frac{9.5(f_c')^{0.5}}{D} \tag{19}$$

but not above 800 lb/in^2 (5516 kPa). Thus, $u_{u,\text{allow}} = 9.5(4,000)^{0.5}/0.875 = 687$ lb/in^2 (4736.9 kPa).

DESIGN OF INTERIOR SPAN OF A ONE-WAY SLAB

A floor slab that is continuous over several spans carries a live load of 120 lb/ft^2 (5745 N/m^2) and a dead load of 40 lb/ft^2 (1915 N/m^2), exclusive of its own weight. The clear spans are 16 ft (4.9

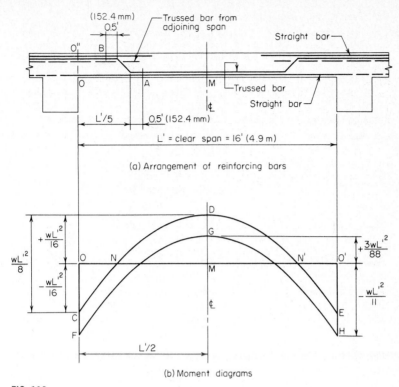

(a) Arrangement of reinforcing bars

$L' = $ clear span = 16' (4.9 m)

(b) Moment diagrams

FIG. 112

m). Design the interior span, using $f'_c = 3000$ lb/in² (20,685 kPa) and $f_y = 50,000$ lb/in² (344,750 kPa).

Calculation Procedure:

1. Find the minimum thickness of the slab as governed by the Code

Refer to Fig. 112. The maximum potential positive or negative moment may be found by applying the type of loading that will induce the critical moment and then evaluating this moment. However, such an analysis is time-consuming. Hence, it is wise to apply the moment equations recommended in the ACI *Code* whenever the span and loading conditions satisfy the requirements given there. The slab is designed by considering a 12-in (304.8-mm) strip as an individual beam, making $b = 12$ in (304.8 mm).

Assuming that $L = 17$ ft (5.2 m), we know the minimum thickness of the slab is $t_{min} = L/35 = 17(12)/35 = 5.8$ in (147.32 mm).

2. Assuming a slab thickness, compute the ultimate load on the member

Tentatively assume $t = 6$ in (152.4 mm). Then the beam weight = $(6/12)(150$ lb/ft³ $= 75$ lb/lin ft (1094.5 N/m). Also, $w_u = 1.5(40 + 75) + 1.8(120) = 390$ lb/lin ft (5691.6 N/m).

3. Compute the shearing stress associated with the assumed beam size

From the *Code* for an interior span, $V_u = \frac{1}{2}w_uL' = \frac{1}{2}(390)(16) = 3120$ lb (13,877.8 N); $d = 6 - 1 = 5$ in (127 mm); $v_u = 3120/[12(5)] = 52$ lb/in² (358.54 kPa); $v_c = 93$ lb/in² (641.2 kPa). This is acceptable.

4. Compute the two critical moments

Apply the appropriate moment equations. Compare the computed moments with the moment capacity of the assumed beam size to ascertain whether the size is adequate. Thus, $M_{u,neg} = (\frac{1}{11})w_uL'^2 = (\frac{1}{11})(390)(16)^2(12) = 108,900$ in·lb (12,305.5 N·m), where the value 12 converts the dimension to inches. Then $M_{u,pos} = \frac{1}{16}w_uL'^2 = 74,900$ in·lb (8462.2 N·m). By Eq. 10, $q_{max} = 0.6375(0.85)(87/137) = 0.344$. By Eq. 6, $M_{u,allow} = 0.90(12)(5)^2(3000)(0.344)(0.797) = 222,000$ in·lb (25,081.5 N·m). This is acceptable. The slab thickness will therefore be made 6 in (152.4 mm).

5. Compute the area of reinforcement associated with each critical moment

By Eq. 7, $bdf_c = 12(5)(2.55) = 153.0$ kips (680.54 kN); then $2bf_cM_{u,neg}/\phi = 2(12)(2.55)(108.9)/0.90 = 7405$ kips² (146,505.7 kN²); $A_{s,neg} = [153.0 - (153.0^2 - 7405)^{0.5}]/50 = 0.530$ in² (3.4196 cm²). Similarly, $A_{s,pos} = 0.353$ in² (2.278 cm²).

6. Select the reinforcing bars, and locate the bend points

For positive reinforcement, use no. 4 trussed bars 13 in (330.2 mm) on centers, alternating with no. 4 straight bars 13 in (330.2 mm) on centers, thus obtaining $A_s = 0.362$ in² (2.336 cm²).

For negative reinforcement, supplement the trussed bars over the support with no. 4 straight bars 13 in (330.2 mm) on centers, thus obtaining $A_s = 0.543$ in² (3.502 cm²).

The trussed bars are usually bent upward at the fifth points, as shown in Fig. 112a. The reinforcement satisfies a section of the ACI *Code* which requires that "at least . . . one-fourth the positive moment reinforcement in continuous beams shall extend along the same face of the beam into the support at least 6 in (152.4 mm)."

7. Investigate the adequacy of the reinforcement beyond the bend points

In accordance with the *Code*, $A_{min} = A_t = 0.0020bt = 0.0020(12)(6) = 0.144$ in² (0.929 cm²).

A section of the *Code* requires that reinforcing bars be extended beyond the point at which they become superfluous with respect to flexure a distance equal to the effective depth or 12 bar diameters, whichever is greater. In the present instance, extension $= 12(0.5) = 6$ in (152.4 mm). Therefore, the trussed bars in effect terminate as positive reinforcement at section A (Fig. 112). Then $L'/5 = 3.2$ ft (0.98 m); $AM = 8 - 3.2 - 0.5 = 4.3$ ft (1.31 m).

The conditions immediately to the left of A are $M_u = M_{u,pos} - \frac{1}{2}w_u(AM)^2 = 74,900 - \frac{1}{2}(390)(4.3)^2(12) = 31,630$ in·lb (3573.56 N·m); $A_{s,pos} = 0.181$ in² (1.168 cm²); $q = 0.181(50)/[12(5)(3)] = 0.0503$. By Eq. 5, $M_{u,allow} = 0.90(0.181)(50,000)(5)(0.970) = 39,500$ in·lb (4462.7 N·m). This is acceptable.

Alternatively, Eq. 11 may be applied to obtain the following conservative approximation: $M_{u,allow} = 74,900(0.181)/0.353 = 38,400$ in·lb (4338.43 N·m).

The trussed bars in effect terminate as negative reinforcement at B, where $O''B = 3.2 - 0.33 - 0.5 = 2.37$ ft (72.23 m). The conditions immediately to the right of B are $|M_u| = M_{u,neg} - 12(3120 \times 2.37 - \frac{1}{2} \times 390 \times 2.37^2) = 33,300$ in·lb (3762.23 N·m). Then $A_{s,neg} = 0.362$ in² (2.336 cm²). As a conservative approximation, $M_{u,allow} = 108,900(0.362)/0.530 = 74,400$ in·lb (8405.71 N·m). This is acceptable.

8. Locate the point at which the straight bars at the top may be discontinued

9. Investigate the bond stresses

In accordance with Eq. 19, $u_{u,allow} = 800$ lb/in² (5516 kPa).

If *CDE* in Fig. 112b represents the true moment diagram, the bottom bars are subjected to bending stress in the interval *NN'*. Manifestly, the maximum bond stress along the bottom occurs at these boundary points (points of contraflexure), where the shear is relatively high and the straight bars alone are present. Thus $MN = 0.354L'$; V_u at N/V_u at support $= 0.354L'/(0.5L') = 0.71$; V_u at $N = 0.71(3120) = 2215$ lb (9852.3 N). By Eq. 18, $u_u = V_u/(\phi\Sigma ojd) = 2215/[0.85(1.45)(0.875)(5)] = 411$ lb/in² (2833.8 kPa). This is acceptable. It is apparent that the maximum bond stress in the top bars has a smaller value.

ANALYSIS OF A TWO-WAY SLAB BY THE YIELD-LINE THEORY

The slab in Fig. 113a is simply supported along all four edges and is isotropically reinforced. It supports a uniformly distributed ultimate load of w_u lb/ft² (kPa). Calculate the ultimate unit moment m_u for which the slab must be designed.

Calculation Procedure:

1. Draw line GH perpendicular to AE at E; express distances b and c in terms of a

Consider a slab to be reinforced in orthogonal directions. If the reinforcement in one direction is identical with that in the other direction, the slab is said to be *isotropically reinforced;* if the

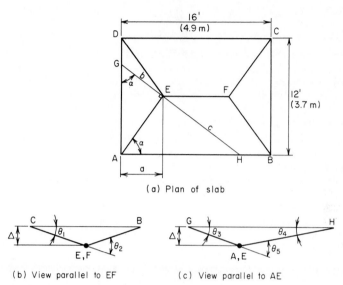

(a) Plan of slab

(b) View parallel to EF   (c) View parallel to AE

FIG. 113   Analysis of two-way slab by mechanism method.

reinforcements differ, the slab is described as *orthogonally anisotropic*. In the former case, the capacity of the slab is identical in all directions; in the latter case, the capacity has a unique value in every direction. In this instance, assume that the slab size is excessive with respect to balanced design, the result being that the failure of the slab will be characterized by yielding of the steel.

In a steel beam, a plastic hinge forms at a *section;* in a slab, a plastic hinge is assumed to form along a *straight line,* termed a *yield line*. It is plausible to assume that by virtue of symmetry of loading and support conditions the slab in Fig. 113a will fail by the formation of a central yield line *EF* and diagonal yield lines such as *AE*, the ultimate moment at these lines being positive. The ultimate *unit* moment m_u is the moment acting on a unit length.

Although it is possible to derive equations that give the location of the yield lines, this procedure is not feasible because the resulting equations would be unduly cumbersome. The procedure followed in practice is to assign a group of values to the distance a and to determine the corresponding values of m_u. The true value of m_u is the highest one obtained. Either the static or mechanism method of analysis may be applied; the latter will be applied here.

Expressing the distances b and c in terms of a gives $\tan \alpha = 6/a = AE/b = c/(AE)$; $b = aAE/6$; $c = 6AE/a$.

2. Find the rotation of the plastic hinges

Allow line *EF* to undergo a virtual displacement Δ after the collapse load is reached. During the virtual displacement, the portions of the slab bounded by the yield lines and the supports rotate as planes. Refer to Fig. 113b and c: $\theta_1 = \Delta/6$; $\theta_2 = 2\theta_1 = \Delta/3 = 0.333\Delta$; $\theta_3 = \Delta/b$; $\theta_4 = \Delta/c$; $\theta_5 = \Delta(1/b + 1/c) = [\Delta/(AE)](6/a + a/6)$.

3. Select a trial value of a, and evaluate the distances and angles

Using $a = 4.5$ ft (1.37 m) as the trial value, we find $AE = (a^2 + 6^2)^{0.5} = 7.5$ ft (2.28 m); $b = 5.63$ ft (1.716 m); $c = 10$ ft (3.0 m); $\theta_5 = (\Delta/7.5)(6/4.5 + 4.5/6) = 0.278\Delta$.

4. Develop an equation for the external work W_E performed by the uniform load on a surface that rotates about a horizontal axis

In Fig. 114, consider that the surface ABC rotates about axis AB through an angle θ while carrying a uniform load of w lb/ft^2 (kPa). For the elemental area dA, the deflection, total load, and external work are $\delta = x\theta$; $dW = w\,dA$; $dW_E = \delta\,dW = x\theta w\,dA$. The total work for the surface is $W_E = w\theta \int x\,dA$, or

$$W_E = w\theta Q \qquad (20)$$

where Q = static moment of total area, with respect to the axis of rotation.

5. Evaluate the external and internal work for the slab

Using the assumed value, we see $a = 4.5$ ft (1.37 m), $EF = 16 - 9 = 7$ ft (2.1 m). The external work for the two triangles is $2w_u(\Delta/4.5)(\frac{1}{2})(12)(4.5)^2 = 18w_u\Delta$. The external work for the two trapezoids is $2w_u(\Delta/6)(\frac{1}{2})(16 + 2 \times 7)(6)^2 = 60w_u\Delta$. Then $W_E = w_u\Delta(18 + 60) = 78w_u\Delta$; $W_I = m_u(7\theta_2 + 4 \times 7.5\theta_5) = 10.67m_u\Delta$.

6. Find the value of m_u corresponding to the assumed value of a

Equate the external and internal work to find this value of m_u. Thus, $10.67m_u\Delta = 78w_u\Delta$; $m_u = 7.31w_u$.

7. Determine the highest value of m_u

Assign other trial values to a, and find the corresponding values of m_u. Continue this procedure until the highest value of m_u is obtained. This is the true value of the ultimate unit moment.

Design of Flexural Members by the Working-Stress Method

As demonstrated earlier, the analysis or design of a composite beam by the working-stress method is most readily performed by transforming the given beam to an equivalent homogeneous beam. In the case of a reinforced-concrete member, the transformation is made by replacing the reinforcing steel with a strip of concrete having an area nA_s and located at the same distance from the neutral axis as the steel. This substitute concrete is assumed capable of sustaining tensile stresses.

The following symbols, shown in Fig. 115, are to be added to the notational system given earlier: kd = distance from extreme compression fiber to neutral axis, in (mm); jd = distance

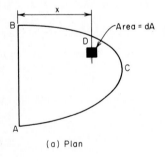

(a) Plan

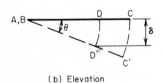

(b) Elevation

FIG. 114

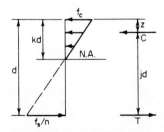

FIG. 115 Stress and resultant forces.

between action lines of C and T, in (mm); z = distance from extreme compression fiber to action line of C, in (mm).

The basic equations for the working-stress design of a rectangular beam reinforced solely in tension are

$$k = \frac{f_c}{f_c + f_s/n} \tag{21}$$

$$j = 1 - \frac{k}{3} \tag{22}$$

$$M = Cjd = \tfrac{1}{2}f_c kjbd^2 \tag{23}$$

$$M = \tfrac{1}{6}f_c k(3 - k)bd^2 \tag{24}$$

$$M = Tjd = f_s A_s jd \tag{25}$$

$$M = f_s pjbd^2 \tag{26}$$

$$M = \frac{f_s k^2(3 - k)bd^2}{6n(1 - k)} \tag{27}$$

$$p = \frac{f_c k}{2f_s} \tag{28}$$

$$p = \frac{k^2}{2n(1 - k)} \tag{29}$$

$$k = [2pn + (pn)^2]^{0.5} - pn \tag{30}$$

For a given set of values of f_c, f_s, and n, M is directly proportional to the beam property bd^2. Let K denote the constant of proportionality. Then

$$M = Kbd^2 \tag{31}$$

where

$$K = \tfrac{1}{2}f_c kj = f_s pj \tag{32}$$

The allowable flexural stress in the concrete and the value of n, which are functions of the ultimate strength f'_c, are given in the ACI *Code*, as is the allowable flexural stress in the steel. In all instances in the following procedures, the assumption is that the reinforcement is intermediate-grade steel having an allowable stress of 20,000 lb/in^2 (137,900 kPa).

Consider that the load on a beam is gradually increased until a limiting stress is induced. A beam that is so proportioned that the steel and concrete simultaneously attain their limiting stress is said to be in *balanced design*. For each set of values of f'_c and f_s, there is a corresponding set of values of K, k, j, and p associated with balanced design. These values are recorded in Table 6.

TABLE 6 Values of Design Parameters at Balanced Design

f'_c and n	f_c	f_s	K	k	j	p
2500 10	1125	20,000	178	0.360	0.880	0.0101
3000 9	1350	20,000	223	0.378	0.874	0.0128
4000 8	1800	20,000	324	0.419	0.860	0.0188
5000 7	2250	20,000	423	0.441	0.853	0.0248

In Fig. 116, AB represents the stress line of the transformed section for a beam in balanced design. If the area of reinforcement is increased while the width and depth remain constant, the neutral axis is depressed to O', and $A'O'B$ represents the stress line under the allowable load. But if the width is increased while the depth and area of reinforcement remain constant, the neutral axis is elevated to O'', and $AO''B'$ represents the stress line under the allowable load. This analysis leads to these conclusions: If the reinforcement is in excess of that needed for balanced design, the concrete is the first material to reach its limiting stress under a gradually increasing load. If the beam size is in excess of that needed for balanced design, the steel is the first material to reach its limiting stress.

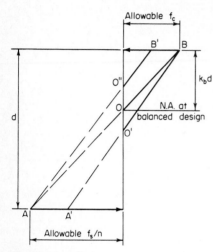

FIG. 116 Stress diagrams.

STRESSES IN A RECTANGULAR BEAM

A beam of 2500-lb/in^2 (17,237.5-kPa) concrete has a width of 12 in (304.8 mm) and an effective depth of 19.5 in (495.3 mm). It is reinforced with one no. 9 and two no. 7 bars. Determine the flexural stresses caused by a bending moment of 62 ft·kips (84.1 kN·m) (a) without applying the basic equations of reinforced-concrete beam design; (b) by applying the basic equations.

Calculation Procedure:

1. Record the pertinent beam data

Thus $f'_c = 2500$ lb/in^2 (17,237.5 kPa); ∴ $n = 10$; $A_s = 2.20$ in^2 (14.194 cm^2); $nA_s = 22.0$ in^2 (141.94 cm^2). Then $M = 62,000(12) = 744,000$ in·lb (84,057.1 N·m).

2. Transform the given section to an equivalent homogeneous section, as in Fig. 117b

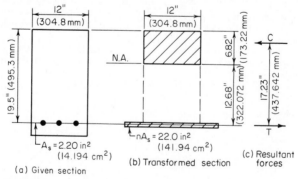

FIG. 117

3. Locate the neutral axis of the member

The neutral axis coincides with the centroidal axis of the transformed section. To locate the neutral axis, set the static moment of the transformed area with respect to its centroidal axis equal to zero: $12(kd)^2/2 - 22.0(19.5 - kd) = 0$; $kd = 6.82$; $d - kd = 12.68$ in (322.072 mm).

4. Calculate the moment of inertia of the transformed section

Then evaluate the flexural stresses by applying the stress equation: $I = (\frac{1}{12})(12)(6.82)^3 + 22.0(12.68)^2 = 4806$ in^4 (200,040.6 cm^4); $f_c = Mkd/I = 744,000(6.82)/4806 = 1060$ lb/in^2 (7308.7 kPa); $f_s = 10(744,000)(12.68)/4806 = 19,600$ lb/in^2

5. Alternatively, evaluate the stresses by computing the resultant forces C and T

Thus $jd = 19.5 - 6.82/3 = 17.23$ in (437.642 mm); $C = T = M/jd = 744,000/17.23 = 43,200$ lb (192,153.6 N). But $C = \frac{1}{2}f_c(6.82)12$; $\therefore f_c = 1060$ lb/in^2 (7308.7 kPa); and $T = 2.20f_s$; $\therefore f_s = 19,600$ lb/in^2 (135,142 kPa). This concludes part a of the solution. The next step constitutes the solution to part b.

6. Compute pn and then apply the basic equations in the proper sequence

Thus $p = A_s/(bd) = 2.20/[12(19.5)] = 0.00940$; $pn = 0.0940$. Then by Eq. 30, $k = [0.188 + (0.094)^2]^{0.5} - 0.094 = 0.350$. By Eq. 22, $j = 1 - 0.350/3 = 0.883$. By Eq. 23, $f_c = 2M/(kjbd^2) = 2(744,000)/[0.350(0.883)(12)(19.5)^2] = 1060$ lb/in^2 (7308.7 kPa). By Eq. 25, $f_s = M/(A_sjd) = 744,000/[2.20(0.883)(19.5)] = 19,600$ lb/in^2 (135,142 kPa).

CAPACITY OF A RECTANGULAR BEAM

The beam in Fig. 118a is made of 2500-lb/in^2 (17,237.5-kPa) concrete. Determine the flexural capacity of the member (a) without applying the basic equations of reinforced-concrete beam design; (b) by applying the basic equations.

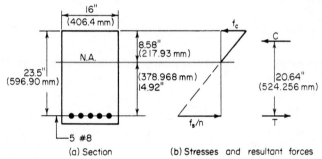

(a) Section   (b) Stresses and resultant forces

FIG. 118

Calculation Procedure:

1. Record the pertinent beam data

Thus, $f_c' = 2500$ lb/in^2 (17,237.5 kPa); $\therefore f_{c,allow} = 1125$ lb/in^2 (7756.9 kPa); $n = 10$; $A_s = 3.95$ in^2 (25.485 cm^2); $nA_s = 39.5$ in^2 (254.85 cm^2).

2. Locate the centroidal axis of the transformed section

Thus, $16(kd)^2/2 - 39.5(23.5 - kd) = 0$; $kd = 8.58$ in (217.93 mm); $d - kd = 14.92$ in (378.968 mm).

3. Ascertain which of the two allowable stresses governs the capacity of the member

For this purpose, assume that $f_c = 1125$ lb/in^2 (7756.9 kPa). By proportion, $f_s = 10(1125)(14.92/8.58) = 19,560$ lb/in^2 (134,866 kPa) $< 20,000$ lb/in^2 (137,900 kPa). Therefore, concrete stress governs.

4. Calculate the allowable bending moment

Thus, $jd = 23.5 - 8.58/3 = 20.64$ in (524.256 mm); $M = Cjd = \frac{1}{2}(1125)(16)(8.58)(20.64) = 1,594,000$ in·lb (180,090.1 N·m); or $M = Tjd = 3.95(19,560)(20.64) = 1,594,000$ in·lb (180,090.1 N·m). This concludes part a of the solution. The next step comprises part b.

5. Compute p and compare with p_b to identify the controlling stress

Thus, from Table 6, $p_b = 0.0101$; then $p = A_s/(bd) = 3.95/[16(23.5)] = 0.0105 > p_b$. Therefore, concrete stress governs.

Applying the basic equations in the proper sequence yields $pn = 0.1050$; by Eq. 30, $k = [0.210 + 0.105^2]^{0.5} - 0.105 = 0.365$; by Eq. 24, $M = (\frac{1}{6})(1125)(0.365)(2.635)(16)(23.5)^2 = 1,593,000$ in·lb (179,977.1 N·m). This agrees closely with the previously computed value of M.

DESIGN OF REINFORCEMENT IN A RECTANGULAR BEAM OF GIVEN SIZE

A rectangular beam of 4000-lb/in² (27,580-kPa) concrete has a width of 14 in (355.6 mm) and an effective depth of 23.5 in (596.9 mm). Determine the area of reinforcement if the beam is to resist a bending moment of (a) 220 ft·kips (298.3 kN·m); (b) 200 ft·kips (271.2 kN·m).

Calculation Procedure:

1. Calculate the moment capacity of this member at balanced design

Record the following values: $f_{c,allow} = 1800$ lb/in² (12,411 kPa); $n = 8$. From Table 6, $j_b = 0.860$; $K_b = 324$ lb/in² (2234.0 kPa); $M_b = K_b b d^2 = 324(14)(23.5)^2 = 2,505,000$ in·lb (283,014.9 N·m).

2. Determine which material will be stressed to capacity under the stipulated moment

For part a, $M = 220,000(12) = 2,640,000$ in·lb (3,579,840 N·m) $> M_b$. This result signifies that the beam size is deficient with respect to balanced design, and the concrete will therefore be stressed to capacity.

3. Apply the basic equations in proper sequence to obtain A_s

By Eq. 24, $k(3 - k) = 6M/(f_c bd^2) = 6(2,640,000)/[1800(14)(23.5)^2] = 1.138$; $k = 0.446$. By Eq. 29, $p = k^2/[2n(1 - k)] = 0.446^2/[16(0.554)] = 0.0224$; $A_s = pbd = 0.0224(14)(23.5) = 7.37$ in² (47.551 cm²).

4. Verify the result by evaluating the flexural capacity of the member

For part b, compute A_s by the exact method and then describe the approximate method used in practice.

5. Determine which material will be stressed to capacity under the stipulated moment

Here $M = 200,000(12) = 2,400,000$ in·lb (3,254,400 N·m) $< M_b$. This result signifies that the beam size is excessive with respect to balanced design, and the steel will therefore be stressed to capacity.

6. Apply the basic equations in proper sequence to obtain A_s

By using Eq. 27, $k^2(3 - k)/(1 - k) = 6nM/(f_s bd^2) = 6(8)(2,400,000)/[20,000(14)(23.5)^2] = 0.7448$; $k = 0.411$. By Eq. 22, $j = 1 - 0.411/3 = 0.863$. By Eq. 25, $A_s = M/(f_s jd) = 2,400,000/[20,000(0.863)(23.5)] = 5.92$ in² (38.196 cm²).

7. Verify the result by evaluating the flexural capacity of this member

The value of j obtained in step 6 differs negligibly from the value $j_b = 0.860$. Consequently, in those instances where the beam size is only moderately excessive with respect to balanced design, the practice is to consider that $j = j_b$ and to solve Eq. 25 directly on this basis. This practice is conservative, and it obviates the need for solving a cubic equation, thus saving time.

DESIGN OF A RECTANGULAR BEAM

A beam on a simple span of 13 ft (3.9 m) is to carry a uniformly distributed load, exclusive of its own weight, of 3600 lb/lin ft (52,538.0 N/m) and a concentrated load of 17,000 lb (75,616 N) applied at midspan. Design the section, using $f'_c = 3000$ lb/in² (20,685 kPa).

Calculation Procedure:

1. Record the basic values associated with balanced design

There are two methods of allowing for the beam weight: (a) to determine the bending moment with an estimated beam weight included; (b) to determine the beam size required to resist the external loads alone and then increase the size slightly. The latter method is used here.

From Table 6, K_b = 223 lb/in^2 (1537.6 kPa); p_b = 0.0128; j_b = 0.874.

2. Calculate the maximum moment caused by the external loads

Thus, the maximum moment M_e = ¼PL + ⅛wL^2 = ¼(17,000)(13)(12) + ⅛(3600)(13)^2(12) = 1,576,000 in·lb (178,056.4 N·m).

3. Establish a trial beam size

Thus, bd^2 = M/K_b = 1,576,000/223 = 7067 in^3 (115,828.1 cm^3). Setting b = (⅔)d, we find b = 14.7 in (373.38 mm), d = 22.0 in (558.8 mm). Try b = 15 in (381 mm) and d = 22.5 in (571.5 mm), producing an overall depth of 25 in (635 mm) if the reinforcing bars may be placed in one row.

4. Calculate the maximum bending moment with the beam weight included; determine whether the trial section is adequate

Thus, beam weight = 15(25)(150)/144 = 391 lb/lin ft (5706.2 N/m); M_w = (⅛)(391)(13)^2(12) = 99,000 in·lb (11,185.0 N·m); M = 1,576,000 + 99,000 = 1,675,000 in·lb (189,241.5 N·m); M_b = $K_b bd^2$ = 223(15)(22.5)^2 = 1,693,000 in·lb (191,275.1 N·m). The trial section is therefore satisfactory because it has adequate capacity.

5. Design the reinforcement

Since the beam size is slightly excessive with respect to balanced design, the steel will be stressed to capacity under the design load. Equation 25 is therefore suitable for this calculation. Thus, A_s = $M/(f_s jd)$ = 1,675,000/[20,000(0.874)(22.5)] = 4.26 in^2 (27.485 cm^2).

An alternative method of calculating A_s is to apply the value of p_b while setting the beam width equal to the dimension actually required to produce balanced design. Thus, A_s = 0.0128(15)(1675)(22.5)/1693 = 4.27 in^2 (27.550 cm^2).

Use one no. 10 and three no. 9 bars, for which A_s = 4.27 in^2 (27.550 cm^2) and b_{min} = 12.0 in (304.8 mm).

6. Summarize the design

Thus, beam size is 15 × 25 in (381 × 635 mm); reinforcement is with one no. 10 and three no. 9 bars.

DESIGN OF WEB REINFORCEMENT

A beam 14 in (355.6 mm) wide with an 18.5-in (469.9-mm) effective depth carries a uniform load of 3.8 kips/lin ft (55.46 N/m) and a concentrated midspan load of 2 kips (8.896 kN). The beam is simply supported, and the clear distance between supports is 13 ft (3.9 m). Using f'_c = 3000 lb/in^2 (20,685 kPa) and an allowable stress f_v in the stirrups of 20,000 lb/in^2 (137,900 kPa), design web reinforcement in the form of vertical U stirrups.

Calculation Procedure:

1. Construct the shearing-stress diagram for half-span

The design of web reinforcement by the working-stress method parallels the design by the ultimate-strength method, given earlier. Let v = nominal shearing stress, lb/in^2 (kPa); v_c = shearing stress resisted by concrete; v' = shearing stress resisted by web reinforcement.

The ACI Code provides two alternative methods of computing the shearing stress that may be resisted by the concrete. The simpler method is used here. This sets

$$v_c = 1.1(f'_c)^{0.5} \tag{33}$$

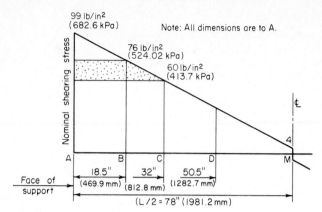

FIG. 119 Shearing-stress diagram.

The equation for nominal shearing stress is

$$v = \frac{V}{bd} \tag{34}$$

The shearing-stress diagram for a half-span is shown in Fig. 119. Establish the region AD within which web reinforcement is required. Thus, $v_c = 1.1(3000)^{0.5} = 60$ lb/in^2 (413.7 kPa). At the face of the support, $V = 6.5(3800) + 1000 = 25,700$ lb (114,313.6 N); $v = 25,700/[14(18.5)] = 99$ lb/in^2 (682.6 kPa).

At midspan, $V = 1000$ lb (4448 N); $v = 4$ lb/in^2 (27.6 kPa); slope of diagram $= -(99 - 4)/78 = -1.22$ lb/(in$^2 \cdot$ in) $(-0.331$ kPa/mm). At distance d from the face of the support, $v = 99 - 18.5(1.22) = 76$ lb/in^2 (524.02 kPa); $v' = 76 - 60 = 16$ lb/in^2 (110.3 kPa); $AC = (99 - 60)/1.22 = 32$ in (812.8 mm); $AD = AC + d = 32 + 18.5 = 50.5$ in (1282.7 mm).

2. Check the beam size for compliance with the Code

Thus, $v_{max} = 5(f'_c)^{0.5} = 274$ lb/in^2 (1889.23 kPa) > 76 lb/in^2 (524.02 kPa). This is acceptable.

3. Select the stirrup size

Use the method given earlier in the ultimate-strength calculation procedure to select the stirrup size, establish the maximum allowable spacing, and devise a satisfactory spacing.

CAPACITY OF A T BEAM

Determine the flexural capacity of the T beam in Fig. 120a, using $f'_c = 3000$ lb/in^2 (20,685 kPa).

Calculation Procedure:

1. Record the pertinent beam values

The neutral axis of a T beam often falls within the web. However, to simplify the analysis, the resisting moment developed by the concrete lying between the neutral axis and the flange is usually disregarded. Let A_f denote the flange area. The pertinent beam values are $f_{c,\text{allow}} = 1350$ lb/in^2 (9308.3 kPa); $n = 9$; $k_b = 0.378$; $nA_s = 9(4.00) = 36.0$ in^2 (232.3 cm^2).

2. Tentatively assume that the neutral axis lies in the web

Locate this axis by taking static moments with respect to the top line. Thus $A_f = 5(16) = 80$ in^2 (516.2 cm^2); $kd = [80(2.5) + 36.0(21.5)]/(80 + 36.0) = 8.40$ in (213.36 mm).

3. Identify the controlling stress

Thus $k = 8.40/21.5 = 0.391 > k_b$; therefore, concrete stress governs.

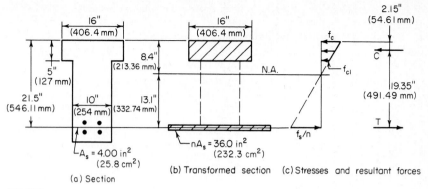

2.15"
(54.61 mm)

16"
(406.4 mm)

16"
(406.4 mm)

8.4"
(213.36 mm)

N.A.

f_c

C

5"
(127 mm)

19.35"
(491.49 mm)

21.5"
(546.11 mm)

10"
(254 mm)

13.1"
(332.74 mm)

f_{cl}

A_s = 4.00 in²
(25.8 cm²)

nA_s = 36.0 in²
(232.3 cm²)

f_s/n

T

(a) Section

(b) Transformed section

(c) Stresses and resultant forces

FIG. 120

4. Calculate the allowable bending moment

Using Fig. 120c, we see $f_{cl} = 1350(3.40)/8.40 = 546$ lb/in² (3764.7 kPa); $C = \frac{1}{2}(80)(1350 + 546) = 75,800$ lb (337,158.4 N). The action line of this resultant force lies at the centroidal axis of the stress trapezoid. Thus, $z = (\%)(1350 + 2 \times 546)/(1350 + 546) = 2.15$ in (54.61 mm); or $z = (\%)(8.40 + 2 \times 3.40)/(8.40 + 3.40) = 2.15$ in (54.61 mm); $M = Cjd = 75,800(19.35) = 1,467,000$ in·lb (165,741 N·m).

5. Alternatively, calculate the allowable bending moment by assuming that the flange extends to the neutral axis

Then apply the necessary correction. Let C_1 = resultant compressive force if the flange extended to the neutral axis, lb (N); C_2 = resultant compressive force in the imaginary extension of the flange, lb (N). Then $C_1 = \frac{1}{2}(1350)(16)(8.40) = 90,720$ lb (403,522.6 N); $C_2 = 90,720(3.40/8.40)^2 = 14,860$ lb (66,097.3 N); $M = 90,720(21.5 - 8.40/3) - 14,860(21.5 - 5 - 3.40/3) = 1,468,000$ in·lb (165,854.7 N·m).

DESIGN OF A T BEAM HAVING CONCRETE STRESSED TO CAPACITY

A concrete girder of 2500-lb/in² (17,237.5-kPa) concrete has a simple span of 22 ft (6.7 m) and is built integrally with a 5-in (127-mm) slab. The girders are spaced 8 ft (2.4 m) on centers; the overall depth is restricted to 20 in (508 mm) by headroom requirements. The member carries a load of 4200 lb/lin ft (61,294.4 N/m), exclusive of the weight of its web. Design the section, using tension reinformcement only.

Calculation Procedure:

1. Establish a tentative width of web

Since the girder is built integrally with the slab that it supports, the girder and slab constitute a structural entity in the form of a T beam. The effective flange width is established by applying the criteria given in the ACI Code, and the bending stress in the flange is assumed to be uniform across a line parallel to the neutral axis. Let A_f = area of flange in² (cm²); b = width of flange, in (mm); b' = width of web, in (mm); t = thickness of flange, in (mm); s = center-to-center spacing of girders.

To establish a tentative width of web, try $b' = 14$ in (355.6 mm). Then the weight of web = $14(15)(150)/144 = 219$, say 220 lb/lin ft (3210.7 N/m); $w = 4200 + 220 = 4420$ lb/lin ft (64,505.0 N/m).

Since two rows of bars are probably required, $d = 20 - 3.5 = 16.5$ in (419.1 mm). The critical shear value is $V = w(0.5L - d) = 4420(11 - 1.4) = 42,430$ lb (188,728.7 N); $v = V/b'd = 42,430/[14(16.5)] = 184$ lb/in² (1268.7 kPa). From the Code, $v_{max} = 5(f_c')^{0.5} = 250$ lb/in² (1723.8 kPa). This is acceptable.

Upon designing the reinforcement, consider whether it is possible to reduce the width of the web.

2. Establish the effective width of the flange according to the Code

Thus, $\frac{1}{4}L = \frac{1}{4}(22)(12) = 66$ in (1676.4 mm); $16t + b' = 16(5) + 14 = 94$ in (2387.6 mm); $s = 8(12) = 96$ in (2438.4 mm); therefore $b = 66$ in (1676.4 mm).

3. Compute the moment capacity of the member at balanced design

Compare the result with the moment in the present instance to identify the controlling stres. With Fig. 120 as a guide, $k_b d = 0.360(16.5) = 5.94$ in (150.876 mm); $A_f = 5(66) = 330$ in^2 (2129.2 cm^2); $f_{c1} = 1125(0.94)/5.94 = 178$ lb/in^2 (1227.3 kPa); $C_b = T_b = \frac{1}{2}(330)(1125 + 178) = 215,000$ lb (956,320 N); $z_b = (\frac{2}{3})(5.94 + 2 \times 0.94)/(5.94 + 0.94) = 1.89$ in (48.0 mm); $jd = 14.61$ in (371.094 mm); $M_b = 215,000(14.61) = 3,141,000$ in·lb (354,870.2 N·m); $M = (\frac{1}{8})(4420)(22)^2(12) = 3,209,000$ in·lb (362,552.8 N·m).

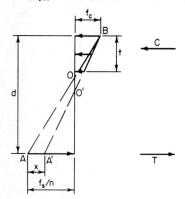

The beam size is slightly deficient with respect to balanced design, and the concrete will therefore be stressed to capacity under the stipulated load. In Fig. 121, let AOB represent the stress line associated with balanced design and $A'O'B$ represent the stress line in the present instance. (The magnitude of AA' is exaggerated for clarity.)

4. Develop suitable equations for the beam

Refer to Fig. 121:

FIG. 121 Stress diagram for T beam.

$$T = T_b + \frac{bt^2x}{2d} \tag{35}$$

where T and T_b = tensile force in present instance and at balanced design, respectively. And

$$M = M_b + \frac{bt^2(3d - 2t)x}{6d} \tag{36}$$

5. Apply the equations from step 4

Thus, $M - M_b = 68,000$ in·lb (7682.6 N·m). By Eq. 36, $x = 68,000(6)(16.5)/[66(25)(49.5 - 10)] = 103$ lb/in^2 (710.2 kPa); $f_s = 20,000 - 10(103) = 18,970$ lb/in^2 (130,798.2 kPa). By Eq. 35, $T = 215,000 + 66(25)(103)/33 = 220,200$ lb (979,449.6 N).

6. Design the reinforcement; establish the web width

Thus $A_s = 220,200/18,970 = 11.61$ in^2 (74.908 cm^2). Use five no. 11 and three no. 10 bars, placed in two rows. Then $A_s = 11.61$ in^2 (74.908 cm^2); $b'_{min} = 14.0$ in (355.6 mm). It is therefore necessary to maintain the 14-in (355.6-mm) width.

7. Summarize the design

Width of web: 14 in (355.6 mm); reinforcement: five no. 11 and three no. 10 bars.

8. Verify the design by computing the capacity of the member

Thus $nA_s = 116.1$ in^2 (749.08 cm^2); $kd = [330(2.5) + 116.1(16.5)]/(330 + 116.1) = 6.14$ in (155.956 mm); $k = 6.14/16.5 = 0.372 > k_b$; therefore, concrete is stressed to capacity. Then $f_s = 10(1125)(10.36)/6.14 = 18,980$ lb/in^2 (130,867.1 kPa); $z = (\frac{2}{3})(6.14 + 2 \times 1.14)/(6.14 + 1.14) = 1.93$ in (49.022 mm); $jd = 14.57$ in (370.078 mm); $M_{allow} = 11.61(18,980)(14.57) = 3,210,000$ in·lb (362,665.8 N·m). This is acceptable.

DESIGN OF A T BEAM HAVING STEEL STRESSED TO CAPACITY

Assume that the girder in the previous calculation procedure carries a total load, including the weight of the web, of 4100 lb/lin ft (59,835.0 N/m). Compute the area of reinforcement.

Calculation Procedure:

1. Identify the controlling stress

Thus, $M = (\%)(4100)(22)^2(12) = 2,977,000$ in·lb (336,341.5 N·m). From the previous calculation procedure, $M_b = 3,141,000$ in·lb (354,870.2 N·m). Since $M_b > M$, the beam size is slightly excessive with respect to balanced design, and the steel will therefore be stressed to capacity under the stipulated load.

2. Compute the area of reinforcement

As an approximation, this area may be found by applying the value of jd associated with balanced design, although it is actually slightly larger. From the previous calculation procedure, $jd = 14.61$ in (371.094 mm). Then $A_s = 2,977,000/[20,000(14.61)] = 10.19$ in² (65.746 cm²).

3. Verify the design by computing the member capacity

Thus, $nA_s = 101.9$ in² (657.46 cm²); $kd = (330 \times 2.5 + 101.9 \times 16.5)/(330 + 101.9) = 5.80$ in (147.32 mm); $z = (\%)(5.80 + 2 \times 0.80)/(5.80 + 0.80) = 1.87$ in (47.498 mm); $jd = 14.63$ in (371.602 mm); $M_{\text{allow}} = 10.19(20,000)(14.63) = 2,982,000$ in·lb (336,906.4 N·m). This is acceptable.

REINFORCEMENT FOR DOUBLY REINFORCED RECTANGULAR BEAM

A beam of 4000-lb/in² (27,580-kPa) concrete that will carry a bending moment of 230 ft·kips (311.9 kN·m) is restricted to a 15-in (381-mm) width and a 24-in (609.6-mm) total depth. Design the reinforcement.

Calculation Procedure:

1. Record the pertinent beam data

In Fig. 122, where the imposed moment is substantially in excess of that corresponding to balanced design, it is necessary to reinforce the member in compression as well as tension. The loss in concrete area caused by the presence of the compression reinforcement may be disregarded.

Since plastic flow generates a transfer of compressive stress from the concrete to the steel, the ACI *Code* provides that "in doubly reinforced beams and slabs, an effective modular ratio of $2n$ shall be used to transform the compression reinforcement and compute its stress, which shall not be taken as greater than the allowable tensile stress." This procedure is tantamount to considering that the true stress in the compression reinforcement is twice the value obtained by assuming a linear stress distribution.

Let A_s = area of tension reinforcement, in² (cm²); A'_s = area of compression reinforcement, in² (cm²); f_s = stress in tension reinforcement, lb/in² (kPa); f'_s = stress in compression reinforcement, lb/in² (kPa); C' = resultant force in compression reinforcement, lb (N); M_1 = moment capacity of member if reinforced solely in tension to produce balanced design; M_2 = incremental moment capacity resulting from use of compression reinforcement.

The data recorded for the beam are $f_c = 1800$ lb/in² (12.411 kPa); $n = 8$; $K_b = 324$ lb/in² (2234.0 kPa); $k_b = 0.419$; $j_b = 0.860$; $M = 230,000(12) = 2,760,000$ in·lb (311,824.8 N·m).

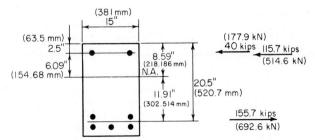

FIG. 122 Doubly reinforced beam.

2. Ascertain whether one row of tension bars will suffice

Assume tentatively that the presence of the compression reinforcement does not appreciably alter the value of j. Then $jd = 0.860(21.5) = 18.49$ in (469.646 mm); $A_s = M/(f_s jd) = 2,760,000/[20,000(18.49)] = 7.46$ in^2 (48.132 cm^2). This area of steel cannot be accommodated in the 15-in (381-mm) beam width, and two rows of bars are therefore required.

3. Evaluate the moments M_1 and M_2

Thus, $d = 24 - 3.5 = 20.5$ in (520.7 mm); $M_1 = K_b bd^2 = 324(15)(20.5)^2 = 2,040,000$ in·lb (230,479.2 N·m); $M_2 = 2,760,000 - 2,040,000 = 720,000$ in·lb (81,345.6 N·m).

4. Compute the forces in the reinforcing steel

For convenience, assume that the neutral axis occupies the same position as it would in the absence of compression reinforcement. For M_1, arm $= j_b d = 0.860(20.5) = 17.63$ in (447.802 mm); for M_2, arm $= 20.5 - 2.5 = 18.0$ in (457.2 mm); $T = 2,040,000/17.63 + 720,000/18.0 = 155,700$ lb (692,553.6 N); $C' = 40,000$ lb (177,920 N).

5. Compute the areas of reinforcement and select the bars

Thus $A_s = T/f_s = 155,700/20,000 = 7.79$ in^2 (50.261 cm^2); $kd = 0.419(20.5) = 8.59$ in (218.186 mm); $d - kd = 11.91$ in (302.514 mm). By proportion, $f_s' = 2(20,000)(6.09)/11.91 = 20,500$ lb/in^2 (141,347.5 kPa); therefore, set $f_s' = 20,000$ lb/in^2 (137,900 kPa). Then, $A_s' = C'/f_s' = 40,000/20,000 = 2.00$ in^2 (12.904 cm^2). Thus tension steel: five no. 11 bars, $A_s = 7.80$ in^2 (50.326 cm^2); compression steel: two no. 9 bars, $A_s = 2.00$ in^2 (12.904 cm^2).

DEFLECTION OF A CONTINUOUS BEAM

The continuous beam in Fig. 123a and b carries a total load of 3.3 kips/lin ft (48.16 kN/m). When it is considered as a T beam, the member has an effective flange width of 68 in (1727.2 mm). Determine the deflection of the beam upon application of full live load, using $f_c' = 2500$ lb/in^2 (17,237.5 kPa) and $f_y = 40,000$ lb/in^2 (275,800 kPa).

Calculation Procedure:

1. Record the areas of reinforcement

At support: $A_s = 4.43$ in^2 (28.582 cm^2) (top); $A_s' = 1.58$ in^2 (10.194 cm^2) (bottom). At center: $A_s = 3.16$ in^2 (20.388 cm^2) (bottom).

2. Construct the bending-moment diagram

Apply the ACI equation for maximum midspan moment. Refer to Fig. 123c: $M_1 = (\frac{1}{8})wL'^2 = (\frac{1}{8})3.3(22)^2 = 200$ ft·kips (271.2 kN·m); $M_2 = (\frac{1}{16})wL'^2 = 100$ ft·kips (135.6 kN·m); $M_3 = 100$ ft·kips (135.6 kN·m).

3. Determine upon what area the moment of inertia should be based

Apply the criterion set forth in the ACI Code to determine whether the moment of inertia is to be based on the transformed gross section or the transformed cracked section. At the support $pf_y = 4.43(40,000)/[14(20.5)] = 617 > 500$. Therefore, use the cracked section.

4. Determine the moment of inertia of the transformed cracked section at the support

Refer to Fig. 123d: $nA_s = 10(4.43) = 44.3$ in^2 (285.82 cm^2); $(n - 1)A_s' = 9(1.58) = 14.2$ in^2 (91.62 cm^2). The static moment with respect to the neutral axis is $Q = -\frac{1}{2}(14y^2) + 44.3(20.5 - y) - 14.2(y - 2.5) = 0$; $y = 8.16$ in (207.264 mm). The moment of inertia with respect to the neutral axis is $I_1 = (\frac{1}{3})14(8.16)^3 + 14.2(8.16 - 2.5)^2 + 44.3(20.5 - 8.16)^2 = 9737$ in^4 (40.53 dm^4).

5. Calculate the moment of inertia of the transformed cracked section at the center

Referring to Fig. 123e and assuming tentatively that the neutral axis falls within the flange, we see $nA_s = 10(3.16) = 31.6$ in^2 (203.88 cm^2). The static moment with respect to the neutral axis is $Q = \frac{1}{2}(68y^2) - 31.6(20.5 - y) = 0$; $y = 3.92$ in (99.568 mm). The neutral axis therefore falls within the flange, as assumed. The moment of inertia with respect to the neutral axis is $I_2 = (\frac{1}{3})68(3.92)^3 + 31.6(20.5 - 3.92)^2 = 10,052$ in^4 (41.840 dm^4).

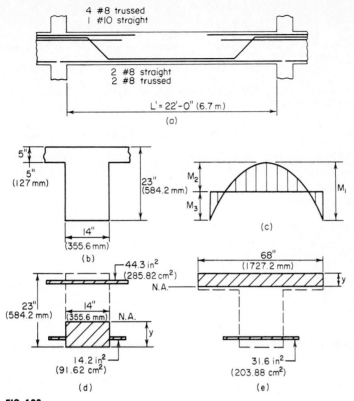

FIG. 123

6. Calculate the deflection at midspan

Use the equation

$$\Delta = \frac{L'^2}{EI}\left(\frac{5M_1}{48} - \frac{M_3}{8}\right) \tag{37}$$

where I = average moment of intertia, in⁴ (dm⁴). Thus, $I = \frac{1}{2}(9737 + 10,052) = 9895$ in⁴ (41.186 dm⁴); $E = 145^{1.5} \times 33(f'_c)^{0.5} = 57,600\ (2500)^{0.5} = 2,880,000$ lb/in² (19,857.6 MPa). Then $\Delta = [22^2 \times 1728/(2880 \times 9895)](5 \times 200/48 - 100/8) = 0.244$ in (6.198 mm).

Where the deflection under sustained loading is to be evaluated, it is necessary to apply the factors recorded in the ACI *Code*.

Design of Compression Members by Ultimate-Strength Method

The notational system is P_u = ultimate axial compressive load on member, lb (N); P_b = ultimate axial compressive load at balanced design, lb (N); P_0 = allowable ultimate axial compressive load in absence of bending moment, lb (N); M_u = ultimate bending moment in member, lb·in (N·m); M_b = ultimate bending moment at balanced design; d' = distance from exterior surface to centroidal axis of adjacent row of steel bars, in (mm); t = overall depth of rectangular section or diameter of circular section, in (mm).

A compression member is said to be *spirally reinforced* if the longitudinal reinforcement is held in position by spiral hooping and *tied* if this reinforcement is held by means of intermittent lateral ties.

The presence of a bending moment in a compression member reduces the ultimate axial load that the member may carry. In compliance with the ACI *Code*, it is necessary to design for a minimum bending moment equal to that caused by an eccentricity of 0.05*t* for spirally reinforced members and 0.10*t* for tied members. Thus, every compression member that is designed by the ultimate-strength method must be treated as a beam column. This type of member is considered to be in balanced design if failure would be characterized by the simultaneous crushing of the concrete, which is assumed to occur when $\epsilon_c = 0.003$, and incipient yielding of the tension steel, which occurs when $f_s = f_y$. The ACI *Code* set $\phi = 0.75$ for spirally reinforced members and $\phi = 0.70$ for tied members.

ANALYSIS OF A RECTANGULAR MEMBER BY INTERACTION DIAGRAM

A short tied member having the cross section shown in Fig. 124*a* is to resist an axial load and a bending moment that induces compression at *A* and tension at *B*. The member is made of 3000-

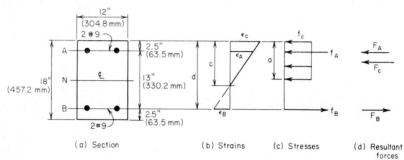

(a) Section (b) Strains (c) Stresses (d) Resultant forces

FIG. 124

lb/in² (20,685-kPa) concrete, and the steel has a yield point of 40,000 lb/in² (275,800 kPa). By starting with $c = 8$ in (203.2 mm) and assigning progressively higher values to *c*, construct the interaction diagram for this member.

Calculation Procedure:

1. Compute the value of c associated with balanced design

An *interaction diagram*, as the term is used here, is one in which every point on the curve represents a set of simultaneous values of the ultimate moment and allowable ultimate axial load. Let ϵ_A and ϵ_B = strain of reinforcement at *A* and *B*, respectively; ϵ_c = strain of extreme fiber of concrete; f_A and f_B = stress in reinforcement at *A* and *B*, respectively, lb/in² (kPa); F_A and F_B = resultant force in reinforcement at *A* and *B*, respectively; F_c = resultant force in concrete, lb (N).

Compression will be considered positive and tension negative. For simplicity, disregard the slight reduction in concrete area caused by the steel at *A*.

Referring to Fig. 124*b*, compute the value of *c* associated with balanced design. Computing P_b and M_b yields $c_b/d = 0.003/(0.003 + f_y/E_s) = 87,000/(87,000 + f_y)$; $c_b = 10.62$ in (269.748 mm). Then $\epsilon_A/\epsilon_B = (10.62 - 2.5)/(15.5 - 10.62) > 1$; therefore, $f_A = f_y$; $a_b = 0.85(10.62) = 9.03$ in (229.362 mm); $F_c = 0.85(3000)(12a_b) = 276,300$ lb (1,228,982.4 N); $F_A = 40,000(2.00) = 80,000$ lb ((355,840 N); $F_B = -80,000$ lb (−355,840 N); $P_b = 0.70(276,300) = 193,400$ lb (860,243.2 N). Also,

$$M_b = 0.70 \left[\frac{F_c(t - a)}{2} + \frac{(F_A - F_B)(t - 2d')}{2} \right] \tag{38}$$

Thus, $M_b = 0.70[276,300(18 - 9.03)/2 + 160,000(6.5)] = 1,596,000$ in·lb (180,316.1 N·m).

When $c > c_b$, the member fails by crushing of the concrete; when $c < c_b$, it fails by yielding of the reinforcement at line *B*.

2. Compute the value of c associated with incipient yielding of the compression steel

Compute the corresponding values of P_u and M_u. Since ϵ_A and ϵ_B are numerically equal, the neutral axis lies at N. Thus, $c = 9$ in (228.6 mm); $a = 0.85(9) = 7.65$ in (194.31 mm); $F_c = 30,600(7.65) = 234,100$ lb (1,041,276.8 N); $F_A = 80,000$ lb (355,840 N); $F_B = -80,000$ lb ($-355,840$ N); $P_u = 0.70 (234,100) = 163,900$ lb (729,027.2 N); $M_u = 0.70(234,100 \times 5.18 + 160,000 \times 6.5) = 1,577,000$ in·lb (178,169.5 N·m).

3. Compute the minimum value of c at which the entire concrete area is stressed to $0.85f_c'$

Compute the corresponding values of P_u and M_u. Thus, $a = t = 18$ in (457.2 mm); $c = 18/0.85 = 21.18$ in (537.972 mm); $f_B = \epsilon_c E_s(c - d)/c = 87,000(21.18 - 15.5)/21.18 = 23,300$ lb/in^2 (160,653.5 kPa); $F_c = 30,600(18) = 550,800$ lb (2,449,958.4 N); $F_A = 80,000$ lb (355,840 N); $F_B = 46,600$ lb (207,276.8 N); $P_u = 0.70(550,800 + 80,000 + 46,600) = 474,200$ lb (2,109,241.6 N); $M_u = 0.70(80,000 - 46,600)6.5 = 152,000$ in·lb (17,192.9 N·m).

4. Compute the value of c at which $M_u = 0$; compute P_0

The bending moment vanishes when F_B reaches 80,000 lb (355,840 N). From the calculation in step 3, $f_B = 87,000(c - d)/c = 40,000$ lb/in^2 (275,800 kPa); therefore, $c = 28.7$ in (728.98 mm); $P_0 = 0.70(550,800 + 160,000) = 497,600$ lb (2,213,324.8 N).

5. Assign other values to c, and compute P_u and M_u

By assigning values to c ranging from 8 to 28.7 in (203. 2 to 728.98 mm), typical calculations are: when $c = 8$ in (203.2 mm), $f_B = -40,000$ lb/in^2 ($-275,800$ kPa); $f_A = 40,000(5.5/7.5) = 29,300$ lb/in^2 (202,023.5 kPa); $a = 6.8$ in (172.72 mm); $F_c = 30,600(6.8) = 208,100$ lb (925,628.8 N); $P_u = 0.70(208,100 + 58,600 - 80,000) = 130,700$ lb (581,353.6 N); $M_u = 0.70 (208,100 \times 5.6 + 138,600 \times 6.5) = 1,446,000$ in·lb (163,369.1 N·m).

When $c = 10$ in (254 mm), $f_A = 40,000$ lb/in^2 (275,800 kPa); $f_B = -40,000$ lb/in^2 ($-275,800$ kPa); $a = 8.5$ in (215.9 mm); $F_c = 30,600(8.5) = 260,100$ lb (1,156,924.8 N); $P_u = 0.70(260,100) = 182,100$ lb (809,980 N); $M_u = 0.70(260,100 \times 4.75 + 160,000 \times 6.5) = 1,593,000$ in·lb (179,997.1 N·m).

When $c = 14$ in (355.6 mm), $f_B = 87,000(14 - 15.5)/14 = -9320$ lb/in^2 ($-64,261.4$ kPa); $a = 11.9$ in (302.26 mm); $F_c = 30,600(11.9) = 364,100$ lb (1,619,516.8 N); $P_u = 0.70(364,100 + 80,000 - 18,600) = 297,900$ lb (1,325,059.2 N); $M_u = 0.70(364,100 \times 3.05 + 98,600 \times 6.5) = 1,226,000$ in·lb (138,513.5 N·m).

6. Plot the points representing computed values of P_u and M_u in the interaction diagram

Figure 125 shows these points. Pass a smooth curve through these points. Note that when $P_u < P_b$, a reduction in M_u is accompanied by a reduction in the allowable load P_u.

AXIAL-LOAD CAPACITY OF RECTANGULAR MEMBER

The member analyzed in the previous calculation procedure is to carry an eccentric longitudinal load. Determine the allowable ultimate load if the eccentricity as measured from N is (a) 9.2 in (233.68 mm); (b) 6 in (152.4 mm).

Calculation Procedure:

1. Evaluate the eccentricity associated with balanced design

Let e denote the eccentricity of the load and e_b the eccentricity associated with balanced design. Then $M_u = P_u e$. In Fig. 125, draw an arbitrary radius vector OD; then $\tan \theta = ED/OE =$ eccentricity corresponding to point D.

Proceeding along the interaction diagram from A to C, we see that the value of c increases and the value of e decreases. Thus, c and e vary in the reverse manner. To evaluate the allowable loads, it is necessary to identify the portion of the interaction diagram to which each eccentricity applies.

From the computations of the previous calculation procedure, $e_b = M_b/P_b = 1,596,000/193,400 = 8.25$ in (209.55 mm). This result discloses that an eccentricity of 9.2 in (233.68 mm)

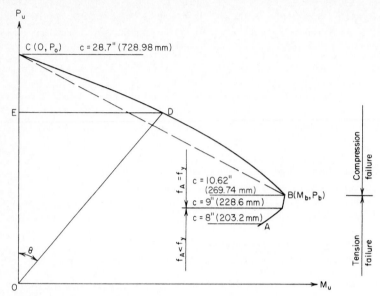

FIG. 125 Interaction diagram.

corresponds to a point on AB and an eccentricity of 6 in (152.4 mm) corresponds to a point on BC.

2. Evaluate P_u when $e = 9.2$ in (233.68 mm)

It was found that $c = 9$ in (228.6 mm) is a significant value. The corresponding value of e is $1,577,000/163,900 = 9.62$ in (244.348 mm). This result discloses that in the present instance $c > 9$ in (228.6 mm) and consequently $f_A = f_y$; $F_A = 80,000$ lb (355,840 N); $F_B = -80,000$ lb $(-355,840$ N); $F_c = 30,600a$; $P_u/0.70 = 30,600a$; $M_u/0.70 = 30,600a(18 - a)/2 + 160,000(6.5)$; $e = M_u/P_u = 9.2$ in (233.68 mm). Solving gives $a = 8.05$ in (204.47 mm), $P_u = 172,400$ lb (766,835.2 N).

3. Evaluate P_u when $e = 6$ in (152.4 mm)

To simplify this calculation, the ACI *Code* permits replacement of curve BC in the interaction diagram with a straight line through B and C. The equation of this line is

$$P_u = P_o - (P_o - P_b)\frac{M_u}{M_b} \qquad (39)$$

By replacing M_u with $P_u e$, the following relation is obtained:

$$P_u = \frac{P_o}{1 + (P_o - P_b)e/M_b} \qquad (39a)$$

In the present instance, $P_o = 497,600$ lb (2,213,324.8 N); $P_b = 193,400$ lb (860,243.2 N); $M_b = 1,596,000$ in·lb (180,316.1 N·m). Thus $P_u = 232,100$ lb (1,032,380 N).

ALLOWABLE ECCENTRICITY OF A MEMBER

The member analyzed in the previous two calculation procedures is to carry an ultimate longitudinal load of 150 kips (667.2 kN) that is eccentric with respect to axis N. Determine the maximum eccentricity with which the load may be applied.

Calculation Procedure:

1. Express P_u in terms of c, and solve for c

From the preceding calcluation procedures, it is seen that the value of c corresponding to the maximum eccentricity lies between 8 and 9 in (203.2 and 228.6 mm), and therefore $f_A < f_y$. Thus $f_B = -40,000$ lb/in^2 ($-275,800$ kPa); $f_A = 40,000(c - 2.5)/(15.5 - c)$; $F_c = 30,600(0.85c) = 26,000c$; $150,000 = 0.70\{26,000c + 80,000[(c - 2.5)/(15.5 - c) - 1]\}$; $c = 8.60$ in (218.44 mm).

2. Compute M_u and evaluate the eccentricity

Thus, $a = 7.31$ in (185.674 mm); $F_c = 223,700$ lb (995,017.6 N); $f_A = 35,360$ lb/in^2 (243,807.2 kPa); $M_u = 0.70(223,700 \times 5.35 + 150,700 \times 6.5) = 1,523,000$ in·lb (172,068.5 N·m); $e = M_u/P_u = 10.15$ in (257.81 mm).

Design of Compression Members by Working-Stress Method

The notational system is as follows: A_g = gross area of section, in^2 (cm^2); A_s = area of tension reinforcement, in^2 (cm^2); A_{st} = total area of longitudinal reinforcement, in^2 (cm^2); D = diameter of circular section, in (mm); $p_g = A_{st}/A_g$; P = axial load on member, lb (N); f_s = allowable stress in longitudinal reinforcement, lb/in^2 (kPa); $m = f_y/(0.85f_c')$.

The working-stress method of designing a compression member is essentially an adaptation of the ultimate-strength method. The allowable ultimate loads and bending moments are reduced by applying an appropriate factor of safety, and certain simplifications in computing the ultimate values are introduced.

The allowable concentric load on a short spirally reinforced column is $P = A_g(0.25f_c' + f_s p_g)$, or

$$P = 0.25f_c'A_g + f_sA_{st} \tag{40}$$

where $f_s = 0.40f_y$, but not to exceed 30,000 lb/in^2 (206,850 kPa).

The allowable concentric load on a short tied column is $P = 0.85A_g(0.25f_c' + f_s p_g)$, or

$$P = 0.2125f_c'A_g + 0.85f_sA_{st} \tag{41}$$

A section of the ACI *Code* provides that p_g may range from 0.01 to 0.08. However, in the case of a circular column in which the bars are to be placed in a single circular row, the upper limit of p_g is often governed by clearance. This section of the *Code* also stipulates that the minimum bar size to be used is no. 5 and requires a minimum of six bars for a spirally reinforced column and four bars for a tied column.

DESIGN OF A SPIRALLY REINFORCED COLUMN

A short circular column, spirally reinforced, is to support a concentric load of 420 kips (1868.16 kN). Design the member, using $f_c' = 4000$ lb/in^2 (27,580 kPa) and $f_y = 50,000$ lb/in^2 (344,750 kPa).

Calculation Procedure:

1. Assume $p_g = 0.025$ and compute the diameter of the section

Thus, $0.25f_c' = 1000$ lb/in^2 (6895 kPa); $f_s = 20,000$ lb/in^2 (137,900 kPa). By Eq. 40, $A_g = 420/(1 + 20 \times 0.025) = 280$ in^2 (1806.6 cm^2). Then $D = (A_g/0.785)^{0.5} = 18.9$ in (130.32 mm). Set $D = 19$ in (131.01 mm), making $A_g = 283$ in^2 (1825.9 cm^2).

2. Select the reinforcing bars

The load carried by the concrete = 283 kips (1258.8 kN). The load carried by the steel = 420 − 283 = 137 kips (609.4 kN). Then the area of the steel is $A_{st} = 137/20 = 6.85$ in^2 (44.196 cm^2). Use seven no. 9 bars, each having an area of 1 in^2 (6.452 cm^2). Then $A_{st} = 7.00$ in^2 (45.164 cm^2). The *Reinforced Concrete Handbook* shows that a 19-in (482.6-mm) column can accommodate 11 no. 9 bars in a single row.

3. Design the spiral reinforcement

The portion of the column section bounded by the outer circumference of the spiral is termed the *core* of the section. Let A_c = core area, in² (cm²); D_c = core diameter, in (mm); a_s = cross-sectional area of spiral wire, in² (cm²); g = pitch of spiral, in (mm); p_s = ratio of volume of spiral reinforcement to volume of core.

The ACI *Code* requires 1.5-in (38.1-mm) insulation for the spiral, with g restricted to a maximum of $D_c/6$. Then $D_c = 19 - 3 = 16$ in (406.4 mm); $A_c = 201$ in² (1296.9 cm²); $D_c/6 = 2.67$ in (67.818 mm). Use a 2.5-in (63.5-mm) spiral pitch. Taking a 1-in (25.4-mm) length of column,

$$p_s = \frac{\text{volume of spiral}}{\text{volume of core}} = \frac{a_s\pi D_c/g}{\pi D_c^2/4}$$

or

$$a_s = \frac{gD_cp_s}{4} \tag{42}$$

The required value of p_s as given by the ACI *Code* is

$$p_s = \frac{0.45(A_g/A_c - 1)f_c'}{f_y} \tag{43}$$

or $p_s = 0.45(283/201 - 1)4/50 = 0.0147$; $a_s = 2.5(16)(0.0147)/4 = 0.147$ in² (0.9484 cm²). Use ½-in (12.7-mm) diameter wire with $a_s = 0.196$ in² (1.2646 cm²).

4. Summarize the design

Thus: column size: 19-in (482.6-mm) diameter; longitudinal reinforcement: seven no. 9 bars; spiral reinforcement: ½-in (12.7-mm) diameter wire, 2.5-in (63.5-mm) pitch.

ANALYSIS OF A RECTANGULAR MEMBER BY INTERACTION DIAGRAM

A short tied member having the cross section shown in Fig. 126 is to resist an axial load and a bending moment that induces rotation about axis N. The member is made of 4000-lb/in² (27,580-kPa) concrete, and the steel has a yield point of 50,000 lb/in² (344,750 kPa). Construct the interaction diagram for this member.

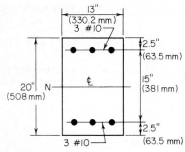

FIG. 126

Calculation Procedure:

1. Compute P_a and M_f

Consider a composite member of two materials having equal strength in tension and compression, the member being subjected to an axial load P and bending moment M that induce the allowable stress in one or both materials. Let P_a = allowable axial load in absence of bending moment, as computed by dividing the allowable ultimate load by a factor of safety; M_f = allowable bending moment in absence of axial load, as computed by dividing the allowable ultimate moment by a factor of safety.

Find the simultaneous allowable values of P and M by applying the interaction equation

$$\frac{P}{P_a} + \frac{M}{M_f} = 1 \tag{44}$$

Alternate forms of this equation are

$$M = M_f\left(1 - \frac{P}{P_a}\right) \qquad P = P_a\left(1 - \frac{M}{M_f}\right) \tag{44a}$$

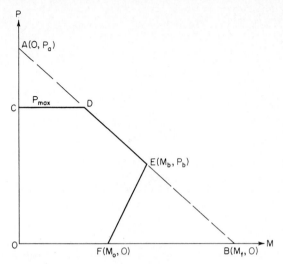

FIG. 127 Interaction diagram.

$$P = \frac{P_a M_f}{M_f + P_a M/P} \tag{44b}$$

Equation 44 is represented by line AB in Fig. 127; it is also valid with respect to a reinforced-concrete member for a certain range of values of P and M. This equation is not applicable in the following instances: (a) If M is relatively small, Eq. 44 yields a value of P in excess of that given by Eq. 41. Therefore, the interaction diagram must contain line CD, which represents the maximum value of P.

(b) If M is relatively large, the section will crack, and the equal-strength assumption underlying Eq. 44 becomes untenable.

Let point E represent the set of values of P and M that will cause cracking in the extreme concrete fiber. And let P_b = axial load represented by point E; M_b = bending moment represented by point E; M_o = allowable bending moment in reinforced-concrete member in absence of axial load, as computed by dividing the allowable ultimate moment by a factor of safety. (M_o differs from M_f in that the former is based on a cracked section and the latter on an uncracked section. The subscript b as used by the ACI *Code* in the present instance does *not* refer to balanced design. However, its use illustrates the analogy with ultimate-strength analysis.) Let F denote the point representing M_o.

For simplicity, the interaction diagram is assumed to be linear between E and F. The interaction equation for a cracked section may therefore be expressed in any of the following forms:

$$M = M_o + \left(\frac{P}{P_b}\right)(M_b - M_o) \qquad P = P_b\left(\frac{M - M_o}{M_b - M_o}\right) \tag{45a}$$

$$P = \frac{P_b M_o}{M_o - M_b + P_b M/P} \tag{45b}$$

The ACI *Code* gives the following approximations: For spiral columns:

$$M_o = 0.12A_{st}f_y D_s \tag{46a}$$

where D_s = diameter of circle through center of longitudinal reinforcement. For symmetric tied columns:

$$M_o = 0.40A_s f_y(d - d') \tag{46b}$$

For unsymmetric tied columns:

$$M_o = 0.40A_s f_y jd \qquad (46c)$$

For symmetric spiral columns:

$$\frac{M_b}{P_b} = 0.43p_g mD_s + 0.14t \qquad (47a)$$

For symmetric tied columns:

$$\frac{M_b}{P_b} = d(0.67p_g m + 0.17) \qquad (47b)$$

For unsymmetric tied columns:

$$\frac{M_b}{P_b} = \frac{p'm(d - d') + 0.1d}{(p' - p)m + 0.6} \qquad (47c)$$

where p' = ratio of area of compression reinforcement to effective area of concrete. The value of P_a is taken as

$$P_a = 0.34f'_c A_g(1 + p_g m) \qquad (48)$$

The value of M_f is found by applying the section modulus of the transformed uncracked section, using a modular ratio of $2n$ to account for stress transfer between steel and concrete engendered by plastic flow. (If the steel area is multiplied by $2n - 1$, allowance is made for the reduction of the concrete area.)

Computing P_a and M_f yields $A_g = 260$ in^2 (1677.5 cm^2); $A_{st} = 7.62$ in^2 (49.164 cm^2); $p_g = 7.62/260 = 0.0293$; $m = 50/[0.85(4)] = 14.7$; $p_g m = 0.431$; $n = 8$; $P_a = 0.34(4)(260)(1.431) = 506$ kips (2250.7 kN).

The section modulus to be applied in evaluating M_f is found thus: $I = (\frac{1}{12})(13)(20)^3 + 7.62(15)(7.5)^2 = 15,100$ in^4 (62.85 dm^4); $S = I/c = 15,100/10 = 1510$ in^3 (24,748.9 cm^3); $M_f = Sf_c = 1510(1.8) = 2720$ in·kips (307.3 kN·m).

2. Compute P_b and M_b

By Eq. 47b, $M_b/P_b = 17.5(0.67 \times 0.431 + 0.17) = 8.03$ in (203.962 mm). By Eq. 44b, $P_b = P_a M_f/(M_f + 8.03P_a) = 506 \times 2720/(2720 + 8.03 \times 506) = 203$ kips (902.9 kN); $M_b = 8.03(203) = 1630$ in·kips (184.2 kN·m).

3. Compute M_o

By Eq. 46b, $M_o = 0.40(3.81)(50)(15) = 1140$ in·kips (128.8 kN·m).

4. Compute the limiting value of P

As established by Eq. 41, $P_{max} = 0.2125(4)(260) + 0.85(20)(7.62) = 351$ kips (1561.2 kN).

5. Construct the interaction diagram

The complete diagram is shown in Fig. 127.

AXIAL-LOAD CAPACITY OF A RECTANGULAR MEMBER

The member analyzed in the previous calculation procedure is to carry an eccentric longitudinal load. Determine the allowable load if the eccentricity as measured from N is (a) 10 in (254 mm); (b) 6 in (152.4 mm).

Calculation Procedure:

1. Evaluate P when e = 10 in (254 mm)

As the preceding calculations show, the eccentricity corresponding to point E in the interaction diagram is 8.03 in (203.962 mm). Consequently, an eccentricity of 10 in (254 mm) corresponds to a point on EF, and an eccentricity of 6 in (152.4 mm) corresponds to a point on ED.

By Eq. 45b, $P = 203(1140)/(1140 - 1630 + 203 \times 10) = 150$ kips (667.2 kN).

2. Evaluate P when e = 6 in (152.4 mm)

By Eq. 44*b*, $P = 506(2720)/(2720 + 506 \times 6) = 239$ kips (1063.1 kN).

Design of Column Footings

A reinforced-concrete footing supporting a single column differs from the usual type of flexural member in the following respects: It is subjected to bending in all directions, the ratio of maximum vertical shear to maximum bending moment is very high, and it carries a heavy load concentrated within a small area. The consequences are as follows: The footing requires two-way reinforcement, its depth is determined by shearing rather than bending stress, the punching-shear stress below the column is usually more critical than the shearing stress that results from ordinary beam action, and the design of the reinforcement is controlled by the bond stress as well as the bending stress.

Since the footing weight and soil pressure are collinear, the former does not contribute to the vertical shear or bending moment. It is convenient to visualize the footing as being subjected to an upward load transmitted by the underlying soil and a downward reaction supplied by the column, this being, of course, an inversion of the true form of loading. The footing thus functions as an overhanging beam. The effective depth of footing is taken as the distance from the top surface to the center of the upper row of bars, the two rows being made identical to avoid confusion.

Refer to Fig. 128, which shows a square footing supporting a square, symmetrically located concrete column. Let P = column load, kips (kN); p = net soil pressure (that caused by the column load alone), lb/ft² (kPa); A = area of footing, ft² (m²); L = side of footing, ft (m); h = side of column, in (mm); d = effective depth

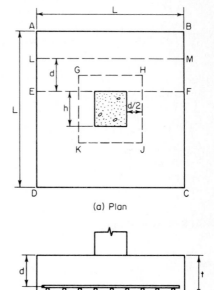

(a) Plan

(b) Elevation

FIG. 128

of footing, ft (m); t = thickness of footing, ft (m); f_b = bearing stress at interface of column, lb/in² (kPa); v_1 = nominal shearing stress under column, lb/in² (kPa); v_2 = nominal shearing stress caused by beam action, lb/in² (kPa); b_o = width of critical section for v_1, ft (m); V_1 and V_2 = vertical shear at critical section for stresses v_1 and v_2, respectively.

In accordance with the ACI *Code*, the critical section for v_1 is the surface *GHJK*, the sides of which lie at a distance $d/2$ from the column faces. The critical section for v_2 is plane *LM*, located at a distance d from the face of the column. The critical section for bending stress and bond stress is plane *EF* through the face of the column. In calculating v_2, f, and u, no allowance is made for the effects of the orthogonal reinforcement.

DESIGN OF AN ISOLATED SQUARE FOOTING

A 20-in (508-mm) square tied column reinforced with eight no. 9 bars carries a concentric load of 380 kips (1690.2 kN). Design a square footing by the working-stress method using these values: the allowable soil pressure is 7000 lb/ft² (335.2 kPa); f'_c = 3000 lb/in² (20,685 kPa); and f_s = 20,000 lb/in² (137,900 kPa).

152

Calculation Procedure:

1. Record the allowable shear, bond, and bearing stresses

From the ACI *Code* table, v_1 = 110 lb/in² (758.5 kPa); v_2 = 60 lb/in² (413.7 kPa); f_b = 1125 lb/in² (7756.9 kPa); u = $4.8(f_c')^{0.5}$/bar diameter = 264/bar diameter.

2. Check the bearing pressure on the footing

Thus, f_b = 380/[20(20)] = 0.95 kips/in² (7.258 MPa) < 1.125 kips/in² (7.7568 MPa). This is acceptable.

3. Establish the length of footing

For this purpose, assume the footing weight is 6 percent of the column load. Then A = 1.06(380)/7 = 57.5 ft² (5.34 m²). Make L = 7 ft 8 in = 7.67 ft (2.338 m); A = 58.8 ft² (5.46 m²).

4. Determine the effective depth as controlled by v_1

Apply

$$(4v_1 + p)d^2 + h(4v_1 + 2p)d = p(A - h^2) \qquad (49)$$

Verify the result after applying this equation. Thus p = 380/58.8 = 6.46 kips/ft² (0.309 MPa); v_1 = 0.11(144) = 15.84 kips/ft² (0.758 MPa); $69.8d^2 + 127.1d$ = 361.8; d = 1.54 ft (0.469 m). Checking in Fig. 128, we see GH = 1.67 + 1.54 = 3.21 ft (0.978 m); V_1 = 6.46(58.8 − 3.21²) = 313 kips (1392.2 kN); v_1 = $V_1/(b_o d)$ = 313/[4(3.21)(1.54)] = 15.83 kips/ft² (0.758 MPa). This is acceptable.

5. Establish the thickness and true depth of footing

Compare the weight of the footing with the assumed weight. Allowing 3 in (76.2 mm) for insulation and assuming the use of no. 8 bars, we see that t = d + 4.5 in (114.3 mm). Then t = 1.54(12) + 4.5 = 23.0 in (584.2 mm). Make t = 24 in (609.6 mm); d = 19.5 in = 1.63 ft (0.496 m). The footing weight = 58.8(2)(0.150) = 17.64 kips (1384.082 kN). The assumed weight = 0.06(380) = 22.8 kips (101.41 kN). This is acceptable.

6. Check v_2

In Fig. 128, AL = (7.67 − 1.67)/2 − 1.63 = 1.37 ft (0.417 m); V_2 = 380(1.37/7.67) = 67.9 kips (302.02 kN); v_2 = $V_2/(Ld)$ = 67,900/[92(19.5)] = 38 lb/in² (262.0 kPa) < 60 lb/in² (413.7 kPa). This is acceptable.

7. Design the reinforcement

In Fig. 128, EA = 3.00 ft (0.914 m); V_{EF} = 380(3.00/7.67) = 148.6 kips (666.97 kN); M_{EF} = 148.6(½)(3.00)(12) = 2675 in·kips (302.22 kN·m); A_s = 2675/[20(0.874)(19.5)] = 7.85 in² (50.648 cm²). Try 10 no. 8 bars each way. Then A_s = 7.90 in² (50.971 cm²); Σo = 31.4 in (797.56 mm); u = $V_{EF}/\Sigma ojd$ = 148,600/[31.4(0.874)(19.5)] = 278 lb/in² (1916.81 kPa); u_{allow} = 264/1 = 264 lb/in² (1820.3 kPa).

The bond stress at EF is slightly excessive. However, the ACI *Code*, in sections based on ultimate-strength considerations, permits disregarding the local bond stress if the average bond stress across the length of embedment is less than 80 percent of the allowable stress. Let L_e denote this length. Then L_e = EA − 3 = 33 in (838.2 mm); $0.80u_{allow}$ = 211 lb/in² (1454.8 kPa); u_{av} = $A_s f_s/(L_e \Sigma o)$ = 0.79(20,000)/[33(3.1)] = 154 lb/in² (1061.8 kPa). This is acceptable.

8. Design the dowels to comply with the Code

The function of the dowels is to transfer the compressive force in the column reinforcing bars to the

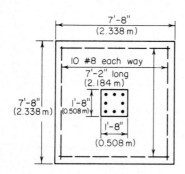

7'-8"
(2.338 m)

10 #8 each way
7'-2" long
(2.184 m)

7'-8"
(2.338 m)

1'-8"
(0.508 m)

1'-8"
(0.508 m)

1'-8"
(0.508 m)

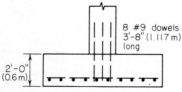

8 #9 dowels
3'-8" (1.117 m)
long

2'-0"
(0.6 m)

FIG. 129

footing. Since this is a tied column, assume the stress in the bars is $0.85(20,000) = 17,000$ lb/in^2 (117,215.0 kPa). Try eight no. 9 dowels with $f_y = 40,000$ lb/in^2 (275,800.0 kPa). Then $u = 264/$ $(9/8) = 235$ lb/in^2 (1620.3 kPa); $L_e = 1.00(17,000)/[235(3.5)] = 20.7$ in (525.78 mm). Since the footing can provide a 21-in (533.4-mm) embedment length, the dowel selection is satisfactory. Also, the length of lap $= 20(9/8) = 22.5$ in (571.5 mm); length of dowels $= 20.7 + 22.5 = 43.2$, say 44 in (1117.6 mm). The footing is shown in Fig. 129.

COMBINED FOOTING DESIGN

An 18 -in (457.2-mm) square exterior column and a 20-in (508.0-mm) square interior column carry loads of 250 kips (1112 kN) and 370 kips (1645.8 kN), respectively. The column centers are 16 ft (4.9 m) apart, and the footing cannot project beyond the face of the exterior column. Design a combined rectangular footing by the working-stress method, using $f'_c = 3000$ lb/in^2 (20,685.0 kPa), $f_s = 20,000$ lb/in^2 (137,900.0 kPa), and an allowable soil pressure of 5000 lb/in^2 (239.4 kPa).

Calculation Procedure:

1. *Establish the length of footing, applying the criterion of uniform soil pressure under total live and dead loads*

In many instances, the exterior column of a building cannot be individually supported because the required footing would project beyond the property limits. It then becomes necessary to use a combined footing that supports the exterior column and the adjacent interior column, the footing being so proportioned that the soil pressure is approximately uniform.

The footing dimensions are shown in Fig. 130a, and the reinforcement is seen in Fig. 131. It is convenient to visualize the combined footing as being subjected to an upward load transmitted by the underlying soil and reactions supplied by the columns. The member thus functions as a beam that overhangs one support. However, since the footing is considerably wider than the columns, there is a transverse bending as well as longitudinal bending in the vicinity of the columns. For simplicity, assume that the transverse bending is confined to the regions bounded by planes AB and EF and by planes GH and NP, the distance m being $h/2$ or $d/2$, whichever is smaller.

In Fig. 130a, let Z denote the location of the resultant of the column loads. Then $x = 370(16)/$ $(250 + 370) = 9.55$ ft (2.910 m). Since Z is to be the centroid of the footing, $L = 2(0.75 + 9.55)$ $= 20.60$ ft (6.278 m). Set $L = 20$ ft 8 in (6.299 m), but use the value 20.60 ft (6.278 m) in the stress calculations.

2. *Construct the shear and bending-moment diagrams*

The net soil pressure per foot of length $= 620/20.60 = 30.1$ kips/lin ft (439.28 kN/m). Construct the diagrams as shown in Fig. 130.

3. *Establish the footing thickness*

Use

$$(Pv_2 + 0.17VL + Pp')d - 0.17Pd^2 = VLp' \qquad (50)$$

where P = aggregate column load, kips (kN); V = maximum vertical shear at a column face, kips (kN); p' = gross soil pressure, kips/ft^2 (MPa).

Assume that the longitudinal steel is centered 3½ in (88.9 mm) from the face of the footing. Then $P = 620$ kips (2757.8 kN); $V = 229.2$ kips (1019.48 kN); $v_2 = 0.06(144) = 8.64$ kips/ft^2 (0.414 MPa); $9260d - 105.4d^2 = 23,608$; $d = 2.63$ ft (0.801 m); $t = 2.63 + 0.29 = 2.92$ ft. Set $t = 2$ ft 11 in (0.889 m); $d = 2$ ft 7½ in (0.800 m).

4. *Compute the vertical shear at distance d from the column face*

Establish the width of the footing. Thus $V = 229.2 - 2.63(30.1) = 150.0$ kips (667.2 kN); $v = V/(Wd)$, or $W = V/(vd) = 150/[8.64(2.63)] = 6.60$ ft (2.012 m). Set $W = 6$ ft 8 in (2.032 m).

5. *Check the soil pressure*

The footing weight $= 20.67(6.67)(2.92)(0.150) = 60.4$ kips (268.66 kN); $p' = (620 + 60.4)/$ $[(20.67)(6.67)] = 4.94$ kips/ft^2 (0.236 MPa) < 5 kips/ft^2 (0.239 MPa). This is acceptable.

154

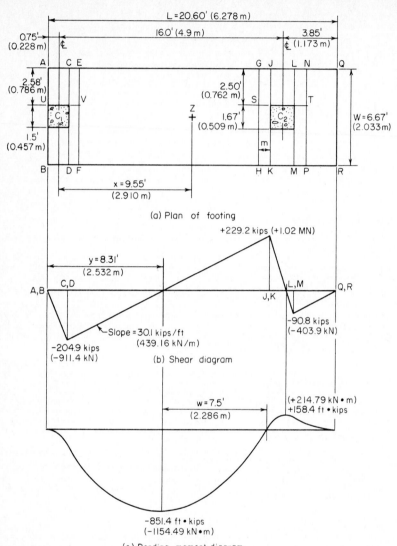

(a) Plan of footing

(b) Shear diagram

(c) Bending-moment diagram

FIG. 130

6. Check the punching shear

Thus, $p = 4.94 - 2.92(0.150) = 4.50$ kips/ft^2 (0.215 MPa). At C1: $b_o = 18 + 31.5 + 2 (18 + 15.8) = 117$ in (2971.8 mm); $V = 250 - 4.50(49.5)(33.8)/144 = 198$ kips (880.7 kN); $v_1 = 198,000/[117(31.5)] = 54$ lb/in^2 (372.3 kPa) < 110 lb/in^2 (758.5 kPa); this is acceptable.

At C2: $b_o = 4(20 + 31.5) = 206$ in (5232.4 mm); $V = 370 - 4.50(51.5)^2/144 = 287$ kips (1276.6 kN); $v_1 = 287,000/[206(31.5)] = 44$ lb/in^2 (303.4 kPa). This is acceptable.

7. Design the longitudinal reinforcement for negative moment

Thus, $M = 851,400$ ft·lb $= 10,217,000$ in·lb (1,154,316.6 N·m); $M_b = 223(80)(31.5)^2 = 17,700,000$ in·lb (1,999,746.0 N·m). Therefore, the steel is stressed to capacity, and $A_s =$

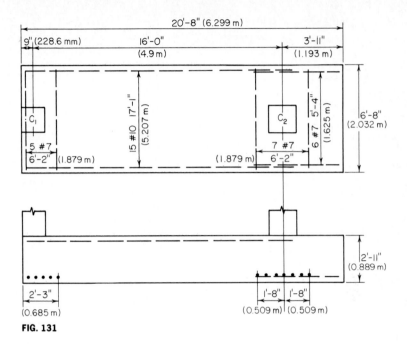

FIG. 131

$10,217,000/[20,000(0.874)(31.5)] = 18.6 \text{ in}^2$ (120.01 cm^2). Try 15 no. 10 bars with $A_s = 19.1 \text{ in}^2$ (123.2 cm^2); $\Sigma o = 59.9$ in (1521.46 mm).

The bond stress is maximum at the point of contraflexure, where $V = 15.81(30.1) - 250 = 225.9$ kips (1004.80 kN); $u = 225,900/[59.9(0.874)(31.5)] = 137 \text{ lb/in}^2$ (944.6 kPa); $u_{\text{allow}} = 3.4(3000)^{0.5}/1.25 = 149 \text{ lb/in}^2$ (1027.4 kPa). This is acceptable.

8. Design the longitudinal reinforcement for positive moment

For simplicity, design for the maximum moment rather than the moment at the face of the column. Then $A_s = 158,400(12)/[20,000(0.874)(31.5)] = 3.45 \text{ in}^2$ (22.259 cm^2). Try six no. 7 bars with $A_s = 3.60 \text{ in}^2$ (23.227 cm^2); $\Sigma o = 16.5$ in (419.10 mm). Take LM as the critical section for bond, and $u = 90,800/[16.5(0.874)(31.5)] = 200 \text{ lb/in}^2$ (1379.0 kPa); $u_{\text{allow}} = 4.8(3000)^{0.5}/0.875 = 302 \text{ lb/in}^2$ (2082.3 kPa). This is acceptable.

9. Design the transverse reinforcement under the interior column

For this purpose, consider member $GNPH$ as an independent isolated footing. Then $V_{ST} = 370(2.50/6.67) = 138.8$ kips (617.38 kN); $M_{ST} = \frac{1}{2}(138.8)(2.50)(12) = 2082$ in·kips (235.22 kN·m). Assume $d = 35 - 4.5 = 30.5$ in (774.7 mm); $A_s = 2,082,000/[20,000(0.874)(30.5)] = 3.91 \text{ in}^2$ (25.227 cm^2). Try seven no. 7 bars; $A_s = 4.20 \text{ in}^2$ (270.098 cm^2); $\Sigma o = 19.2$ in (487.68 mm); $u = 138,800/[19.2(0.874)(30.5)] = 271 \text{ lb/in}^2$ (1868.5 kPa); $u_{\text{allow}} = 302 \text{ lb/in}^2$ (2082.3 kPa). This is acceptable.

Since the critical section for shear falls outside the footing, shearing stress is not a criterion in this design.

10. Design the transverse reinforcement under the exterior column; disregard eccentricity

Thus, $V_{UV} = 250(2.58/6.67) = 96.8$ kips (430.57 kN); $M_{UV} = \frac{1}{2}(96.8)(2.58)(12) = 1498$ in·kips (169.3 kN·m); $A_s = 2.72 \text{ in}^2$ (17.549 cm^2). Try five no. 7 bars; $A_s = 3.00 \text{ in}^2$ (19.356 cm^2); $\Sigma o = 13.7$ in (347.98 mm); $u = 96,800/[13.7(0.874)(31.5)] = 257 \text{ lb/in}^2$ (1772.0 kPa). This is acceptable.

Cantilever Retaining Walls

Retaining walls having a height ranging from 10 to 20 ft (3.0 to 6.1 m) are generally built as reinforced-concrete cantilever members. As shown in Fig. 132, a cantilever wall comprises a vertical stem to retain the soil, a horizontal base to support the stem, and in many instances a key that projects into the underlying soil to augment the resistance to sliding. Adequate drainage is an essential requirement, because the accumulation of water or ice behind the wall would greatly increase the horizontal thrust.

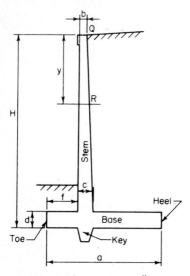

FIG. 132   Cantilever retaining wall.

The calculation of earth thrust in this section is based on Rankine's theory, which is developed in a later calculation procedure. When a live load, termed a *surcharge*, is applied to the retained soil, it is convenient to replace this load with a hypothetical equivalent prism of earth. Referring to Fig. 132, consider a portion QR of the wall, R being at distance y below the top. Take the length of wall normal to the plane of the drawing as 1 ft (0.3 m). Let T = resultant earth thrust on QR; M = moment of this thrust with respect to R; h = height of equivalent earth prism that replaces surcharge; w = unit weight of earth; C_a = coefficient of active earth pressure; C_p = coefficient of passive earth pressure. Then

$$T = \tfrac{1}{2}C_a wy(y + 2h) \tag{51}$$

$$M = (\tfrac{1}{6})C_a wy^2(y + 3h) \tag{52}$$

DESIGN OF A CANTILEVER RETAINING WALL

Applying the working-stress method, design a reinforced-concrete wall to retain an earth bank 14 ft (4.3 m) high. The top surface is horizontal and supports a surcharge of 500 lb/ft² (23.9 kPa). The soil weighs 130 lb/ft³ (20.42 kN/m³), and its angle of internal friction is 35°; the coefficient of friction of soil and concrete is 0.5. The allowable soil pressure is 4000 lb/ft² (191.5 kPa); f_c' = 3000 lb/in² (20,685 kPa) and f_y = 40,000 lb/in² (275,800 kPa). The base of the structure must be set 4 ft (1.2 m) below ground level to clear the frost line.

Calculation Procedure:

1.   Secure a trial section of the wall

Apply these relations: a = 0.60H; $b \geq 8$ in (203.2 mm); $c = d = b + 0.045h$; $f = a/3 - c/2$.

The trial section is shown in Fig. 133a, and the reinforcement is shown in Fig. 134. As the calculation will show, it is necessary to provide a key to develop the required resistance to sliding. The sides of the key are sloped to ensure that the surrounding soil will remain undisturbed during excavation.

2.   Analyze the trial section for stability

The requirements are that there be a factor of safety (FS) against sliding and overturning of at least 1.5 and that the soil pressure have a value lying between 0 and 4000 lb/ft² (0 and 191.5 kPa). Using the equation developed later in this handbook gives h = surcharge/soil weight = 500/130 = 3.85 ft (1.173 m); sin 35° = 0.574; tan 35° = 0.700; C_a = 0.271; C_p = 3.69; $C_a w$ = 35.2 lb/ft³ (5.53 kN/m³); $C_p w$ = 480 lb/ft³ (75.40 kN/m³); T_{AB} = ½(35.2)18(18 + 2 × 3.85) = 8140 lb (36,206.7 N); M_{AB} = (⅙)35.2(18)²(18 + 3 × 3.85) = 56,200 ft·lb (76,207.2 N·m).

The critical condition with respect to stability is that in which the surcharge extends to G. The moments of the stabilizing forces with respect to the toe are computed in Table 7. In Fig. 133c, x = 81,030/21,180 = 3.83 ft (1.167 m); e = 5.50 − 3.83 = 1.67 ft (0.509 m). The fact that the

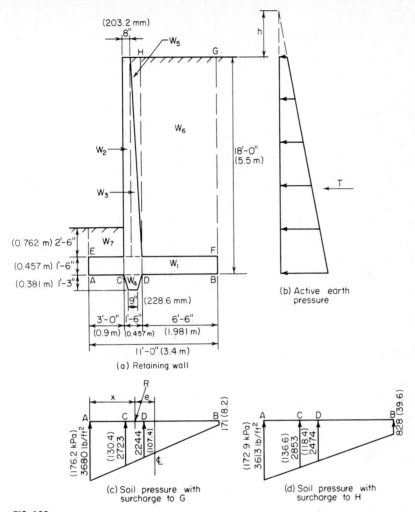

(203.2 mm)
8"

W_5

H

G

h

W_6

W_2

W_3

18'-0"
(5.5 m)

T

(0.762 m) 2'-6"
E

W_7

(0.457 m) 1'-6"

(0.381 m) 1'-3"

A   C W_4 D

W_1

F

B

9" (228.6 mm)

3'-0" 1'-6"   6'-6"
(0.9 m) (0.457 m) (1.981 m)

11'-0" (3.4 m)

(a) Retaining wall

(b) Active earth
pressure

R

x   e

A   C D   B 171 (8.2)

(176.2 kPa)
3680 lb/ft²

(130.4)
2723

2244
(107.4)

₵

(c) Soil pressure with
surcharge to G

A   C D   B 828 (39.6)

(172.9 kPa)
3613 lb/ft²

(136.6)
2853

(118.4)
2474

(d) Soil pressure with
surcharge to H

FIG. 133

resultant strikes the base within the middle third attests to the absence of uplift. By $f = (P/A)(1 \pm 6e_x/d_x \pm 6e_y/d_y)$, $p_a = (21{,}180/11)(1 + 6 \times 1.67/11) = 3680$ lb/ft² (176.2 kPa); $p_b = (21{,}180/11)(1 - 6 \times 1.67/11) = 171$ lb/ft² (8.2 kPa). Check: $x = (11/3)(3680 + 2 \times 171)/(3680 + 171) = 3.83$ ft (1.167 m), as before. Also, $p_c = 2723$ lb/ft² (130.4 kPa); $p_d = 2244$ lb/ft² (107.4 kPa); FS against overturning $= 137{,}230/56{,}200 = 2.44$. This is acceptable.

Lateral displacement of the wall produces sliding of earth on earth to the left of C and of concrete on earth to the right of C. In calculating the passive pressure, the layer of earth lying above the base is disregarded, since its effectiveness is unknown. The resistance to sliding is as follows: friction, A to C (Fig. 133): ½(3680 + 2723)(3)(0.700) = 6720 lb (29,890.6 N); friction, C to B: ½(2723 + 171)(8)(0.5) = 5790 lb (25,753.9 N); passive earth pressure: ½(480)(2.75)² = 1820 lb (8095.4 N). The total resistance to sliding is the sum of these three items, or 14,330 lb (63,739.8 N). Thus, the FS against sliding is 14,330/8140 = 1.76. This is acceptable because it exceeds 1.5. Hence the trial section is adequate with respect to stability.

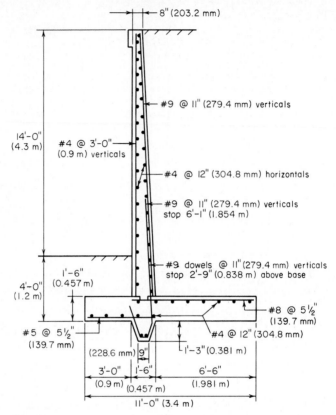

FIG. 134

3. Calculate the soil pressures when the surcharge extends to H

Thus $W_s = 500(6.5) = 3250$ lb (14,456 N); $\Sigma W = 21,180 + 3250 = 24,430$ lb (108,664.6 N); $M_a = 81,030 + 3250(7.75) = 106,220$ ft·lb (144,034.3 N·m); $x = 106,220/24,430 = 4.35$ ft (1.326 m); $e = 1.15$ ft (0.351 m); $p_a = 3613$ lb/ft² (173 kPa); $p_b = 828$ lb/ft² (39.6 kPa); $p_c = 2853$ lb/ft² (136.6 kPa); $p_d = 2474$ lb/ft² (118.5 kPa).

TABLE 7 Stability of Retaining Wall

Force, lb (N)		Arm, ft (m)	Moment, ft·lb (N·m)
W_1 1.5(11)(150)	= 2,480 (11,031.0)	5.50 (1.676)	13,640 (18,495.8)
W_2 0.67(16.5)(150)	= 1,650 (7,339.2)	3.33 (1.015)	5,500 (7,458.0)
W_3 0.5(0.83)(16.5)(150)	= 1,030 (4,581.4)	3.95 (1.204)	4,070 (5,518.9)
W_4 1.25(1.13)(150)	= 210 (934.1)	3.75 (1.143)	790 (1,071.2)
W_5 0.5(0.83)(16.5)(130)	= 890 (3,958.7)	4.23 (1.289)	3,760 (5,098.6)
W_6 6.5(16.5)(130)	= 13,940 (62,005.1)	7.75 (2.362)	108,000 (146,448.0)
W_7 2.5(3)(130)	= 980 (4,359.1)	1.50 (0.457)	1,470 (1993.3)
Total	21,180 (94,208.6)	....	137,230 (186,083.8)
Overturning moment	....		56,200 (76,207.2)
Net moment about A	....		81,030 (109,876.6)

4. Design the stem

At the base of the stem, $y = 16.5$ ft (5.03 m) and $d = 18 - 3.5 = 14.5$ in (368.30 mm); $T_{EF} = 7030$ lb (31,269.4 N); $M_{EF} = 538,000$ in·lb (60,783.24 N·m). The allowable shear at a distance d above the base is $V_{allow} = vbd = 60(12)(14.5) = 10,440$ lb (46,437.1 N). This is acceptable. Also, $M_b = 223(12)(14.5)^2 = 563,000$ in·lb (63,607.74 N·m); therefore, the steel is stressed to capacity, and $A_s = 538,000/[20,000(0.874)(14.5)] = 2.12$ in^2 (13.678 cm^2). Use no. 9 bars 5½ in (139.70 mm) on centers. Thus, $A_s = 2.18$ in^2 (14.065 cm^2); $\Sigma o = 7.7$ in (195.58/mm); $u = 7030/[7.7(0.874)(14.5)] = 72$ lb/in^2 (496.5 kPa); $u_{allow} = 235$ lb/in^2 (1620.3 kPa). This is acceptable.

Alternate bars will be discontinued at the point where they become superfluous. As the following calculations demonstrate, the theoretical cutoff point lies at $y = 11$ ft 7 in (3.531 m), where $M = 218,400$ in·lb (24,674.8 N·m); $d = 4.5 + 10(11.58/16.5) = 11.52$ in (292.608 mm); $A_s = 218,400/[20,000 (0.874)(11.52)] = 1.08$ in^2 (6.968 cm^2). This is acceptable. Also, $T = 3930$ lb (17,480.6 N); $u = 101$ lb/in^2 (696.4 kPa). This is acceptable. From the ACI *Code*, anchorage $= 12(9/8) = 13.5$ in (342.9 mm).

The alternate bars will therefore be terminated at 6 ft 1 in (1.854 m) above the top of the base. The *Code* requires that special precautions be taken where more than half the bars are spliced at a point of maximum stress. To circumvent this requirement, the short bars can be extended into the footing; therefore only the long bars require splicing. For the dowels, $u_{allow} = 0.75(235) = 176$ lb/in^2 (1213.5 kPa); length of lap $= 1.00(20,000)/[176(3.5)] = 33$ in (838.2 mm).

5. Design the heel

Let V and M denote the shear and bending moment, respectively, at section D. Case 1: surcharge extending to G—downward pressure $p = 16.5(130) + 1.5(150) = 2370$ lb/ft^2 (113.5 kPa); $V = 6.5[2370 - \frac{1}{2}(2244 + 171)] = 7560$ lb (33,626.9 N); $M = 12(6.5)^2 [\frac{1}{2} \times 2370 - \frac{1}{6}(2244 + 2 \times 171)] = 383,000$ in·lb (43,271.3 N·m).

Case 2: surcharge extending to H—$p = 2370 + 500 = 2870$ lb/ft^2 (137.4 kPa); $V = 6.5[2870 - \frac{1}{2}(2474 + 828)] = 7920$ lb (35,228.1 N) $< V_{allow}$; $M = 12(6.5)^2 [\frac{1}{2} \times 2870 - \frac{1}{6}(2474 + 2 \times 828)] = 379,000$ in·lb (42,819.4 N·m); $A_s = 2.12(383/538) = 1.51$ in^2 (9.742 cm^2).

To maintain uniform bar spacing throughout the member, use no. 8 bars 5½ in (139.7 mm) on centers. In the heel, tension occurs at the *top* of the slab, and $A_s = 1.72$ in^2 (11.097 cm^2); $\Sigma o = 6.9$ in (175.26 mm); $u = 91$ lb/in^2 (627.4 kPa); $u_{allow} = 186$ lb/in^2 (1282.5 kPa). This is acceptable.

6. Design the toe

For this purpose, assume the absence of backfill on the toe, but disregard the minor modification in the soil pressure that results. Let V and M denote the shear and bending moment, respectively, at section C (Fig. 133). The downward pressure $p = 1.5(150) = 225$ lb/ft^2 (10.8 kPa).

Case 1: surcharge extending to G (Fig. 133)—$V = 3[\frac{1}{2}(3680 + 2723) - 225] = 8930$ lb (39,720.6 N); $M = 12(3)^2 [(\frac{1}{6})(2723 + 2 \times 3680) - \frac{1}{2}(225)] = 169,300$ in·lb (19,127.5 N·m).

Case 2: surcharge extending to H (Fig. 133)—$V = 9020$ lb (40,121.0 N) $< V_{allow}$; $M = 169,300$ in·lb (19,127.5 N·m); $A_s = 2.12(169,300/538,000) = 0.67$ in^2 (4.323 cm^2). Use no. 5 bars 5½ in (139.7 mm) on centers. Then $A_s = 0.68$ in^2 (4.387 cm^2); $\Sigma o = 4.3$ in (109.22 mm); $u = 166$ lb/in^2 (1144.4 kPa); $u_{allow} = 422$ lb/in^2 (2909.7 kPa). This is acceptable.

The stresses in the key are not amenable to precise evaluation. Reinforcement is achieved by extending the dowels and short bars into the key and bending them.

In addition to the foregoing reinforcement, no. 4 bars are supplied to act as temperature reinforcement and spacers for the main bars, as shown in Fig. 134.

Prestressed Concrete

Prestressed-concrete construction is designed to enhance the suitability of concrete as a structural material by inducing prestresses opposite in character to the stresses resulting from gravity loads. These prestresses are created by the use of steel wires or strands, called *tendons*, that are incorporated in the member and subjected to externally applied tensile forces. This prestressing of the steel may be performed either before or after pouring of the concrete. Thus, two methods of prestressing a concrete beam are available: *pretensioning* and *posttensioning*.

In pretensioning, the tendons are prestressed to the required amount by means of hydraulic jacks, their ends are tied to fixed abutments, and the concrete is poured around the tendons. When hardening of the concrete has advanced to the required state, the tendons are released. The tendons now tend to contract longitudinally to their original length and to expand laterally to their original diameter, both these tendencies being opposed by the surrounding concrete. As a result of the longitudinal restraint, the concrete exerts a tensile force on the steel and the steel exerts a compressive force on the concrete. As a result of the lateral restraint, the tendons are deformed to a wedge shape across a relatively short distance at each end of the member. It is within this distance, termed the *transmission length*, that the steel becomes bonded to the concrete and the two materials exert their prestressing forces on each other. However, unless greater precision is warranted, it is assumed for simplicity that the prestressing forces act at the end sections.

The tendons may be placed either in a straight line or in a series of straight-line segments, being deflected at designated points by means of holding devices. In the latter case, prestressing forces between steel and concrete occur both at the ends and at these deflection points.

In posttensioning, the procedure usually consists of encasing the tendons in metal or rubber hoses, placing these in the forms, and then pouring the concrete. When the concrete has hardened, the tendons are tensioned and anchored to the ends of the concrete beam by means of devices called *end anchorages*. If the hoses are to remain in the member, the void within the hose is filled with grout. Posttensioning has two important advantages compared with pretensioning: It may be performed at the job site, and it permits the use of parabolic tendons.

The term *at transfer* refers to the instant at which the prestressing forces between steel and concrete are developed. (In posttensioning, where the tendons are anchored to the concrete one at a time, in reality these forces are developed in steps.) Assume for simplicity that the tendons are straight and that the resultant prestressing force in these tendons lies below the centroidal axis of the concrete section. At transfer, the member cambers (deflects upward), remaining in contact with the casting bed only at the ends. Thus, the concrete beam is compelled to resist the prestressing force and to support its own weight simultaneously.

At transfer, the prestressing force in the steel diminishes because the concrete contracts under the imposed load. The prestressing force continues to diminish as time elapses as a result of the relaxation of the steel and the shrinkage and plastic flow of the concrete subsequent to transfer. To be effective, prestressed-concrete construction therefore requires the use of high-tensile steel in order that the reduction in prestressing force may be small in relation to the initial force. In all instances, we assume that the ratio of final to initial prestressing force is 0.85. Moreover, to simplify the stress calculations, we also assume that the full initial prestressing force exists at transfer and that the entire reduction in this force occurs during some finite interval following transfer.

Therefore, two loading states must be considered in the design: the initial state, in which the concrete sustains the initial prestressing force and the beam weight; and the final state, in which the concrete sustains the final prestressing force, the beam weight, and all superimposed loads. Consequently, the design of a prestressed-concrete beam differs from that of a conventional type in that designers must consider two stresses at each point, the initial stress and the final stress, and these must fall between the allowable compressive and tensile stresses. A beam is said to be in *balanced design* if the critical initial and final stresses coincide precisely with the allowable stresses.

The term *prestress* designates the stress induced by the *initial* prestressing force. The terms *prestress shear* and *prestress moment* refer to the vertical shear and bending moment, respectively, that the initial prestressing force induces in the concrete at a given section.

The *eccentricity* of the prestressing force is the distance from the action line of this resultant force to the centroidal axis of the section. Assume that the tendons are subjected to a uniform prestress. The locus of the centroid of the steel area is termed the *trajectory* of the steel or of the prestressing force.

The sign convention is as follows: The eccentricity is positive if the action line of the prestressing force lies below the centroidal axis. The trajectory has a positive slope if it inclines downward to the right. A load is positive if it acts downward. The vertical shear at a given section is positive if the portion of the beam to the left of this section exerts an upward force on the concrete. A bending moment is positive if it induces compression above the centroidal axis and tension below it. A compressive stress is positive; a tensile stress, negative.

The notational system is as follows. *Cross-sectional properties:* A = gross area of section, in^2 (cm^2) A_s = area of prestressing steel, in^2 (cm^2); d = effective depth of section at ultimate strength, in (mm); h = total depth of section, in (mm); I = moment of inertia of gross area, in^4 (cm^4); y_b = distance from centroidal axis to bottom fiber, in (mm); S_b = section modulus with respect to bottom fiber $= I/y_b$, in^3 (cm^3); $k_b =$ distance from centroidal axis to lower kern point, in (mm); $k_t =$ distance from centroidal axis to upper kern point, in (mm). *Forces and moments:* $F_i =$ initial prestressing force, lb (N); $F_f =$ final prestressing force, lb (N); $\eta = F_f/F_i$; $e =$ eccentricity of prestressing force, in (mm); $e_{con} =$ eccentricity of prestressing force having concordant trajectory; $\theta =$ angle between trajectory (or tangent to trajectory) and horizontal line; $m =$ slope of trajectory; $w =$ vertical load exerted by curved tendons on concrete in unit distance; $w_w =$ unit beam weight; $w_s =$ unit superimposed load; $w_{DL} =$ unit dead load; $w_{LL} =$ unit live load; $w_u =$ unit ultimate load; $V_p =$ prestress shear; $M_p =$ prestress moment; $M_w =$ bending moment due to beam weight; $M_s =$ bending moment due to superimposed load; $C_u =$ resultant compressive force at ultimate load; $T_u =$ resultant tensile force at ultimate load. *Stresses:* $f'_c =$ ultimate compressive strength of concrete, lb/in^2 (kPa); $f'_{ci} =$ compressive strength of concrete at transfer; $f'_s =$ ultimate strength of prestressing steel; $f_{su} =$ stress in prestressing steel at ultimate load; $f_{bp} =$ stress in bottom fiber due to initial prestressing force; $f_{bw} =$ bending stress in bottom fiber due to beam weight; $f_{bs} =$ bending stress in bottom fiber due to superimposed loads; $f_{bi} =$ stress in bottom fiber at initial state $= f_{bp} + f_{bw}$; $f_{bf} =$ stress in bottom fiber at final state $= \eta f_{bp} + f_{bw} + f_{bs}$; $f_{cai} =$ initial stress at centroidal axis. *Camber:* $\Delta_p =$ camber due to initial prestressing force, in (mm); $\Delta_w =$ camber due to beam weight; $\Delta_i =$ camber at initial state; $\Delta_f =$ camber at final state.

The symbols that refer to the bottom fiber are transformed to their counterparts for the top fiber by replacing the subscript b with t. For example, f_{ti} denotes the stress in the top fiber at the initial state.

DETERMINATION OF PRESTRESS SHEAR AND MOMENT

The beam in Fig. 135a is simply supported at its ends and prestressed with an initial force of 300 kips (1334.4 kN). At section C, the eccentricity of this force is 8 in (203.2 mm), and the slope of

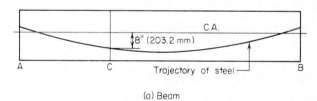

(a) Beam

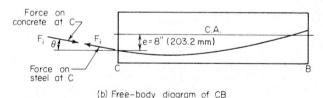

(b) Free-body diagram of CB

FIG. 135

the trajectory is 0.014. (In the drawing, vertical distances are exaggerated in relation to horizontal distances.) Find the prestress shear and prestress moment at C.

Calculation Procedure:

1. *Analyze the prestressing forces*

If the composite concrete-and-steel member is regarded as a unit, the prestressing forces that the steel exerts on the concrete are purely internal. Therefore, if a beam is simply supported, the prestressing force alone does not induce any reactions at the supports.

Refer to Fig. 135b, and consider the forces acting on the beam segment CB solely as a result of F_i. The left portion of the beam exerts a tensile force F_i on the tendons. Since CB is in equilibrium, the left portion also induces compressive stresses on the concrete at C, these stresses having a resultant that is numerically equal to and collinear with F_i.

2. Express the prestress shear and moment in terms of F_i

Using the sign convention described, express the prestress shear and moment in terms of F_i and θ. (The latter is positive if the slope of the trajectory is positive.) Thus $V_p = -F_i \sin \theta$; $M_p = -F_i e \cos \theta$.

3. Compute the prestress shear and moment

Since θ is minuscule, apply these approximations: $\sin \theta = \tan \theta$, and $\cos \theta = 1$. Then

$$V_p = -F_i \tan \theta \tag{53}$$

Or, $V_p = -300,000(0.014) = -4200$ lb $(-18,681.6$ N).
    Also,

$$M_p = -F_i e \tag{54}$$

Or, $M_p = -300,000(8) = -2,400,000$ in$\cdot$lb $(-271,152$ N$\cdot$m).

STRESSES IN A BEAM WITH STRAIGHT TENDONS

A 12×18 in (304.8×457.2 mm) rectangular beam is subjected to an initial prestressing force of 230 kips (1023.0 kN) applied 3.3 in (83.82 mm) below the center. The beam is on a simple span of 30 ft (9.1 m) and carries a superimposed load of 840 lb/lin ft (12,258.9 N/m). Determine the initial and final stresses at the supports and at midspan. Construct diagrams to represent the initial and final stresses along the span.

Calculation Procedures:

1. Compute the beam properties

Thus, $A = 12(18) = 216$ in^2 (1393.6 cm^2); $S_b = S_t = (\%)(12)(18)^2 = 648$ in^3 (10,620.7 cm^3); $w_w = (216/144)(150) = 225$ lb/lin ft (3,283.6 N/m).

2. Calculate the prestress in the top and bottom fibers

Since the section is rectangular, apply $f_{bp} = (F_i/A)(1 + 6e/h) = (230,000/216)(1 + 6 \times 3.3/18) = +2236$ lb/in^2 (+15,417.2 kPa); $f_{tp} = (F_i/A)(1 - 6e/h) = -106$ lb/in^2 (−730.9 kPa).
    For convenience, record the stresses in Table 8 as they are obtained.

TABLE 8   Stresses in Prestressed-Concrete Beam

	At support		At midspan	
	Bottom fiber	Top fiber	Bottom fiber	Top fiber
(a) Initial prestress, lb/in^2 (kPa)	+2,236 (+15,417.2)	−106 (−730.9)	+2,236 (+15,417.2)	−106 (−730.9)
(b) Final prestress, lb/in^2 (kPa)	+1,901 (+13,107.4)	−90 (−620.6)	+1,901 (+13,107.4)	−90 (−620.6)
(c) Stress due to beam weight, lb/in^2 (kPa)	. . . . . .	. . . . . .	−469 (−3,233.8)	+469 (3,233.8)
(d) Stress due to superimposed load, lb/in^2 (kPa)	. . . . . .	. . . . . .	−1,750 (−12,066.3)	+1,750 (+12,066.3)
Initial stress: (a) + (c)	+2,236 (+15,417.2)	−106 (−730.9)	+1,767 (+12,183.5)	+363 (+2,502.9)
Final stress: (b) + (c) + (d)	+1,901 (+13,107.4)	−90 (−620.6)	−318 (−2,192.6)	+2,129 (+14,679.5)

3. Determine the stresses at midspan due to gravity loads

Thus $M_s = (\%)(840)(30)^2(12) = 1{,}134{,}000$ in·lb (128,119.32 N·m); $f_{bs} = -1{,}134{,}000/648 = -1750$ lb/in^2 ($-12{,}066.3$ kPa); $f_{ts} = +1750$ lb/in^2 (12,066.3 kPa). By proportion, $f_{bw} = -1750(225/840) = -469$; $f_{tw} = +469$ lb/in^2 ($+3233.8$ kPa).

4. Compute the initial and final stresses at the supports

Thus, $f_{bi} = +2236$ lb/in^2 ($+15{,}417.2$ kPa); $f_{ti} = -106$ lb/in^2 (-730.9 kPa); $f_{bf} = 0.85(2236) = +1901$ lb/in^2 ($+13{,}107.4$ kPa); $f_{tf} = 0.85(-106) = -90$ lb/in^2 (-620.6 kPa).

5. Determine the initial and final stresses at midspan

Thus $f_{bi} = +2236 - 469 = +1767$ lb/in^2 ($+12{,}183.5$ kPa); $f_{ti} = -106 + 469 = +363$ lb/in^2 ($+2502.9$ kPa); $f_{bf} = +1901 - 469 - 1750 = -318$ lb/in^2 (-2192.6 kPa); $f_{tf} = -90 + 469 + 1750 = +2129$ lb/in^2 ($+14{,}679.5$ kPa).

6. Construct the initial-stress diagram

In Fig. 136a, construct the initial-stress diagram $A_t A_b BC$ at the support and the initial-stress diagram $M_t M_b DE$ at midspan. Draw the parabolic arcs BD and CE. The stress diagram at an inter-

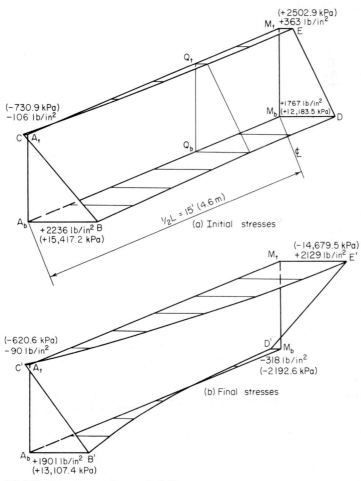

FIG. 136 Isometric stress diagrams for half-span.

mediate section Q is obtained by passing a plane normal to the longitudinal axis. The offset from a reference line through B to the arc BD represents the value of f_{bw} at that section.

7. Construct the final-stress diagram

Construct Fig. 136b in an analogous manner. The offset from a reference line through B' to the arc $B'D'$ represents the value of $f_{bw} + f_{bs}$ at the given section.

8. Alternatively, construct composite stress diagrams for the top and bottom fibers

The diagram pertaining to the bottom fiber is shown in Fig. 137. The difference between the ordinates to DE and AB represents f_{bi}, and the difference between the ordinates to FG and AC represents f_{bf}.

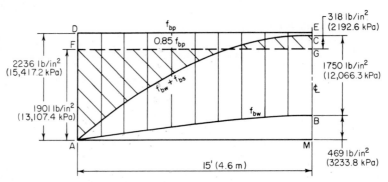

FIG. 137 Stresses in bottom fiber along half-span.

This procedure illustrates the following principles relevant to a beam with straight tendons carrying a uniform load: At transfer, the critical stresses occur at the supports; under full design load, the critical stresses occur at midspan if the allowable final stresses exceed η times the allowable initial stresses in absolute value.

The primary objective in prestressed-concrete design is to maximize the capacity of a given beam by maximizing the absolute values of the prestresses at the section having the greatest superimposed-load stresses. The three procedures that follow, when taken as a unit, illustrate the manner in which the allowable prestresses may be increased numerically by taking advantage of the beam-weight stresses, which are opposite in character to the prestresses. The next procedure will also demonstrate that when a beam is not in balanced design, there is a range of values of F_i that will enable the member to carry this maximum allowable load. In summary, the objective is to maximize the capacity of a given beam and to provide the minimum prestressing force associated with this capacity.

DETERMINATION OF CAPACITY AND PRESTRESSING FORCE FOR A BEAM WITH STRAIGHT TENDONS

An 8×10 in (203.2 $\times$ 254 mm) rectangular beam, simply supported on a 20-ft (6.1-m) span, is to be prestressed by means of straight tendons. The allowable stresses are: *initial*, $+ 2400$ and -190 lb/in² ($+16,548$ and -1310.1 kPa); *final*, $+ 2250$ and -425 lb/in² ($+15,513.8$ and -2930.3 kPa). Evaluate the allowable unit superimposed load, the maximum and minimum prestressing force associated with this load, and the corresponding eccentricities.

Calculation Procedure:

1. Compute the beam properties

Here $A = 80$ in² (516.16 cm²); $S = 133$ in² (858.1 cm²); $w_w = 83$ lb/lin ft (1211.3 N/m).

2. Compute the stresses at midspan due to the beam weight

Thus, $M_w = (\frac{1}{8})(83)(20)^2(12) = 49,800$ in·lb (5626.4 N·m); $f_{bw} = -49,800/133 = -374$ lb/in² (-2578.7 kPa); $f_{tw} = +374$ lb/in² (2578.7 kPa).

3. Set the critical stresses equal to their allowable values to secure the allowable unit superimposed load

Use Fig. 136 or 137 as a guide. At support: $f_{bi} = +2400$ lb/in² ($+16,548$ kPa); $f_{ti} = -190$ lb/in² (-1310.1 kPa); at midspan, $f_{bf} = 0.85(2400) - 374 + f_{bs} = -425$ lb/in² (-2930.4 kPa); $f_{tf} = 0.85(-190) + 374 + f_{ts} = +2250$ lb/in² ($+15,513.8$ kPa). Also, $f_{bs} = -2091$ lb/in² ($-14,417.4$ kPa); $f_{ts} = +2038$ lb/in² ($+14,052$ kPa).

Since the superimposed-load stresses at top and bottom will be numerically equal, the latter value governs the beam capacity. Or $w_s = w_w f_{ts}/f_{tw} = 83(2038/374) = 452$ lb/lin ft (6596.4 N/m).

4. Find $F_{i,max}$ and its eccentricity

The value of w_s was found by setting the critical value of f_{ti} and of f_{tf} equal to their respective allowable values. However, since S_b is excessive for the load w_s, there is flexibility with respect to the stresses at the bottom. The designer may set the critical value of either f_{bi} or f_{bf} equal to its allowable value or produce some intermediate condition. As shown by the calculations in step 3, f_{bf} may vary within a range of $2091 - 2038 = 53$ lb/in² (365.4 kPa). Refer to Fig. 138, where the lines represent the stresses indicated.

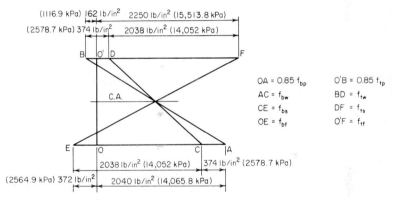

FIG. 138 Stresses at midspan under maximum prestressing force.

Points B and F are fixed, but points A and E may be placed anywhere within the 53-lb/in² (365.4-kPa) range. To maximize F_i, place A at its limiting position to the right; i.e., set the critical value of f_{bi} rather than that of f_{bf} equal to the allowable value. Then $f_{cai} = F_{i,max}/A = \frac{1}{2}(2400 - 190) = +1105$ lb/in² ($+7619.0$ kPa); $F_{i,max} = 1105(80) = 88,400$ lb (393,203.2 N); $f_{bp} = 1105 + 88,400e/133 = +2400$; $e = 1.95$ in (49.53 mm).

5. Find $F_{i,min}$ and its eccentricity

For this purpose, place A at its limiting position to the left. Then $f_{bp} = 2,400 - (53/0.85) = +2338$ lb/in² ($+16,120.5$ kPa); $f_{cai} = +1074$ lb/in² ($+7405.2$ kPa); $F_{i,min} = 85,920$ lb (382,172.2 N); $e = 1.96$ in (49.78 mm).

6. Verify the value of $F_{i,max}$ by checking the critical stresses

At support: $f_{bi} = +2400$ lb/in² ($+16,548.0$ kPa); $f_{ti} = -190$ lb/in² (-1310.1 kPa). At midspan: $f_{bf} = +2040 - 374 - 2038 = -372$ lb/in² (-2564.9 kPa); $f_{tf} = -162 + 374 + 2038 = +2250$ lb/in² ($+15,513.8$ kPa).

7. *Verify the value of $F_{i,min}$ by checking the critical stresses*

At support: $f_{bi} = +2338$ lb/in² (16,120.5 kPa); $f_{ti} = -190$ lb/in² (-1310.1 kPa). At midspan: $f_{bf} = 0.85(2338) - 374 - 2038 = -425$ lb/in² (-2930.4 kPa); $f_{tf} = +2250$ lb/in² (+15,513.8 kPa).

BEAM WITH DEFLECTED TENDONS

The beam in the previous calculation procedure is to be prestressed by means of tendons that are deflected at the quarter points of the span, as shown in Fig. 139a. Evaluate the allowable unit superimposed load, the magnitude of the prestressing force, the eccentricity e_1 in the center interval, and the maximum and minimum allowable values of the eccentricity e_2 at the supports. What increase in capacity has been obtained by deflecting the tendons?

Calculation Procedure:

1. *Compute the beam-weight stresses at B*

In the composite stress diagram, Fig. 139b, the difference between an ordinate to EFG and the corresponding ordinate to AHJ represents the value of f_{ti} at the given section. It is apparent that

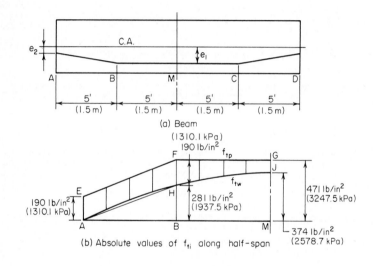

(a) Beam

(b) Absolute values of f_{ti} along half-span

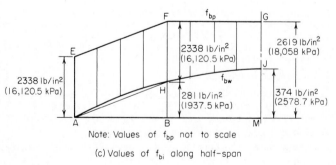

Note: Values of f_{bp} not to scale

(c) Values of f_{bi} along half-span

FIG. 139

if AE does not exceed HF, then f_{ti} does not exceed HF in absolute value anywhere along the span. Therefore, for the center interval BC, the critical stresses at transfer occur at the boundary sections B and C. Analogous observations apply to Fig. 139c.

Computing the beam-weight stresses at B yields $f_{bw} = (\frac{3}{4})(-374) = -281$ lb/in^2 (-1937.5 kPa); $f_{tw} = +281$ lb/in^2 ($+1937.5$ kPa).

2.   Tentatively set the critical stresses equal to their allowable values to secure the allowable unit superimposed load

Thus, at B: $f_{bi} = f_{bp} - 281 = +2400$; $f_{ti} = f_{tp} + 281 = -190$; $f_{bp} = +2681$ lb/in^2 ($+18,485.5$ kPa); $f_{tp} = -471$ lb/in^2 (-3247.5 kPa).

At M: $f_{bf} = 0.85(2681) - 374 + f_{bs} = -425$; $f_{tf} = 0.85(-471) + 374 + f_{ts} = +2250$; $f_{bs} = -2330$ lb/in^2 ($-16,065.4$ kPa); $f_{ts} = +2277$ lb/in^2 ($+15,699.9$ kPa). The latter value controls.

Also, $w_s = 83(2277/374) = 505$ lb/lin ft (7369.9 N/m); $505/452 = 1.12$. The capacity is increased 12 percent.

When the foregoing calculations are compared with those in the previous calculation procedure, the effect of deflecting the tendons is to permit an increase of 281 lb/in^2 (1937.5 kPa) in the absolute value of the prestress at top and bottom. The accompanying increase in f_{ts} is 0.85(281) $= 239$ lb/in^2 (1647.9 kPa).

3.   Find the minimum prestressing force and the eccentricity e_1

Examination of Fig. 138 shows that f_{cat} is not affected by the form of trajectory used. Therefore, as in the previous calculation procedure, $F_i = 85,920$ lb (382,172.2 N); $f_{tp} = 1074 - 85,920e_1/133 = -471$; $e_1 = 2.39$ in (60.706 mm).

Although it is not required, the value of $f_{bp} = 1074 + 1074 - (-471) = +2619$ lb/in^2 ($+18,058$ kPa), or $f_{bp} = 2681 - 53/0.85 = +2619$ lb/in^2 ($+18,058$ kPa).

4.   Establish the allowable range of values of e_2

At the supports, the tendons may be placed an equal distance above or below the center. Then $e_{2,\max} = 1.96$ in (23.44 mm); $e_{2,\min} = -1.96$ in (-23.44 mm).

BEAM WITH CURVED TENDONS

The beam in the second previous calculation procedure is to be prestressed by tendons lying in a parabolic arc. Evaluate the allowable unit superimposed load, the magnitude of the prestressing force, the eccentricity of this force at midspan, and the increase in capacity accruing from the use of curved tendons.

Calculation Procedure:

1.   Tentatively set the initial and final stresses at midspan equal to their allowable values to secure the allowable unit superimposed load

Since the prestressing force has a parabolic trajectory, lines EFG in Fig. 139b and c will be parabolic in the present case. Therefore, it is possible to achieve the full allowable initial stresses at midspan. Thus, $f_{bi} = f_{bp} - 374 = +2400$; $f_{ti} = f_{tp} + 374 = -190$; $f_{bp} = +2774$ lb/in^2 ($+19,126.7$ kPa); $f_{tp} = -564$ lb/in^2 (-3888.8 kPa); $f_{bf} = 0.85(2774) - 374 + f_{bs} = -425$; $f_{tf} = 0.85(-564) + 374 + f_{ts} = +2250$; $f_{bs} = -2409$ lb/in^2 ($-16,610.1$ kPa); $f_{ts} = +2356$ lb/in^2 ($+16,244.6$ kPa). The latter value controls.

Also, $w_s = 83(2356/374) = 523$ lb/lin ft (7632.6 N/m); $523/452 = 1.16$. Thus the capacity is increased 16 percent.

When the foregoing calculations are compared with those in the earlier calculation procedure, the effect of using parabolic tendons is to permit an increase of 374 lb/in^2 (2578.7 kPa) in the absolute value of the prestress at top and bottom. The accompanying increase in f_{ts} is 0.85(374) $= 318$ lb/in^2 (2192.6 kPa).

2.   Find the minimum prestressing force and its eccentricity at midspan

As before, $F_i = 85,920$ lb (382,172.2 N); $f_{tp} = 1074 - 85,920e/133 = -564$; $e = 2.54$ in (64.516 mm).

DETERMINATION OF SECTION MODULI

A beam having a cross-sectional area of 500 in^2 (3226 cm^2) sustains a beam-weight moment equal to 3500 in·kips (395.4 kN·m) at midspan and a superimposed moment that varies parabolically from 9000 in·kips (1016.8 kN·m) at midspan to 0 at the supports. The allowable stresses are: *initial*, +2400 and −190 lb/in^2 (+16,548 and −1310.1 kPa); *final*, + 2250 and −200 lb/in^2 (+15,513.8 and −1379 kPa). The member will be prestressed by tendons deflected at the quarter points. Determine the section moduli corresponding to balanced design, the magnitude of the prestressing force, and its eccentricity in the center interval. Assume that the calculated eccentricity is attainable (i.e., that the centroid of the tendons will fall within the confines of the section while satisfying insulation requirements).

Calculation Procedure:

1. Equate the critical initial stresses, and the critical final stresses, to their allowable values

Let M_w and M_s denote the indicated moments at midspan; the corresponding moments at the quarter point are three-fourths as large. The critical initial stresses occur at the quarter point, while the critical final stresses occur at midspan. After equating the stresses to their allowable values, solve the resulting simultaneous equations to find the section moduli and prestresses. Thus: *stresses in bottom fiber*, $f_{bi} = f_{bp} - 0.75M_w/S_b = +2400$; $f_{bf} = 0.85f_{bp} - M_w/S_b - M_s/S_b = -200$. Solving gives $S_b = (M_s + 0.3625M_w)/2240 = 4584$ in^3 (75,131.7 cm^3) and $f_{bp} = +2973$ lb/in^2 (+20,498.8 kpa); *stresses in top fiber*, $f_{ti} = f_{tp} + 0.75(M_w/S_t) = -190$; $f_{tf} = 0.85f_{tp} + M_w/S_t + M_s/S_t = +2250$. Solving yields $S_t = (M_s + 0.3625M_w)/2412 = 4257$ in^3 (69,772.2 cm^3) and $f_{tp} = -807$ lb/in^2 (−5564.2 kPa).

2. Evaluate F_i and e

In this instance, e denotes the eccentricity in the center interval. Thus $f_{bp} = F_i/A + F_ie/S_b = +2973$; $f_{tp} = F_i/A - F_ie/S_t = -807$; $F_i = (2973S_b - 807S_t)A/(S_b + S_t) = 576,500$ lb (2,564,272.0 N); $e = 2973S_b/F_i - S_b/A = 14.47$ in (367.538 mm).

3. Alternatively, evaluate F_i by assigning an arbitrary depth to the member

Thus, set $h = 10$ in (254 mm); $y_b = S_th/(S_b + S_t) = 4.815$ in (122.301 mm); $f_{cai} = f_{bp} - (f_{bp} - f_{tp})y_b/h = 2973 - (2973 + 807)0.4815 = +1153$ lb/in^2 (+7949.9 kPa); $F_i = 1153(500) = 576,500$ lb (2,564,272.0 N).

EFFECT OF INCREASE IN BEAM SPAN

Consider that the span of the beam in the previous calculation procedure increases by 10 percent, thereby causing the midspan moment due to superimposed load to increase by 21 percent. Show that the member will be adequate with respect to flexure if all cross-sectional dimensions are increased by 7.2 percent. Compute the new eccentricity in the center interval, and compare this with the original value.

Calculation Procedure:

1. Calculate the new section properties and bending moments

Thus $A = 500(1.072)^2 = 575$ in^2 (3709.9 cm^2); $S_b = 4584(1.072)^3 = 5647$ in^3 (92,554.3 cm^3); $S_t = 4257(1.072)^3 = 5244$ in^3 (85,949.2 cm^3); $M_s = 9000(1.21) = 10,890$ in·kips (1230.4 kN·m); $M_w = 3500(1.072)^2(1.21) = 4867$ in·kips (549.9 kN·m).

2. Compute the required section moduli, prestresses, prestressing force, and its eccentricity in the central interval, using the same sequence as in the previous calculation procedure

Thus $S_b = 5649$ in^3 (92,587.1 cm^3); $S_t = 5246$ in^3 (85,981.9 cm^3). Both these values are acceptable. Then $f_{bp} = + 3046$ lb/in^2 (+21,002.2 kPa); $f_{tp} = -886$ lb/in^2 (−6108.9 kPa); $F_i = 662,800$ lb (2,948,134.4 N); $e = 16.13$ in (409.7 mm). The eccentricity has increased by 11.5 percent.

In practice, it would be more efficient to increase the vertical dimensions more than the horizontal dimensions. Nevertheless, as the span increases, the eccentricity increases more rapidly than the depth.

EFFECT OF BEAM OVERLOAD

The beam in the second previous calculation procedure is subjected to a 10 percent overload. How does the final stress in the bottom fiber compare with that corresponding to the design load?

Calculation Procedure:

1. Compute the value of f_{bs} under design load

Thus, $f_{bs} = -M_s/S_b = -9,000,000/4584 = -1963 \text{ lb/in}^2 \ (-13,534.8 \text{ kPa})$.

2. Compute the increment or f_{bs} caused by overload and the revised value of f_{bf}

Thus, $\Delta f_{bs} = 0.10(-1963) = -196 \text{ lb/in}^2 \ (-1351.4 \text{ kPa})$; $f_{bf} = -200 - 196 = -396 \text{ lb/in}^2$ (-2730.4 kPa). Therefore, a 10 percent overload virtually doubles the tensile stress in the member.

PRESTRESSED-CONCRETE BEAM DESIGN GUIDES

On the basis of the previous calculation procedures, what conclusions may be drawn that will serve as guides in the design of prestressed-concrete beams?

Calculation Procedure:

1. Evaluate the results obtained with different forms of tendons

The capacity of a given member is increased by using deflected rather than straight tendons, and the capacity is maximized by using parabolic tendons. (However, in the case of a pretensioned beam, an economy analysis must also take into account the expense incurred in deflecting the tendons.)

2. Evaluate the prestressing force

For a given ratio of y_b/y_t, the prestressing force that is required to maximize the capacity of a member is a function of the cross-sectional area and the allowable stresses. It is independent of the form of the trajectory.

3. Determine the effect of section moduli

If the section moduli are in excess of the minimum required, the prestressing force is minimized by setting the critical values of f_{bf} and f_{ti} equal to their respective allowable values. In this manner, points A and B in Fig. 138 are placed at their limiting positions to the left.

4. Determine the most economical short-span section

For a short-span member, an I section is most economical because it yields the required section moduli with the minimum area. Moreover, since the required values of S_b and S_t differ, the area should be disposed unsymmetrically about middepth to secure these values.

5. Consider the calculated value of e

Since an increase in span causes a greater increase in the theoretical eccentricity than in the depth, the calculated value of e is not attainable in a long-span member because the centroid of the tendons would fall beyond the confines of the section. For this reason, long-span members are generally constructed as T sections. The extensive flange area elevates the centroidal axis, thus making it possible to secure a reasonably large eccentricity.

6. Evaluate the effect of overload

A relatively small overload induces a disproportionately large increase in the tensile stress in the beam and thus introduces the danger of cracking. Moreover, owing to the presence of many variable quantities, there is not a set relationship between the beam capacity at allowable final stress and the capacity at incipient cracking. It is therefore imperative that every prestressed-concrete beam be subjected to an ultimate-strength analysis to ensure that the beam provides an adequate factor of safety.

KERN DISTANCES

The beam in Fig. 140 has the following properties: $A = 850$ in^2 (5484.2 cm^2); $S_b = 11,400$ in^3 (186,846.0 cm^3); $S_t = 14,400$ in^3 (236,016.0 cm^3). A prestressing force of 630 kips (2802.2 kN) is applied with an eccentricity of 24 in (609.6 mm) at the section under investigation. Calculate f_{bp} and f_{tp} by expressing these stresses as functions of the kern distances of the section.

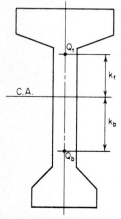

FIG. 140   Kern points.

Calculation Procedure:

1.   Consider the prestressing force to be applied at each kern point, and evaluate the kern distances

Let Q_b and Q_t denote the points at which a compressive force must be applied to induce a zero stress in the top and bottom fiber, respectively. These are referred to as the *kern points* of the section, and the distances k_b and k_t from the centroidal axis to these points are called the *kern distances*.

Consider the prestressing force to be applied at each kern point in turn. Set the stresses f_{tp} and f_{bp} equal to zero to evaluate the kern distances k_b and k_t, respectively. Thus $f_{tp} = F_i/A - F_i k_b/S_t = 0$, Eq. *a*; $f_{bp} = F_i/A - F_i k_t/S_b = 0$, Eq. *b*. Then

$$k_b = \frac{S_t}{a} \quad \text{and} \quad k_t = \frac{S_b}{A} \tag{55}$$

And, $k_b = 14,400/850 = 16.9$ in (429.26 mm); $k_t = 11,400/850 = 13.4$ in (340.36 mm).

2.   Express the stresses f_{bp} and f_{tp} associated with the actual eccentricity as functions of the kern distances

By combining the stress equations with Eqs. *a* and *b*, the following equations are obtained:

$$f_{bp} = \frac{F_i(k_t + e)}{S_b} \quad \text{and} \quad f_{tp} = \frac{F_i(k_b - e)}{S_t} \tag{56}$$

Substituting numerical values gives $f_{bp} = 630,000(13.4 + 24)/11,400 = +2067$ lb/in^2 (+14,252.0 kPa); $f_{tp} = 630,000(16.9 - 24)/14,400 = -311$ lb/in^2 (-2144.3 kPa).

3.   Alternatively, derive Eq. 56 by considering the increase in prestress caused by an increase in eccentricity

Thus, $\Delta f_{bp} = F_i \Delta e/S_b$; therefore, $f_{bp} = F_i(k_t + e)/S_b$.

MAGNEL DIAGRAM CONSTRUCTION

The data pertaining to a girder having curved tendons are $A = 500$ in^2 (3226.0 cm^2); $S_b = 5000$ in^3 (81,950 cm^3); $S_t = 5340$ in^3 (87,522.6 cm^3); $M_w = 3600$ in·kips (406.7 kN·m); $M_s = 9500$ in·kips (1073.3 kN·m). The allowable stresses are: *initial*, + 2400 and - 190 lb/in^2 (+16,548 and -1310.1 kPa); *final*, + 2250 and - 425 lb/in^2 (+15,513.8 and -2930.4 kPa). (*a*) Construct the Magnel diagram for this member. (*b*) Determine the minimum prestressing force and its eccentricity by referring to the diagram. (*c*) Determine the prestressing force if the eccentricity is restricted to 18 in (457.2 mm).

Calculation Procedure:

1.   Set the initial stress in the bottom fiber at midspan equal to or less than its allowable value, and solve for the reciprocal of F_i

In this situation, the superimposed load is given, and the sole objective is to minimize the prestressing force. The Magnel diagram is extremely useful for this purpose because it brings into sharp focus the relationship between F_i and e. In this procedure, let f_{bi}, f_{bf} and so forth represent the *allowable* stresses.

Thus,

$$\frac{1}{F_i} \geq \frac{k_t + e}{M_w + f_{bi}S_b} \qquad (57a)$$

2. Set the final stress in the bottom fiber at midspan equal to or algebraically greater than its allowable value, and solve for the reciprocal of F_i

Thus

$$\frac{1}{F_i} \leq \frac{\eta(k_t + e)}{M_w + M_s + f_{bf}S_b} \qquad (57b)$$

3. Repeat the foregoing procedure with respect to the top fiber

Thus,

$$\frac{1}{F_i} \geq \frac{e - k_b}{M_w - f_{ti}S_t} \qquad (57c)$$

and

$$\frac{1}{F_i} \leq \frac{\eta(e - k_b)}{M_w + M_s - f_{tf}S_t} \qquad (57d)$$

4. Substitute numerical values, expressing F_i in thousands of kips

Thus, $1/F_i \geq (10 + e)/15.60$, Eq. a; $1/F_i \leq (10 + e)/12.91$, Eq. b; $1/F_i \geq (e - 10.68)/4.61$, Eq. c; $1/F_i \leq (e - 10.68)/1.28$, Eq. d.

5. Construct the Magnel diagram

In Fig. 141, consider the foregoing relationships as equalities, and plot the straight lines that represent them. Each point on these lines represents a set of values of $1/F_i$ and e at which the designated stress equals its allowable value.

When the section moduli are in excess of those corresponding to balanced design, as they are

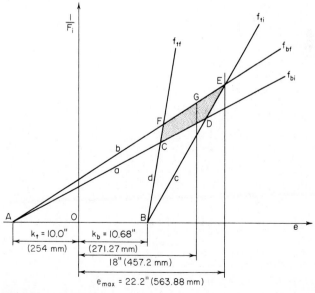

FIG. 141 Magnel diagram.

in the present instance, line b makes a greater angle with the e axis than does a, and line d makes a greater angle than does c. From the sense of each inequality, it follows that $1/F_i$ and e may have any set of values represented by a point within the quadrilateral $CDEF$ or on its circumference.

6. To minimize F_i, determine the coordinates of point E at the intersection of lines b and c

Thus, $1/F_i = (10 + e)/12.91 = (e - 10.68)/4.61$; so $e = 22.2$ in (563.88 mm); $F_i = 401$ kips (1783.6 kN).

The Magnel diagram confirms the third design guide presented earlier in the section.

7. For the case where e is restricted to 18 in (457.2 mm), minimize F_i by determining the ordinate of point G on line b

Thus, in Fig. 141, $1/F_i = (10 + 18)/12.91$; $F_i = 461$ kips (2050.5 kN).

The Magnel diagram may be applied to a beam having deflected tendons by substituting for M_w in Eqs. 57a and 57c the beam-weight moment at the deflection point.

CAMBER OF A BEAM AT TRANSFER

The following pertain to a simply supported prismatic beam: $L = 36$ ft (11.0 m); $I = 40,000$ in⁴ (166.49 dm⁴); $f'_{ci} = 4000$ lb/in² (27,580.0 kPa); $w_w = 340$ lb/lin ft (4961.9 N/m); $F_i = 430$ kips (1912.6 kN); $e = 8.8$ in (223.5 mm) at midspan. Calculate the camber of the member at transfer under each of these conditions: (a) the tendons are straight across the entire span; (b) the tendons are deflected at the third points, and the eccentricity at the supports is zero; (c) the tendons are curved parabolically, and the eccentricity at the supports is zero.

Calculation Procedure:

1. Evaluate E_c at transfer, using the ACI Code

Review the moment-area method of calculating beam deflections, which is summarized earlier. Consider an upward displacement (camber) as positive, and let the symbols Δ_p, Δ_w, and Δ_t, defined earlier, refer to the camber at midspan.

Thus, using the ACI *Code*, $E_c = (145)^{1.5}(33)(4000)^{0.5} = 3,644,000$ lb/in² (25,125.4 MPa).

2. Construct the prestress-moment diagrams associated with the three cases described

See Fig. 142. By symmetry, the elastic curve corresponding to F_i is horizontal at midspan. Consequently, Δ_p equals the deviation of the elastic curve at the support from the tangent to this curve at midspan.

3. Using the literal values shown in Fig. 142, develop an equation for Δ_p by evaluating the tangential deviation; substitute numerical values

Thus, case a:

$$\Delta_p = \frac{ML^2}{8E_cI} \qquad (58)$$

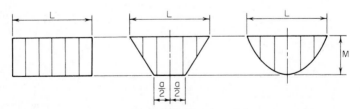

(a) Straight tendons   (b). Deflected tendons   (c) Parabolic tendons

FIG. 142 Prestress-moment diagrams.

or $\Delta_p = 430,000(8.8)(36)^2(144)/[8(3,644,000)(40,000)] = 0.61$ in (15.494 mm). For case b:

$$\Delta_p = \frac{M(2L^2 + 2La - a^2)}{24E_cI} \tag{59}$$

or $\Delta_p = 0.52$ in (13.208 mm). For case c:

$$\Delta_p = \frac{5ML^2}{48E_cI} \tag{60}$$

or $\Delta_p = 0.51$ in (12.954 mm).

4. Compute Δ_w

Thus, $\Delta_w = -5w_wL^4/(384E_cI) = -0.09$ in (-2.286 mm).

5. Combine the foregoing results to obtain Δ_i

Thus: case a, $\Delta_i = 0.61 - 0.09 = 0.52$ in (13.208 mm); case b, $\Delta_i = 0.52 - 0.09 = 0.43$ in (10.922 mm); case c, $\Delta_i = 0.51 - 0.09 = 0.42$ in (10.688 mm).

DESIGN OF A DOUBLE-T ROOF BEAM

The beam in Fig. 143 was selected for use on a simple span of 40 ft (12.2 m) to carry the following loads: roofing, 12 lb/ft^2 (574.5 N/m^2) snow, 40 lb/ft^2 (1915.1 N/m^2); total, 52 lb/ft^2 (2489.6 N/

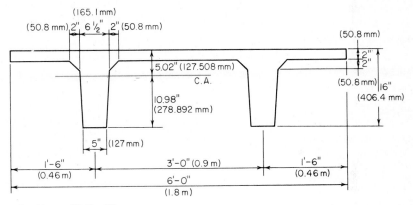

FIG. 143 Double-T roof beam.

m^2). The member will be pretensioned with straight seven-wire strands, $\frac{7}{16}$ in (11.11 mm) diameter, having an area of 0.1089 in^2 (0.70262 cm^2) each and an ultimate strength of 248,000 lb/in^2 (1,709,960.0 kPa). The concrete strengths are $f_c' = 5000$ lb/in^2 (34,475.0 kPa) and $f_{ci}' = 4000$ lb/in^2 (27,580.0 kPa). The allowable stresses are: *initial*, $+2400$ and -190 lb/in^2 ($+16,548.0$ and -1310.1 kPa); *final*, $+2250$ and -425 lb/in^2 ($+15,513.8$ and -2930.4 kPa). Investigate the adequacy of this section, and design the tendons. Compute the camber of the beam after the concrete has hardened and all dead loads are present. For this calculation, assume that the final value of E_c is one-third of that at transfer.

Calculation Procedure:

1. Compute the properties of the cross section

Let f_{bf} and f_{tf} denote the respective stresses at *midspan* and f_{bi} and f_{ti} denote the respective stresses *at the support*. Previous calculation procedures demonstrated that where the section moduli are excessive, the minimum prestressing force is obtained by setting f_{bf} and f_{ti} equal to their allowable values.

Thus $A = 316$ in² (2038.8 cm²); $I = 7240$ in⁴ (30.14 dm⁴); $y_b = 10.98$ in (278.892 mm); $y_t = 5.02$ in (127.508 mm); $S_b = 659$ in³ (10,801.0 cm³); $S_t = 1442$ in³ (23,614 cm³); $w_w = (316/144)150 = 329$ lb/lin ft (4801.4 N/m).

2. Calculate the total midspan moment due to gravity loads and the corresponding stresses

Thus $w_s = 52(6) = 312$ lb/lin ft (4553.3 N/m); $w_w = 329$ lb/lin ft (4801.4 N/m); and $M_w + M_s = (\frac{1}{8})(641)(40^2)(12) = 1,538,000$ in·lb (173,763.2 N·m); $f_{bw} + f_{bs} = -1,538,000/659 = -2334$ lb/in² ($-16,092.9$ kPa); $f_{tw} + f_{ts} = +1,538,000/1442 = +1067$ lb/in² ($+7357.0$ kPa).

3. Determine whether the section moduli are excessive

Do this by setting f_{bf} and f_{ti} equal to their allowable values and computing the corresponding values of f_{bi} and f_{tf}. Thus, $f_{bf} = 0.85f_{bp} - 2334 = -425$; therefore, $f_{bp} = +2246$ lb/in² ($+15,486.2$ kPa); $f_{ti} = f_{tp} = -190$ lb/in² (-1310.1 kPa); $f_{bi} = f_{bp} = +2246 < 2400$ lb/in² ($+16,548.0$ kPa). This is acceptable. Also, $f_{tf} = 0.85(-190) + 1067 = +905 < 2250$ lb/in² ($+15,513.8$ kPa); this is acceptable. The section moduli are therefore excessive.

4. Find the minimum prestressing force and its eccentricity

Refer to Fig. 144. Thus, $f_{bp} = +2246$ lb/in² ($+15,486.2$ kPa); $f_{tp} = -190$ lb/in² (-1310.1 kPa); slope of $AB = 2246 - (-190)/16 = 152.3$ lb/(in²·in) (41.33 MPa/m); $F_i/A = CD = 2246 -$

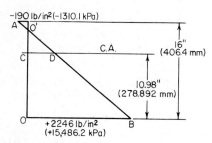

10.98(152.3) = 574 lb/in² (3957.7 kPa); $F_i = 574(316) = 181,400$ lb (806,867.2 N); slope of $AB = F_i e/I = 152.3$; $e = 152.3(7240)/181,400 = 6.07$ in (154.178 mm).

5. Determine the number of strands required, and establish their disposition

In accordance with the ACI *Code*, allowable initial force per strand = 0.1089(0.70)(248,000) = 18,900 lb (84,067.2 N); number required = 181,400/18,900 = 9.6. Therefore, use 10 strands (5 in each web) stressed to 18,140 lb (80,686.7 N) each.

Referring to the ACI *Code* for the minimum clear distance between the strands, we find the allowable center-to-center spacing = $4(\frac{7}{16}) = 1\frac{3}{4}$ in (44.45 mm). Use a 2-in (50.8-mm) spacing. In Fig. 145, locate the centroid of the steel, or $y = (2 \times 2 + 1 \times 4)/5 = 1.60$ in (40.64 mm); $v = 10.98 - 6.07 - 1.60 = 3.31$ in (84.074 mm); set $v = 3\frac{5}{16}$ in (84.138 mm).

FIG. 144 Prestress diagram.

6. Calculate the allowable ultimate moment of the member in accordance with the ACI Code

Thus, $A_s = 10(0.1089) = 1.089$ in² (7.0262 cm²); $d = y_t + e = 5.02 + 6.07 = 11.09$ in (281.686 mm); $p = A_s/(bd) = 1.089/[72(11.09)] = 0.00137$.

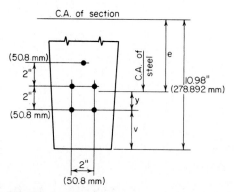

FIG. 145 Location of tendons.

Compute the steel stress and resultant tensile force at ultimate load:

$$f_{su} = f_s'\left(1 - \frac{0.5\,p f_s'}{f_c'}\right) \tag{61}$$

Or, $f_{su} = 248{,}000(1 - 0.5 \times 0.00137 \times 248{,}000/5000) = 240{,}00$ lb/in^2 (1,654,800 kPa); $T_u = A_s f_{su} = 1.089(240{,}000) = 261{,}400$ lb (1,162,707.2 N).
Compute the depth of the compression block. This depth, a, is found from $C_u = 0.85(5000)(72a) = 261{,}400$ lb (1,162,707.2 N); $a = 0.854$ in (21.6916 mm); $jd = d - a/2 = 10.66$ in (270.764 mm); $M_u = \phi T_u jd = 0.90(261{,}400)(10.66) = 2{,}500{,}000$ in·lb (282,450.0 N·m).
Calculate the steel index to ascertain that it is below the limit imposed by the ACI *Code*, or $q = pf_{su}/f_c' = 0.00137\,(240{,}000)/5000 = 0.0658 < 0.30$. This is acceptable.

7. Calculate the required ultimate-moment capacity as given by the ACI Code

Thus, $w_{DL} = 329 + 12(6) = 401$ lb/lin ft (5852.2 N/m); $w_{LL} = 40(6) = 240$ lb/lin ft (3502.5 N/m); $w_u = 1.5w_{DL} + 1.8w_{LL} = 1034$ lb/lin ft (15,090.1 N/m); M_u required $= (\frac{1}{8})(1034)(40)^2(12) = 2{,}480{,}000 < 2{,}500{,}000$ in·lb (282,450.0 N·m). The member is therefore adequate with respect to its ultimate-moment capacity.

8. Calculate the maximum and minimum area of web reinforcement in the manner prescribed in the ACI Code

Since the maximum shearing stress does not vary linearly with the applied load, the shear analysis is performed at ultimate-load conditions. Let A_v = area of web reinforcement placed perpendicular to the longitudinal axis; V_c' = ultimate-shear capacity of concrete; V_p' = vertical component of F_f at the given section; V_u' = ultimate shear at given section; s = center-to-center spacing of stirrups; f_{pc}' = stress due to F_f, evaluated at the centroidal axis, or at the junction of the web and flange when the centroidal axis lies in the flange.
Calculate the ultimate shear at the critical section, which lies at a distance $d/2$ from the face of the support. Then distance from midspan to the critical section $= \frac{1}{2}(L - d) = 19.54$ ft (5.955 m); $V_u' = 1034(19.54) = 20{,}200$ lb (89,849.6 N).
Evaluate V_c' by solving the following equations and selecting the smaller value:

$$V_{ci}' = 1.7b'd(f_c')^{0.5} \tag{62}$$

where d = effective depth, in (mm); b' = width of web at centroidal axis, in (mm); $b' = 2(5 + 1.5 \times 10.98/12) = 12.74$ in (323.596 mm); $V_{ci}' = 1.7(12.74)(11.09)(5000)^{0.5} = 17{,}000$ lb (75,616.0 N). Also,

$$V_{cw}' = b'd(3.5f_c'^{0.5} + 0.3f_{pc}') + V_p' \tag{63}$$

where d = effective depth or 80 percent of the overall depth, whichever is greater, in (mm). Thus, $d = 0.80(16) = 12.8$ in (325.12 mm); $V_p' = 0$. From step 4, $f_{pc} = 0.85(574) = +488$ lb/in^2 (3364.8 kPa); $V_{cw}' = 12.74(12.8)(3.5 \times 5000^{0.5} + 0.3 \times 488) = 64{,}300$ lb (286,006.4 N); therefore, $V_c' = 17{,}000$ lb (75,616.0 N).
Calculate the maximum web-reinforcement area by applying the following equation:

$$A_v = \frac{s(V_u' - \phi V_c')}{\phi d f_y} \tag{64}$$

where d = effective depth at section of maximum moment, in (mm). Use $f_y = 40{,}000$ lb/in^2 (275,800.0 kPa), and set $s = 12$ in (304.8 mm). Then $A_v = 12(20{,}200 - 0.85 \times 17{,}000)/[0.85(11.09)(40{,}000)] = 0.184$ in^2/ft (3.8949 cm^2/m). This is the area required at the ends.
Calculate the minimum web-reinforcement area by applying

$$A_v = \frac{A_s f_s'}{80 f_y} \frac{s}{(b'd)^{0.5}} \tag{65}$$

or $A_v = (1.089/80)(248{,}000/40{,}000)12/(12.74 \times 11.09)^{0.5} = 0.085$ in^2/ft (1.7993 cm^2/m).

9. Calculate the camber under full dead load

From the previous procedure, $E_c = (\frac{1}{8})(3.644)(10)^6 = 1.215 \times 10^6$ lb/in^2 (8.377 $\times 10^6$ kPa); $E_c I = 1.215(10)^6(7240) = 8.8 \times 10^9$ lb·in^2 (25.25 $\times 10^6$ N·m^2); $\Delta_{DL} = -5(401)(40)^4(1728)/$

$[384(8.8)(10)^9] = -2.62$ in (-66.548 mm). By Eq. 58, $\Delta_p = 0.85(181,400)(6.07)(40)^2(144)/$ $[8(8.8)(10)^9] = 3.06$ in (77.724 mm); $\Delta = 3.06 - 2.62 = 0.44$ in (11.176 mm).

DESIGN OF A POSTTENSIONED GIRDER

The girder in Fig. 146 has been selected for use on a 90-ft (27.4-m) simple span to carry the following superimposed loads: dead load, 1160 lb/lin ft (16,928.9 N/m), live load, 1000 lb/lin ft

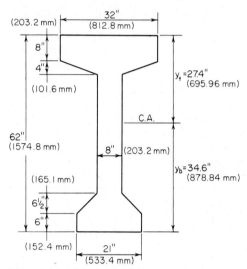

FIG. 146

(14,593.9 N/m). The girder will be posttensioned with Freyssinet cables. The concrete strengths are $f_c' = 5000$ lb/in² (34,475 kPa) and $f_{ci}' = 4000$ lb/in² (27,580 kPa). The allowable stresses are: *initial*, $+2400$ and -190 lb/in² ($+16,548$ and -1310.1 kPa); *final*, $+2250$ and -425 lb/in² ($+15,513.8$ and -2930.4 kPa). Complete the design of this member, and calculate the camber at transfer.

Calculation Procedure:

1. Compute the properties of the cross section

Since the tendons will be curved, the initial stresses at midspan may be equated to the allowable values. The properties of the cross section are $A = 856$ in² (5522.9 cm²); $I = 394,800$ in⁴ (1643 dm⁴); $y_b = 34.6$ in (878.84 mm); $y_t = 27.4$ in (695.96 mm); $S_b = 11,410$ in³ (187,010 cm³); $S_t = 14,410$ in³ (236,180 cm³); $w_w = 892$ lb/lin ft (13,017.8 N/m).

2. Calculate the stresses at midspan caused by gravity loads

Thus $f_{bw} = -950$ lb/in² (-6550.3 kPa); $f_{bs} = -2300$ lb/in² ($-15,858.5$ kPa); $f_{tw} = +752$ lb/in² ($+5185.0$ kPa); $f_{ts} = +1820$ lb/in² ($+12,548.9$ kPa).

3. Test the section adequacy

To do this, equate f_{bf} and f_{ti} to their allowable values and compute the corresponding values of f_{bi} and f_{tf}. Thus $f_{bf} = 0.85 f_{bp} - 950 - 2300 = -425$; $f_{ti} = f_{tp} + 752 = -190$; therefore, $f_{bp} = +3324$ lb/in² ($+22,919.0$ kPa) and $f_{tp} = -942$ lb/in² (-6495.1 kPa); $f_{bi} = +3324 - 950 = +2374 < 2400$ lb/in² (16,548.0 kPa). This is acceptable. And $f_{tf} = 0.85(-942) + 752 + 1820 = +1771 < 2250$ lb/in² (15,513.8 kPa). This is acceptable. The section is therefore adequate.

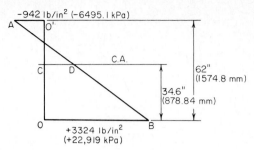

FIG. 147   Prestress diagram.

4.   Find the minimum prestressing force and its eccentricity at midspan

Do this by applying the prestresses found in step 3. Refer to Fig. 147. Slope of AB = [3324 − (−942)]/62 = 68.8 lb/(in²·in) (18.68 kPa/mm); F_i/A = CD = 3324 − 34.6(68.8) = 944 lb/in² (6508.9 kPa); F_i = 944(856) = 808,100 lb (3,594,428.8 N); slope of AB = $F_i e/I$ = 68.8; e = 68.8(394,800)/808,100 = 33.6 in (853.44 mm). Since y_b = 34.6 in (878.84 mm), this eccentricity is excessive.

5.   Select the maximum feasible eccentricity; determine the minimum prestressing force associated with this value

Try e = 34.6 − 3.0 = 31.6 in (802.64 mm). To obtain the minimum value of F_i, equate f_{bf} to its allowable value. Check the remaining stresses. As before, f_{bp} = +3324 lb/in² (+22,919 kPa). But f_{bp} = $F_i/856$ + 31.6F_i/11,410 = +3324; therefore F_i = 844,000 lb (3754.1 kN). Also, f_{tp} = −865 lb/in² (−5964.2 kPa); f_{bi} = +2374 lb/in² (+16,368.7 kPa); f_{ti} = −113 lb/in² (−779.1 kPa); f_{tf} = +1837 lb/in² (+12,666.1 kPa).

6.   Design the tendons, and establish their pattern at midspan

Refer to a table of the properties of Freyssinet cables, and select 12/0.276 cables. The designation indicates that each cable consists of 12 wires of 0.276-in (7.0104-mm) diameter. The ultimate strength is 236,000 lb/in² (1,627,220 kPa). Then A_s = 0.723 in² (4.6648 cm²) per cable. Outside diameter of cable = 1⅝ in (41.27 mm). Recommended final prestress = 93,000 lb (413,664 N) per cable; initial prestress = 93,000/0.85 = 109,400 lb (486,611.2 N) per cable. Therefore, use eight cables at an initial prestress of 105,500 lb (469,264.0 N) each.

A section of the ACI *Code* requires a minimum cover of 1½ in (38.1 mm) and another section permits the ducts to be bundled at the center. Try the tendon pattern shown in Fig. 148. Thus, y = [6(2.5) + 2(4.5)]/8 = 3.0 in (76.2 mm). This is acceptable.

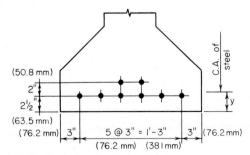

FIG. 148   Location of tendons at midspan.

7.   Establish the trajectory of the prestressing force

Construct stress diagrams to represent the initial and final stresses in the bottom and top fibers along the entire span.

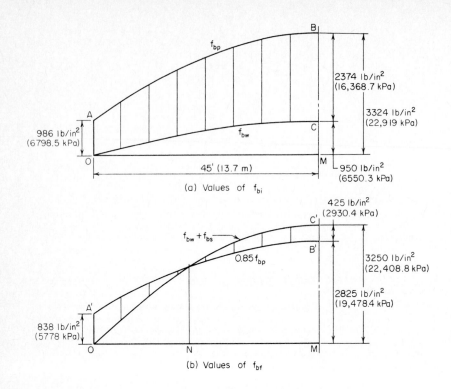

986 lb/in²
(6798.5 kPa)

2374 lb/in²
(16,368.7 kPa)

3324 lb/in²
(22,919 kPa)

45' (13.7 m)

950 lb/in²
(6550.3 kPa)

(a) Values of f_{bi}

425 lb/in²
(2930.4 kPa)

3250 lb/in²
(22,408.8 kPa)

2825 lb/in²
(19,478.4 kPa)

838 lb/in²
(5778 kPa)

(b) Values of f_{bf}

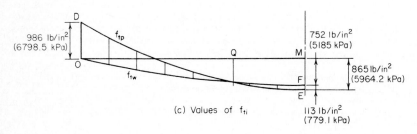

986 lb/in²
(6798.5 kPa)

752 lb/in²
(5185 kPa)

865 lb/in²
(5964.2 kPa)

113 lb/in²
(779.1 kPa)

(c) Values of f_{ti}

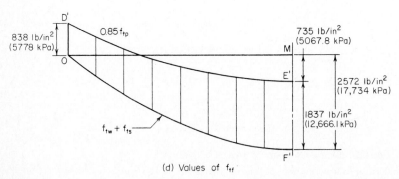

838 lb/in²
(5778 kPa)

735 lb/in²
(5067.8 kPa)

2572 lb/in²
(17,734 kPa)

1837 lb/in²
(12,666.1 kPa)

(d) Values of f_{tf}

FIG. 149

178

For convenience, set $e = 0$ at the supports. The prestress at the ends is therefore $f_{bp} = f_{tp} = 844,000/856 = +986$ lb/in² ($+6798.5$ kPa). Since e varies parabolically from maximum at midspan to zero at the supports, it follows that the prestresses also vary parabolically.

In Fig. 149a, draw the parabolic arc AB with summit at B to represent the absolute value of f_{bp}. Draw the parabolic arc OC in the position shown to represent f_{bw}. The vertical distance between the arcs at a given section represents the value of f_{bi}; this value is maximum at midspan.

In Fig. 149b, draw $A'B'$ to represent the absolute value of the final prestress; draw OC' to represent the absolute value of $f_{bw} + f_{bs}$. The vertical distance between the arcs represents the value of f_{bf}. This stress is compressive in the interval ON and tensile in the interval NM.

Construct Fig. 149c and d in an analogous manner. The stress f_{ti} is compressive in the interval OQ.

8. Calculate the allowable ultimate moment of the member

The midspan section is critical in this respect. Thus, $d = 62 - 3 = 59.0$ in (1498.6 mm); $A_s = 8(0.723) = 5.784$ in² (37.3184 cm²); $p = A_s/(bd) = 5.784/[32(59.0)] = 0.00306$.

Apply Eq. 61, or $f_{su} = 236,000(1 - 0.5 \times 0.00306 \times 236,000/5000) = 219,000$ lb/in² (1,510,005.0 kPa). Also, $T_u = A_s f_{su} = 5.784(219,000) = 1,267,000$ lb (5,635,616.0 N). The concrete area under stress $= 1,267,000/[0.85(5,000)] = 298$ in² (1922.7 cm²). This is the shaded area in Fig. 150, as the following calculation proves: $32(9.53) - 4.59(1.53) = 305 - 7 = 298$ in² (1922.7 cm²).

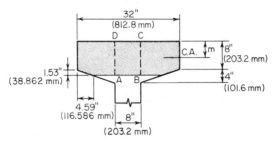

FIG. 150 Concrete area under stress at ultimate load.

Locate the centroidal axis of the stressed area, or $m = [305(4.77) - 7(9.53 - 0.51)]/298 = 4.67$ in (118.618 mm); $M_u = \phi T_u jd = 0.90(1,267,000)(59.0 - 4.67) = 61,950,000$ in·lb (6,999,111.0 N·m).

Calculate the steel index to ascertain that it is below the limit imposed by the ACI *Code*. Refer to Fig. 150. Or, area of $ABCD = 8(9.53) = 76.24$ in² (491.900 cm²). The steel area A_{sr} that is required to balance the force on this web strip is $A_{sr} = 5.784(76.24)/298 = 1.48$ in² (9.549 cm²); $q = A_{sr}f_{su}/(b'\,df_c') = 1.48(219,000)/[8(59.0)(5000)] = 0.137 < 0.30$. This is acceptable.

9. Calculate the required ultimate-moment capacity as given by the ACI Code

Thus, $w_u = 1.5(892 + 1160) + 1.8(1000) = 4878$ lb/lin ft (71,189.0 N/m); M_u required $= (\frac{1}{8})(4878)(90)^2(12) = 59,270,000$ in·lb (6,696,324.6 N·m). This is acceptable. The member is therefore adequate with respect to its ultimate-moment capacity.

10. Design the web reinforcement

Follow the procedure given in step 8 of the previous calculation procedure.

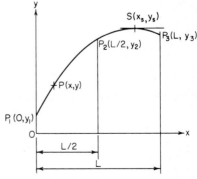

FIG. 151 Parabolic arc.

11. Design the end block

This is usually done by applying isobar charts to evaluate the tensile stresses caused by the concentrated prestressing forces. Refer to Winter et al.—*Design of Concrete Structures*, McGraw-Hill.

12. Compute the camber at transfer

Referring to earlier procedures in this section, we see that $E_cI = 3.644(10)^6(394,800) = 1.44 \times 10^{12}$ lb·in^2 $(4.132 \times 10^9$ N·m$^2)$. Also, $\Delta_w = -5(892)(90)^4(1728)/[384(1.44)(10)^{12}] = -0.91$ in $(-23.11$ mm). Apply Eq. 60, or $\Delta_p = 5(844,000)(31.6)(90)^2(144)/[48(1.44)(10)^{12}] = 2.25$ in (57.15 mm); $\Delta_i = 2.25 - 0.91 = 1.34$ in (34.036 mm).

PROPERTIES OF A PARABOLIC ARC

Figure 151 shows the literal values of the coordinates at the ends and at the center of the parabolic arc $P_1P_2P_3$. Develop equations for y, dy/dx, and d^2y/dx^2 at an arbitrary point P. Find the slope of the arc at P_1 and P_3 and the coordinates of the summit S. (This information is required for the analysis of beams having parabolic trajectories.)

Calculation Procedure:

1. Select a slope for the arc

Let m denote the slope of the arc.

2. Present the results

The equations are

$$y = 2(y_1 - 2y_2 + y_3)\left(\frac{x}{L}\right)^2 - (3y_1 - 4y_2 + y_3)\frac{x}{L} + y_1 \tag{66}$$

$$m = \frac{dy}{dx} = 4(y_1 - 2y_2 + y_3)\left(\frac{x}{L^2}\right) - \frac{3y_1 - 4y_2 + y_3}{L} \tag{67}$$

$$\frac{dm}{dx} = \frac{d^2y}{dx^2} = \frac{4}{L^2}(y_1 - 2y_2 + y_3) \tag{68}$$

$$m_1 = \frac{-(3y_1 - 4y_2 + y_3)}{L} \tag{69a}$$

$$m_3 = \frac{y_1 - 4y_2 + 3y_3}{L} \tag{69b}$$

$$x_s = \frac{(L/4)(3y_1 - 4y_2 + y_3)}{y_1 - 2y_2 + y_3} \tag{70a}$$

$$y_s = \frac{-(1/8)(3y_1 - 4y_2 + y_3)^2}{y_1 - 2y_2 + y_3} + y_1 \tag{70b}$$

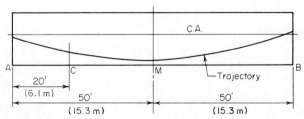

FIG. 152

ALTERNATIVE METHODS OF ANALYZING A BEAM WITH PARABOLIC TRAJECTORY

The beam in Fig. 152 is subjected to an initial prestressing force of 860 kips (3825.3 kN) on a parabolic trajectory. The eccentricities at the left end, midspan, and right end, respectively, are $e_a = 1$ in (25.4 mm); $e_m = 30$ in (762.0 mm); $e_b = -3$ in (-76.2 mm). Evaluate the prestress shear and prestress moment at section C (a) by applying the properties of the trajectory at C; (b) by considering the prestressing action of the steel on the concrete in the interval AC.

Calculation Procedure:

1. Compute the eccentricity and slope of the trajectory at C

Use Eqs. 66 and 67. Let m denote the slope of the trajectory. This is positive if the trajectory slopes downward to the right. Thus $e_a - 2e_m + e_b = 1 - 60 - 3 = -62$ in (1574.8 mm); $3e_a - 4e_m + e_b = 3 - 120 - 3 = -120$ in (-3048 mm); $e_c = 2(-62)(20/100)^2 + 120(20/100) + 1 = 20.04$ in (509.016 mm); $m_c = 4(-62/12)(20/100^2) - (-120/12 \times 100) = 0.0587$.

2. Compute the prestress shear and moment at C

Thus $V_{pc} = -m_c F_i = -0.0587(860,000) = -50,480$ lb (-224,535.0 N); $M_{pc} = -F_i e = -860,000(20.04) = -17,230,000$ in·lb (-1,946,645.4 N·m). This concludes the solution to part a.

3. Evaluate the vertical component w of the radial force on the concrete in a unit longitudinal distance

An alternative approach to this problem is to analyze the forces that the tendons exert on the concrete in the interval AC, namely, the prestressing force transmitted at the end and the radial forces resulting from curvature of the tendons.

Consider the component w to be positive if directed downward. In Fig. 153, $V_{pr} - V_{pq} = -F_i(m_r - m_q)$; therefore, $\Delta V_p/\Delta x = -F_i \Delta m/\Delta x$. Apply Eq. 68: $dV_p/dx = -F_i\, dm/dx = -(4F_i/L^2)(e_a - 2e_m + e_b)$; but $dV_p/dx = -w$. Therefore,

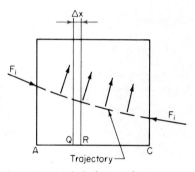

$$w = F_i \frac{dm}{dx} = \left(\frac{4F_i}{L^2}\right)(e_a - 2e_m + e_b) \quad (71)$$

FIG. 153 Free-body diagram of concrete.

This result discloses that when the trajectory is parabolic, w is uniform across the span. The radial forces are always directed toward the center of curvature, since the tensile forces applied at their ends tend to straighten the tendons. In the present instance, $w = (4F_i/100^2)(-62/12) = -0.002067F_i$ lb/lin ft (-0.00678F_i N/m).

4. Find the prestress shear at C

By Eq. 69a, $m_a = -[-120/(100 \times 12)] = 0.1$; $V_{pa} = -0.1F_i$; $V_{pc} = V_{pa} - 20w = F_i(-0.1 + 20 \times 0.002067) = -0.0587F_i = -50,480$ lb (-224,535.0 N).

5. Find the prestress moment at C

Thus, $M_{pc} = M_{pa} + V_{pa}(240) - 20w(120) = F_i(-1 - 0.1 \times 240 + 20 \times 0.002067 \times 120) = -20.04F_i = -17,230,000$ in·lb (1,946,645.4 N·m).

PRESTRESS MOMENTS IN A CONTINUOUS BEAM

The continuous prismatic beam in Fig. 154 has a prestressing force of 96 kips (427.0 kN) on a parabolic trajectory. The eccentricities are $e_a = -0.40$ in (-10.16 mm); $e_d = +0.60$ in (15.24 mm); $e_b = -1.20$ in (-30.48 mm); $e_e = +0.64$ in (16.256 mm); $e_c = -0.60$ in (-15.24 mm). Construct the prestress-moment diagram for this member, indicating all significant values.

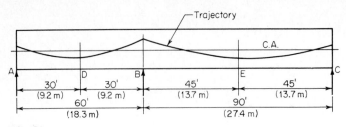

FIG. 154

Calculation Procedure:

1. Find the value of $wL^2/4$ for each span by applying Eq. 71

Refer to Fig. 155. Since members AB and BC are constrained to undergo an identical rotation at B, there exists at this section a bending moment M_{kb} in addition to that resulting from the eccentricity of F_i. The moment M_{kb} induces reactions at the supports. Thus, at every section of the beam there is a moment caused by continuity of the member as well as the moment $-F_ie$. The moment M_{kb} is termed the *continuity moment*; its numerical value is directly proportional to the distance from the given section to the end support.

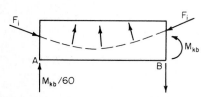

FIG. 155 Free-body diagram of concrete.

The continuity moment may be evaluated by adopting the second method of solution in the previous calculation procedure, since this renders the continuous member amenable to analysis by the theorem of three moments or moment distribution.

Determine $wL^2/4$ for each span: span AB, $w_1L_1^2/4 = F_i(-0.40 - 1.20 - 1.20) = -2.80F_i$ in·lb $(-0.3163F_i$ N·m); span BC, $w_2L_2^2/4 = F_i(-1.20 - 1.28 - 0.60) = -3.08F_i$ in·lb $(-0.3479F_i$ N·m).

2. Determine the true prestress moment at B in terms of F_i

Apply the theorem of three moments; by subtraction, find M_{kb}. Thus, $M_{pa}L_1 + 2M_{pb}(L_1 + L_2) + M_{pc}L_2 = -w_1L_1^3/4 - w_2L_2^3/4$. Substitute the value of L_1 and L_2, in feet (meters), and divide each term by F_i, or $0.40(60) + (2M_{pb} \times 150)/F_i + 0.60(90) = 2.80(60) + 3.08(90)$. Solving gives $M_{pb} = 1.224F_i$ in·lb $(0.1383F_i$ N·m). Also, $M_{kb} = M_{pb} - (-F_ie_b) = F_i(1.224 - 1.20) = 0.024F_i$. Thus, the continuity moment at B is positive.

3. Evaluate the prestress moment at the supports and at midspan

Using foot-pounds (newton-meters) in the moment evaluation yields $M_{pa} = 0.40(96,000)/12 = 3200$ ft·lb $(4339.2$ N·m); $M_{pb} = 1.224(96,000)/12 = 9792$ ft·lb $(13,278$ N·m); $M_{pc} = 0.60(96,000)/12 = 4800$ ft·lb $(6508.0$ N·m); $M_{pd} = -F_ie_d + M_{kd} = F_i(-0.60 + \frac{1}{2} \times 0.024)/12 = -4704$ ft·lb $(-6378$ N·m); $M_{pe} = F_i(-0.64 + \frac{1}{2} \times 0.024)/12 = -5024$ ft·lb $(-6812$ N·m).

4. Construct the prestress-moment diagram

Figure 156 shows this diagram. Apply Eq. 70 to locate and evaluate the maximum negative moments. Thus, $AF = 25.6$ ft $(7.80$ m); $BG = 49.6$ ft $(15.12$ m); $M_{pf} = -4947$ ft·lb $(-6708$ N·m); $M_{pg} = -5151$ ft·lb $(-6985$ N·m).

PRINCIPLE OF LINEAR TRANSFORMATION

For the beam in Fig. 154, consider that the parabolic trajectory of the prestressing force is displaced thus: e_a and e_c are held constant as e_b is changed to -2.0 in $(-50.80$ mm), the eccentricity at any intermediate section being decreased algebraically by an amount directly proportional to the distance from that section to A or C. Construct the prestress-moment diagram.

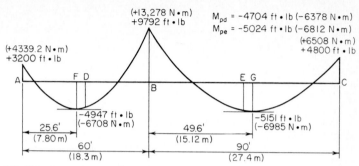

FIG. 156 Prestress-moment diagram.

Calculation Procedure:

1. Compute the revised eccentricities

The modification described is termed a *linear transformation* of the trajectory. Two methods are presented. Steps 1 through 4 comprise method 1; the remaining steps comprise method 2.

The revised eccentricities are $e_a = -0.40$ in (-10.16 mm); $e_d = +0.20$ in (5.08 mm); $e_b = -2.00$ in (-50.8 mm); $e_e = +0.24$ in (6.096 mm); $e_c = -0.60$ in (-15.24 mm).

2. Find the value of $wL^2/4$ for each span

Apply Eq. 71: span AB, $w_1L_1^2/4 = F_i(-0.40 - 0.40 - 2.00) = -2.80F_i$; span BC, $w_2L_2^2/4 = F_i(-2.00 - 0.48 - 0.60) = -3.08F_i$.

These results are identical with those obtained in the previous calculation procedure. The change in e_b is balanced by an equal change in $2e_d$ and $2e_e$.

3. Determine the true prestress moment at B by applying the theorem of three moments; then find M_{kb}

Refer to step 2 in the previous calculation procedure. Since the linear transformation of the trajectory has not affected the value of w_1 and w_2, the value of M_{pb} remains constant. Thus, $M_{kb} = M_{pb} - (-F_i e_b) = F_i(1.224 - 2.0) = -0.776F_i$.

4. Evaluate the prestress moment at midspan

Thus, $M_{pd} = -F_i e_d + M_{kd} = F_i(-0.20 - \frac{1}{2} \times 0.776)/12 = -4704$ ft·lb (-6378.6 N·m); $M_{pe} = F_i(-0.24 - \frac{1}{2} \times 0.776)/12 = -5024$ ft·lb (-6812.5 N·m).

These results are identical with those in the previous calculation procedure. The change in the eccentricity moment is balanced by an accompanying change in the continuity moment. Since three points determine a parabolic arc, the prestress moment diagram coincides with that in Fig. 156. This constitutes the solution by method 1.

5. Evaluate the prestress moments

Do this by replacing the prestressing system with two hypothetical systems that jointly induce eccentricity moments identical with those of the true system.

Let e denote the original eccentricity of the prestressing force at a given section and Δe the change in eccentricity that results from the linear transformation. The final eccentricity moment is $-F_i(e + \Delta e) = -(F_i e + F_i \Delta e)$.

Consider the beam as subjected to two prestressing forces of 96 kips (427.0 kN) each. One has the parabolic trajectory described in the previous calculation procedure; the other has the linear trajectory shown in Fig. 157, where $e_a = 0$, $e_b = -0.80$ in (-20.32 mm), and $e_c = 0$. Under the latter prestressing system, the tendons exert three forces on the concrete—one at each end and one at the deflection point above the interior support caused by the change in direction of the prestressing force.

The horizontal component of the prestressing force is considered equal to the force itself; it therefore follows that the force acting at the deflection point has no horizontal component.

Since the three forces that the tendons exert on the concrete are applied directly at the sup-

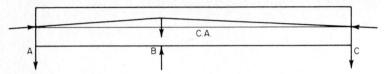

FIG. 157 Hypothetical prestressing system and forces exerted on concrete.

ports, their vertical components do not induce bending. Similarly, since the forces at A and C are applied at the centroidal axis, their horizontal components do not induce bending. Consequently, the prestressing system having the trajectory shown in Fig. 157 does not cause any prestress moments whatsoever. The prestress moments for the beam in the present instance are therefore identical with those for the beam in the previous calculation procedure.

The second method of analysis is preferable to the first because it is general. The first method demonstrates the equality of prestress moments before and after the linear transformation where the trajectory is parabolic; the second method demonstrates this equality without regard to the form of trajectory.

In this calculation procedure, the extremely important *principle of linear transformation* for a two-span continuous beam was developed. This principle states: The prestress moments remain constant when the trajectory of the prestressing force is transformed linearly. The principle is frequently applied in plotting a trial trajectory for a continuous beam.

Two points warrant emphasis. First, in a linear transformation, the eccentricities at the end supports remain constant. Second, the hypothetical prestressing systems introduced in step 5 are equivalent to the true system solely with respect to bending stresses; the axial stress F_i/A under the hypothetical systems is double that under the true system.

CONCORDANT TRAJECTORY OF A BEAM

Referring to the beam in the second previous calculation procedure, transform the trajectory linearly to obtain a concordant trajectory.

Calculation Procedure:

1. Calculate the eccentricities of the concordant trajectory

Two principles apply here. First, in a continuous beam, the prestress moment M_p consists of two elements, a moment $-F_ie$ due to eccentricity and a moment M_k due to continuity. The continuity moment varies linearly from zero at the ends to its maximum numerical value at the interior support. Second, in a linear transformation, the change in $-F_ie$ is offset by a compensatory change in M_k, with the result that M_p remains constant.

It is possible to transform a given trajectory linearly to obtain a new trajectory having the characteristic that $M_k = 0$ along the entire span, and therefore $M_p = -F_ie$. The latter is termed a *concordant trajectory*. Since M_p retains its original value, the concordant trajectory corresponding to a given trajectory is found simply by equating the final eccentricity to $-M_p/F_i$.

Refer to Fig. 154, and calculate the eccentricities of the concordant trajectory. As before, $e_a = -0.40$ in $(-10.16$ mm) and $e_c = -0.60$ in $(-15.24$ mm). Then $e_d = 4704(12)/96,000 = +0.588$ in $(+14.9352$ mm); $e_b = -9792(12)/96,000 = -1.224$ in $(-31.0896$ mm); $e_e = 5024(12)/96,000 = +0.628$ in $(15.9512$ mm).

2. Analyze the eccentricities

All eccentricities have thus been altered by an amount directly proportional to the distance from the adjacent end support to the given section, and the trajectory has undergone a linear transformation. The advantage accruing from plotting a concordant trajectory is shown in the next calculation procedure.

DESIGN OF TRAJECTORY TO OBTAIN ASSIGNED PRESTRESS MOMENTS

The prestress moments shown in Fig. 156 are to be obtained by applying an initial prestressing force of 72 kips (320.3 kN) with an eccentricity of -2 in $(-50.8$ mm) at B. Design the trajectory.

Calculation Procedure:

1. Plot a concordant trajectory

Set $e = M_p/F_i$, or $e_a = -3200(12)/72,000 = -0.533$ in $(-13.5382$ mm$)$; $e_d = +0.784$ in $(19.9136$ mm$)$; $e_b = -1.632$ in $(-41.4528$ mm$)$; $e_e = +0.837$ in $(21.2598$ mm$)$; $e_c = -0.800$ in $(-20.32$ mm$)$.

2. Set e_b = desired eccentricity, and transform the trajectory linearly

Thus, $e_a = -0.533$ in $(-13.5382$ mm$)$; $e_c = -0.800$ in $(-20.32$ mm$)$; $e_d = +0.784 - \frac{1}{2}(2.000 - 1.632) = +0.600$ in $(+15.24$ mm$)$; $e_e = +0.837 - 0.184 = +0.653$ in $(+16.5862$ mm$)$.

EFFECT OF VARYING ECCENTRICITY AT END SUPPORT

For the beam in Fig. 154, consider that the parabolic trajectory in span AB is displaced thus: e_b is held constant as e_a is changed to -0.72 in $(-18.288$ mm$)$, the eccentricity at every intermediate section being decreased algebraically by an amount directly proportional to the distance from that section to B. Compute the prestress moment at the supports and at midspan caused by a prestressing force of 96 kips (427.0 kN).

Calculation Procedure:

1. Apply the revised value of e_a; repeat the calculations of the earlier procedure

Thus, $M_{pa} = 5760$ ft·lb $(7810.6$ N·m$)$; $M_{pd} = -3680$ ft·lb $(-4990.1$ N·m$)$; $M_{pb} = 9280$ ft·lb $(12,583.7$ N·m$)$; $M_{pe} = -5280$ ft·lb $(-7159.7$ N·m$)$; $M_{pc} = 4800$ ft·lb $(6508.8$ N·m$)$.

The change in prestress moment caused by the displacement of the trajectory varies linearly across each span. Figure 158 compares the original and revised moments along span AB. This constitutes method 1.

2. Replace the prestressing system with two hypothetical systems that jointly induce eccentricity moments identical with those of the true system

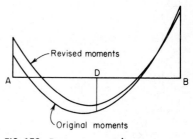

FIG. 158   Prestress-moment diagrams.

This constitutes method 2. For this purpose, consider the beam to be subjected to two prestressing forces of 96 kips (427.0 kN) each. One has the parabolic trajectory described in the earlier procedure; the other has a trajectory that is linear in each span, the eccentricities being $e_a = -0.72 - (-0.40) = -0.32$ in $(-8.128$ mm$)$, $e_b = 0$, and $e_c = 0$.

3. Evaluate the prestress moments induced by the hypothetical system having the linear trajectory

The tendons exert a force on the concrete at A, B, and C, but only the force at A causes bending moment.

Thus, $M_{pa} = -F_i e_a = -96,000(-0.32)/12 = 2560$ ft·lb $(3471.4$ N·m$)$. Also, $M_{pa}L_1 + 2M_{pb}(L_1 + L_2) + M_{pc}L_2 = 0$. But $M_{pc} = 0$; therefore, $M_{pb} = -512$ ft·lb $(-694.3$ N·m$)$; $M_{pd} = \frac{1}{2}(2560 - 512) = 1024$ ft·lb $(1388.5$ N·m$)$; $M_{pe} = \frac{1}{2}(-512) = -256$ ft·lb $(-347.1$ N·m$)$.

4. Find the true prestress moments by superposing the two hypothetical systems

Thus $M_{pa} = 3200 + 2560 = 5760$ ft·lb $(7810.6$ N·m$)$; $M_{pd} = -4704 + 1024 = -3680$ ft·lb $(-4990.1$ N·m$)$; $M_{pb} = 9792 - 512 = 9280$ ft·lb $(12,583.7$ N·m$)$; $M_{pe} = -5024 - 256 = -5280$ ft·lb $(-7159.7$ N·m$)$; $M_{pc} = 4800$ ft·lb $(6508.8$ N·m$)$.

DESIGN OF TRAJECTORY FOR A TWO-SPAN CONTINUOUS BEAM

A T beam that is continuous across two spans of 120 ft (36.6 m) each is to carry a uniformly distributed live load of 880 lb/lin ft (12,842.6 N/m). The cross section has these properties: $A = 1440$ in^2 (9290.8 cm^2); $I = 752,000$ in^4 (3130.05 dm^4); $y_b = 50.6$ in (1285.24 mm); $y_t = 23.4$ in

(594.36 mm). The allowable stresses are: *initial*, $+2400$ and -60 lb/in^2 ($+16,548.0$ and -413.7 kPa); *final*, $+2250$ and -60 lb/in^2 ($+15,513.8$ and 413.7 kPa). Assume that the minimum possible distance from the extremity of the section to the centroidal axis of the prestressing steel is 9 in (228.6 mm). Determine the magnitude of the prestressing force, and design the parabolic trajectory (*a*) using solely prestressed reinforcement; (*b*) using a combination of prestressed and non-prestressed reinforcement.

Calculation Procedure:

1. Compute the section moduli, kern distances, and beam weight

For part *a*, an exact design method consists of these steps: First, write equations for the prestress moment, beam-weight moment, maximum and minimum potential superimposed-load moment, expressing each moment in terms of the distance from a given section to the adjacent exterior support. Second, apply these equations to identify the sections at which the initial and final stresses are critical. Third, design the prestressing system to restrict the critical stresses to their allowable range. Whereas the exact method is not laborious when applied to a prismatic beam carrying uniform loads, this procedure adopts the conventional, simplified method for illustrative purposes. This consists of dividing each span into a suitable number of intervals and analyzing the stresses at each boundary section.

For simplicity, set the eccentricity at the ends equal to zero. The trajectory will be symmetric about the interior support, and the vertical component w of the force exerted by the tendons on the concrete in a unit longitudinal distance will be uniform across the entire length of member. Therefore, the prestress-moment diagram has the same form as the bending-moment diagram of a nonprestressed prismatic beam continuous over two equal spans and subjected to a uniform load across its entire length. It follows as a corollary that the prestress moments at the boundary sections previously referred to have specific *relative* values, although their absolute values are functions of the prestressing force and its trajectory.

The following steps constitute a methodical procedure: Evaluate the relative prestress moments, and select a trajectory having ordinates directly proportional to these moments. The trajectory thus fashioned is concordant. Compute the prestressing force required to restrict the stresses to the allowable range. Then transform the concordant trajectory linearly to secure one that lies entirely within the confines of the section. Although the number of satisfactory concordant trajectories is infinite, the one to be selected is that which requires the minimum prestressing force. Therefore, the selection of the trajectory and the calculation of F_i are blended into one operation.

Divide the left span into five intervals, as shown in Fig. 159. (The greater the number of intervals chosen, the more reliable are the results.)

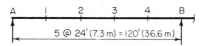

FIG. 159   Division of span into intervals.

Computing the moduli, kern distances, and beam weight gives $S_b = 14,860$ in^3 (243,555.4 cm^3); $S_t = 32,140$ in^3 (526,774.6 cm^3); $k_b = 22.32$ in (566.928 mm); $k_t = 10.32$ in (262.128 mm); $w_w = 1500$ lb/lin ft (21,890.9 N/m).

2. Record the bending-moment coefficients C_1, C_2, and C_3

Use Table 9 to record these coefficients at the boundary sections. The subscripts refer to these conditions of loading: 1, load on entire left span and none on right span; 2, load on entire right span and none on left span; 3, load on entire length of beam.

To obtain these coefficients, refer to the AISC *Manual*, case 29, which represents condition 1. Thus, $R_1 = (\%_6)wL$; $R_3 = -(\%_6)wL$. At section 3, for example, $M_1 = (\%_6)wL(0.6L) - \frac{1}{2}w(0.6L)^2 = [7(0.6) - 8(0.36)]wL^2/16 = 0.0825wL^2$; $C_1 = M_1/(wL^2) = +0.0825$.

To obtain condition 2, interchange R_1 and R_3. At section 3, $M_2 = -(\%_6)wL(0.6L) = -0.0375wL^2$; $C_2 = -0.0375$; $C_3 = C_1 + C_2 = +0.0825 - 0.0375 = +0.0450$.

These moment coefficients may be applied without appreciable error to find the maximum

TABLE 9 Calculations for Two-Span Beam: Part *a*

Section	1	2	3	4	B
1 C_1	+0.0675	+0.0950	+0.0825	+0.0300	−0.0625
2 C_2	−0.0125	−0.0250	−0.0375	−0.0500	−0.0625
3 C_3	+0.0550	+0.0700	+0.0450	−0.0200	−0.1250
4 f_{bw}, lb/in²	−959	−1,221	−785	+349	+2,180
(kPa)	(−6,611)	(−8,418)	(−5,412)	(+2406)	(+15,029)
5 f_{bs1}, lb/in² (kPa)	−691	−972	−844	−307	+640
	(−4,764)	(−6,701)	(−5,819)	(−2,116)	(+4,412)
6 f_{bs2}, lb/in² (kPa)	+128	+256	+384	+512	+640
	(+882)	(+1,765)	(+2,647)	(+3,530)	(+4,412)
7 f_{tw}, lb/in²	+444	+565	+363	−161	−1,008
(kPa)	(+3,060)	(+3895)	(+2,503)	(−1,110)	(−6,949)
8 f_{ts1}, lb/in²	+319	+450	+390	+142	−296
(kPa)	(+2,199)	(+3,102)	(+2689)	(+979)	(−2,041)
9 f_{ts2}, lb/in²	−59	−118	−177	−237	−296
(kPa)	(−407)	(−813)	(−1,220)	(−1,634)	(−2,041)
10 e_{con}, in	+17.19	+21.87	+14.06	−6.25	−39.05
(mm)	(+436.6)	(+555.5)	(+357.1)	(−158.8)	(−991.9)
11 f_{bp}, lb/in²	+2,148	+2,513	+1,903	+318	−2,243
(kPa)	(+14,808)	(+17,325)	(+13,119)	(+2,192)	(−15,463)
12 f_{tp}, lb/in²	+185	+16	+298	+1,031	+2,215
(kPa)	(+128)	(+110)	(+2,054)	(+7,108)	(+15,270)
$0.85f_{bp}$, 13 lb/in²	+1,826	+2,136	+1,618	+270	−1,906
(kPa)	(+12,588)	(+14,726)	(+11,154)	(+1,861)	(−13,140)
$0.85f_{tp}$, 14 lb/in²	+157	+14	+253	+876	+1,883
(kPa)	(+1,082)	(+97)	(+1,744)	(+6,039)	(+12,981)

At midspan: $C_3 = +0.0625$ and $e_{con} = +19.53$ in (496.1 mm)

and minimum potential live-load bending moments at the respective sections. The values of C_3 also represent the relative eccentricities of a concordant trajectory.

Since the gravity loads induce the maximum positive moment at section 2 and the maximum negative moment at section B, the prestressing force and its trajectory will be designed to satisfy the stress requirements at these two sections. (However, the stresses at all boundary sections will be checked.) The Magnel diagram for section 2 is similar to that in Fig. 141, but that for section B is much different.

3. Compute the value of C_3 at midspan

Thus, $C_3 = +0.0625$.

4. Apply the moment coefficients to find the gravity-load stresses

Record the results in Table 9. Thus $M_w = C_3(1500)(120)^2(12) = 259,200,000C_3$ in·lb (29.3C_3 kN·m); $f_{bw} = −259,200,000C_3/14,860 = −17,440C_3$; $f_{bs1} = −10,230C_1$; $f_{bs2} = −10,230C_2$; $f_{tw} = 8065C_3$; $f_{ts1} = 4731C_1$; $f_{ts2} = 4731C_2$.

Since S_t far exceeds S_b, it is manifest that the prestressing force must be designed to confine the bottom-fiber stresses to the allowable range.

5. Consider that a concordant trajectory has been plotted; express the eccentricity at section B relative to that at section 2

Thus, $e_b/e_2 = −0.1250/+0.0700 = −1.786$; therefore, $e_b = −1.786e_2$.

6. Determine the allowable range of values of f_{bp} at sections 2 and B

Refer to Fig. 160. At section 2, $f_{bp} \leq +3621$ lb/in² (+24,966.8 kPa), Eq. *a*; $0.85f_{bp} \geq 1221 + 972 − 60$; therefore, $f_{bp} \geq +2509$ lb/in² (+17,299.5 kPa), Eq *b*. At section B, $f_{bp} \geq −2240$

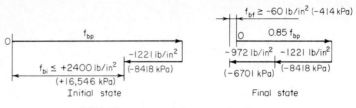

(a) Limiting values of f_{bp} at section 2

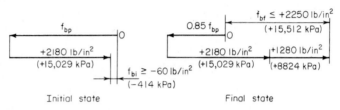

(b) Limiting values of f_{bp} at section B

FIG. 160

lb/in^2 ($-15{,}444.8$ kPa), Eq. c; $0.85f_{bp} \leq -(2180 + 1280) + 2250$; $f_{bp} \leq -1424$ lb/in^2 (-9818.5 kPa), Eq. d.

7. Substitute numerical values in Eq. 56, expressing e_b in terms of e_2

The values obtained are $1/F_i \geq (k_t + e_2)/(3621 S_b)$, Eq. a'; $1/F_i \leq (k_t + e_2)/(2509 S_b)$, Eq. b'; $1/F_i \geq (1.786e_2 - k_t)/(2240 S_b)$, Eq. c'; $1/F_i \leq (1.786e_2 - k_t)/(1424 S_b)$, Eq. d'.

8. Obtain the composite Magnel diagram

Considering the relations in step 7 as equalities, plot the straight lines representing them to obtain the composite Magnel diagram in Fig. 161. The slopes of the lines have these relative values: $m_a = 1/3621$; $m_b = 1/2509$; $m_c = 1.786/2240 = 1/1254$; $m_d = 1.786/1424 = 1/797$. The shaded area bounded by these lines represents the region of permissible sets of values of e_2 and $1/F_i$.

9. Calculate the minimum allowable value of F_i and the corresponding value of e_2

In the composite Magnel diagram, this set of values is represented by point A. Therefore, consider Eqs. b' and c' as equalities, and solve for the unknowns. Or, $(10.32 + e_2)/2509 = (1.786e_2 - 10.32)/2240$; solving gives $e_2 = 21.87$ in (555.5 mm) and $F_i = 1{,}160{,}000$ lb (5,159,680.0 N).

10. Plot the concordant trajectory

Do this by applying the values of C_3 appearing in Table 9; for example, $e_1 = +21.87(0.0550)/0.0700 = +17.19$ in (436.626 mm). At midspan, $e_m = +21.87(0.0625)/0.0700 = +19.53$ in (496.062 mm).

Record the eccentricities on line 10 of the table. It is apparent that this concordant trajectory is satisfactory in the respect that it may be linearly transformed to one falling within the confines of the section; this is proved in step 14.

11. Apply Eq. 56 to find f_{bp} and f_{tp}

Record the results in Table 9. For example, at section 1, $f_{bp} = 1{,}160{,}000(10.32 + 17.19)/14{,}860 = +2148$ lb/in^2 ($+14{,}810.5$ kPa); $f_{tp} = 1{,}160{,}000(22.32 - 17.19)/32{,}140 = +185$ lb/in^2 ($+1275.6$ kPa).

12. Multiply the values of f_{bp} and f_{tp} by 0.85, and record the results

These results appear in Table 9.

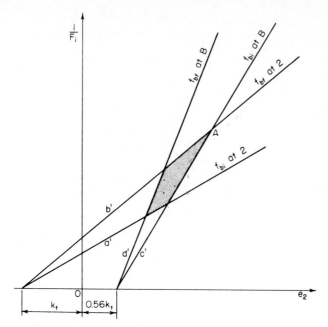

FIG. 161 Composite Magnel diagram.

13. Investigate the stresses at every boundary section

In calculating the final stresses, apply the live-load stress that produces a more critical condition. Thus, at section 1, $f_{bi} = -959 + 2148 = +1189$ lb/in^2 (+8198.2 kPa); $f_{bf} = -959 - 691 + 1826 = +176$ lb/in^2 (+1213.6 kPa); $f_{ti} = +444 + 185 = +629$ lb/in^2 (+4337.0 kPa); $f_{tf} = +444 + 319 + 157 = +920$ lb/in^2 (+6343.4 kPa). At section 2: $f_{bi} = -1221 + 2513 = +1292$ lb/in^2 (+8908.3 kPa); $f_{bf} = -1221 - 972 + 2136 = -57$ lb/in^2 (−393.0 kPa); $f_{ti} = +565 + 16 = +581$ lb/in^2 (+4006.0 kPa); $f_{tf} = +565 + 450 + 14 = +1029$ lb/in^2 (+7095.0 kPa). At section 3: $f_{bi} = -785 + 1903 = +1118$ lb/in^2 (+7706.8 kPa); $f_{bf} = -785 - 844 + 1618 = -11$ lb/in^2 (−75.8 kPa); $f_{ti} = +363 + 298 = +661$ lb/in^2 (+4558.0 kPa); $f_{tf} = +363 + 390 + 253 = +1006$ lb/in^2 (+6936.4 kPa). At section 4: $f_{bi} = +349 + 318 = +667$ lb/in^2 (+4599.0 kPa); $f_{bf} = +349 - 307 + 270 = +312$ lb/in^2 (+2151.2 kPa); or $f_{bf} = +349 + 512 + 270 = +1131$ lb/in^2 (7798.2 kPa); $f_{ti} = -161 + 1031 = +870$ lb/in^2 (+5998.7 kPa); $f_{tf} = -161 - 237 + 876 = +478$ lb/in^2 (+3295.8 kPa), or $f_{tf} = -161 + 142 + 876 = +857$ lb/in^2 (+5909.0 kPa). At section B: $f_{bi} = +2180 - 2243 = -63$ lb/in^2 (−434.4 kPa); $f_{bf} = +2180 + 1280 - 1906 = +1554$ lb/in^2 (+10,714.8 kPa); $f_{ti} = -1008 + 2215 = +1207$ lb/in^2 (+8322.3 kPa); $f_{tf} = -1008 - 592 + 1883 = +283$ lb/in^2 (+1951.3 kPa).

In all instances, the stresses lie within the allowable range.

14. Establish the true trajectory by means of a linear transformation

The imposed limits are $e_{max} = y_b - 9 = 41.6$ in (1056.6 mm), $e_{min} = -(y_t - 9) = -14.4$ in (−365.76 mm).

Any trajectory that falls between these limits and that is obtained by linearly transforming the concordant trajectory is satisfactory. Set $e_b = -14$ in (−355.6 mm), and compute the eccentricity at midspan and the maximum eccentricity.

Thus, $e_m = +19.53 + \frac{1}{2}(39.05 - 14) = +32.06$ in (814.324 mm). By Eq. 70b, $e_s = -(\frac{1}{8})$ $(-4 \times 32.06 - 14)^2/(-2 \times 32.06 - 14) = +32.4$ in (+823.0 mm) < 41.6 in (1056.6 mm). This is acceptable. This constitutes the solution to part a of the procedure. Steps 15 through 20 constitute the solution to part b.

15. *Assign eccentricities to the true trajectory, and check the maximum eccentricity*

The preceding calculation shows that the maximum eccentricity is considerably below the upper limit set by the beam dimensions. Refer to Fig. 161. If the restrictions imposed by line c' are removed, e_2 may be increased to the value corresponding to a maximum eccentricity of 41.6 in (1056.6 mm), and the value of F_i is thereby reduced. This revised set of values will cause an excessive initial tensile stress at B, but the condition can be remedied by supplying nonprestressed reinforcement over the interior support. Since the excess tension induced by F_i extends across a comparatively short distance, the savings accruing from the reduction in prestressing force will more than offset the cost of the added reinforcement.

Assigning the following eccentricities to the true trajectory and checking the maximum eccentricity by applying Eq. 70b, we get $e_a = 0$; $e_m = +41$ in (1041.4 mm); $e_b = -14$ in (-355.6 mm); $e_s = -(\%)(-4 \times 41 - 14)^2/(-2 \times 41 - 14) = +41.3$ in (1049.02 mm). This is acceptable.

16. *To analyze the stresses, obtain a hypothetical concordant trajectory by linearly transforming the true trajectory.*

Let y denote the upward displacement at B. Apply the coefficients C_3 to find the eccentricities of the hypothetical trajectory. Thus, $e_m/e_b = (41 - \frac{1}{2}y)/(-14 - y) = +0.0625/-0.1250$; $y = 34$ in (863.6 mm); $e_a = 0$; $e_m = +24$ in (609.6 mm); $e_b = -48$ in (-1219.2 mm); $e_1 = -48$ ($+0.0550)/-0.1250 = +21.12$ in (536.448 mm); $e_2 = +26.88$ in (682.752 mm); $e_3 = +17.28$ in (438.912 mm); $e_4 = -7.68$ in (-195.072 mm).

17. *Evaluate F_i by substituting in relation (b') of step 7*

Thus, $F_i = 2509(14,860)/(10.32 + 26.88) = 1,000,000$ lb (4448 kN). Hence, the introduction of nonprestressed reinforcement served to reduce the prestressing force by 14 percent.

18. *Calculate the prestresses at every boundary section; then find the stresses at transfer and under design load*

Record the results in Table 10. (At sections 1 through 4, the final stresses were determined by applying the values on lines 5 and 8 in Table 9. The slight discrepancy between the final stress at 2 and the allowable value of -60 lb/in^2 (-413.7 kPa) arises from the degree of precision in the calculations.)

With the exception of f_{bi} at B, all stresses at the boundary sections lie within the allowable range.

TABLE 10 Calculations for Two-Span Continuous Beam: Part b

Section	1	2	3	4	B
e_{con}, in (mm)	+21.12	+26.88	+17.28	−7.68	−48.00
	(536.4)	(+682.8)	(+438.9)	(−195.1)	(−1,219.2)
f_{bp}, lb/in^2 (kPa)	+2,116	+2,503	+1,857	+178	−2,535
	(+14,588)	(+17,256)	(+12,802)	(+1,227)	(−17,476)
f_{tp}, lb/in^2 (kPa)	+37	−142	+157	+933	+2188
	(+255)	(−979)	(+1,082)	(+6,660)	(+15,084)
$0.85f_{bp}$, lb/in^2 (kPa)	+1,799	+2,128	+1,578	+151	−2,155
	(+12,402)	(+14,670)	(+10,879)	(+1,041)	(−14,857)
$0.85f_{tp}$, lb/in^2 (kPa)	+31	−121	+133	+793	+1,860
	(+214)	(−834)	(+917)	(+5,467)	(+12,823)
f_{bi}, lb/in^2 (kPa)	+1,157	+1,282	+1,072	+527	−355
	(+7,976)	(+8,838)	(+7,390)	(+3,633)	(−2,447)
f_{bf}, lb/in^2 (kPa)	+149	−65	−51	+193	+1,305
	(+1,027)	(−448)	(−352)	(+1,331)	(+8,997)
f_{ti}, lb/in^2 (kPa)	+481	+423	+520	+772	+1,180
	(+3,316)	(+2,916)	(+3,585)	(+5,322)	(+8,135)
f_{tf}, lb/in^2 (kPa)	+794	+894	+886	+774	+260
	(+5,474)	(+6,163)	(+6108)	(+5,336)	(+1,792)

19.   Locate the section at which $f_{bi} = -60$ lb/in^2 $(-413.7$ kPa)

Since f_{bp} and f_{bw} vary parabolically across the span, their sum f_{bi} also varies in this manner. Let x denote the distance from the interior support to a given section. Apply Eq. 66 to find the equation for f_{bi}, using the initial-stress values at sections B, 3, and 1. Or, $-355 - 2 \times 1072 + 1157 = -1342$ (-9253.1 kPa); $3(-355) - 4(1072) + 1157 = -4196$ ($-28,931.4$ kPa); $f_{bi} = -2684(x/96)^2 + 4196x/96 - 355$. When $f_{bi} = -60$ (-413.7), $x = 7.08$ ft (2.15 m). The tensile stress at transfer is therefore excessive in an interval of only 14.16 ft (4.32 m).

20.   Design the nonprestressed reinforcement over the interior support

As in the preceding procedures, the member must be investigated for ultimate-strength capacity. The calculation pertaining to any quantity that varies parabolically across the span may be readily checked by verifying that the values at uniformly spaced sections have equal "second differences." For example, with respect to the values of f_{bi} recorded in Table 10, the verification is:

$$+1157 \qquad +1282 \qquad +1072 \qquad +527 \qquad -355$$

$$-125 \qquad +210 \qquad +545 \qquad +882$$

$$+335 \qquad +335 \qquad +337$$

The values on the second and third lines represent the differences between successive values on the preceding line.

REACTIONS FOR A CONTINUOUS BEAM

With reference to the beam in the previous calculation procedure, compute the reactions at the supports caused by the initial prestressing force designed in part a.

Calculation Procedure:

1.   Determine what causes the reactions at the supports

As shown in Fig. 155, the reactions at the supports result from the continuity at B, and $R_a = M_{kb}/L$.

2.   Compute the continuity moment at B; then find the reactions

Thus, $M_p = -F_i e + M_k = -F_i e_{con}$; $M_k = F_i(e - e_{con}) = 1160(-14 + 39.05) = 29,060$ in·kips (3283 kN·m). $R_a = 29,060/[120(12)] = 20.2$ kips (89.8 kN); $R_B = -40.4$ kips (-179.8 kN).

Design of Highway Bridges

Where a bridge is supported by steel trusses, the stresses in the truss members are determined by applying the rules formulated in the truss calculation procedures given earlier in this handbook.

The following procedures show the design of a highway bridge supported by concrete or steel girders. Except for the deviations indicated, the *Standard Specifications for Highway Bridges*, published by the American Association of State Highway and Transportation Officials (AASHTO), are applied.

The AASHTO *Specification* recognizes two forms of truck loading: the H loading, and the HS loading. Both are illustrated in the *Specification*. For a bridge of relatively long span, it is necessary to consider the possibility that several trucks will be present simultaneously. To approximate this condition, the AASHTO *Specification* offers various lane loadings, and it requires that the bridge be designed for the lane loading if this yields greater bending moments and shears than does the corresponding truck loading.

In designing the bridge members, it is necessary to modify the wheel loads to allow for the effects of dynamic loading and the lateral distribution of loads resulting from the rigidity of the floor slab.

The basic notational system is: DF = factor for lateral distribution of wheel loads; IF = impact factor; P = resultant of group of concentrated loads.

The term *live load* as used in the following material refers to the wheel load after correction for distribution but before correction for impact.

DESIGN OF A T-BEAM BRIDGE

A highway bridge consisting of a concrete slab and concrete girders is to be designed for these conditions: loading, HS20-44; clear width, 28 ft (8.5 m); effective span, 54 ft (16.5 m); concrete strength, 3000 lb/in² (20,685 kPa); reinforcement, intermediate grade. The slab and girders will be poured monolithically, and the slab will include a ¾ in (19.05 mm) wearing surface. In addition, the design is to make an allowance of 15 lb/ft² (718 N/m²) for future paving. Design the slab and the cross section of the interior girders.

Calculation Procedure:

1. Record the allowable stresses and modular ratio given in the AASHTO Specification

Refer to Fig. 162, which shows the spacing of the girders and the dimensions of the members. The sizes were obtained by a trial-and-error method. Values from the *Specification* are: $n = 10$

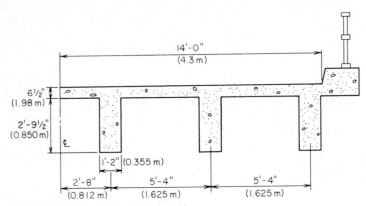

FIG. 162 Transverse section of T-beam bridge.

in stress calculations; $f_c = 0.4f'_c = 1200$ lb/in² (8274 kPa); for beams with web reinforcement, $v_{max} = 0.075f'_c = 225$ lb/in² (1551.4 kPa); $f_s = 20,000$ lb/in² (137.9 MPa); $u = 0.10f'_c = 300$ lb/in² (2068.5 kPa).

2. Compute the design coefficients associated with balanced design

Thus, $k = 1200/(1200 + 2000) = 0.375$, using Eq. 21. Using Eq. 22, $j = 1 - 0.125 = 0.875$. By Eq. 32, $K = \frac{1}{2}(1200)(0.375)(0.875) = 197$ lb/in² (1358.3 kPa).

3. Establish the wheel loads and critical spacing associated with the designated vehicular loading

As shown in the AASHTO *Specification*, the wheel-load system comprises two loads of 16 kips (71.2 kN) each and one load of 4 kips (17.8 kN). Since the girders are simply supported, an axle spacing of 14 ft (4.3 m) will induce the maximum shear and bending moment in these members.

4. Verify that the slab size is adequate and design the reinforcement

The AASHTO *Specification* does not present moment coefficients for the design of continuous members. The positive and negative reinforcement will be made identical, using straight bars for both. Apply a coefficient of ¹/₁₀ in computing the dead-load moment. The *Specification* provides that the span length S of a slab continuous over more than two supports be taken as the clear distance between supports.

In computing the effective depth, disregard the wearing surface, assume the use of No. 6 bars, and allow 1 in (25.4 mm) for insulation, as required by AASHTO. Then, $d = 6.5 - 0.75 - 1.0 - 0.38 = 4.37$ in (110.998 mm); $w_{DL} = (6.5/12)(150) + 15 = 96$ lb/lin ft (1401 N/m); $M_{DL} = (\frac{1}{10})w_{DL}S^2 = (\frac{1}{10})(96)(4.17)^2 = 167$ ft·lb (226 N·m); $M_{LL} = 0.8(S + 2)P_{20}/32$, by AASHTO, or $M_{LL} = 0.8(6.17)(16,000)/32 = 2467$ ft·lb (3345 N·m). Also by AASHTO, IF = 0.30; $M_{total} =$

12(167 + 1.30 × 2467) = 40,500 in·lb (4.6 kN·m). The moment corresponding to balanced design is $M_b = K_b bd^2 = 197(12)(4.37)^2 = 45,100$ in·lb (5.1 kN·m). The concrete section is therefore excessive, but a 6-in (152.4-mm) slab would be inadequate. The steel is stressed to capacity at design load. Or, $A_s = 40,500/(20,000 × 0.875 × 4.37) = 0.53$ in² (3.4 cm²). Use No. 6 bars 10 in (254 mm) on centers, top and bottom.

The transverse reinforcement resists the tension caused by thermal effects and by load distribution. By AASHTO, $A_t = 0.67(0.53) = 0.36$ in² (2.3 cm²). Use five No. 5 bars in each panel, for which $A_t = 1.55/4.17 = 0.37$ in² (2.4 cm²).

5. Calculate the maximum live-load bending moment in the interior girder caused by the moving-load group

The method of positioning the loads to evaluate this moment is described in an earlier calculation procedure in this handbook. The resultant, Fig. 163, has this location: $d = [16(14) + 4(28)]/(16$

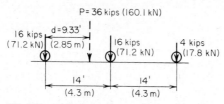

FIG. 163 Load group and its resultant.

+ 16 + 4) = 9.33 ft (2.85 m). Place the loads in the position shown in Fig. 164a. The maximum live-load bending moment occurs under the center load.

The AASHTO prescribes a distribution factor of $S/6$ in the present instance, where S denotes the spacing of girders. However, a factor of $S/5$ will be applied here. Then DF = 5.33/5 = 1.066; 16 × 1.066 = 17.06 kips (75.9 kN); 4 × 1.066 = 4.26 kips (18.9 kN); $P = 2(17.06) + 4.26 = 38.38$ kips (170.7 kN); $R_L = 38.38(29.33)/54 = 20.85$ kips (92.7 kN). The maximum live-load moment is $M_{LL} = 20.85(29.34) - 17.06(14) = 372.8$ ft·kips (505 kN·m).

6. Calculate the maximum live-load shear in the interior girder caused by the moving-load group

Place the loads in the position shown in Fig. 164b. Do not apply lateral distribution to the load at the support. Then, $V_{LL} = 16 + 17.06(40/54) + 4.26(26/54) = 30.69$ kips (136.5 kN).

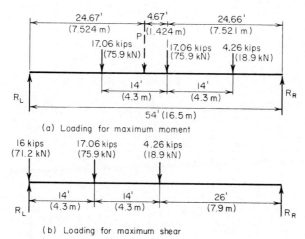

(a) Loading for maximum moment

(b) Loading for maximum shear

FIG. 164

7. Verify that the size of the girder is adequate and design the reinforcement

Thus, $w_{DL} = 5.33(96) + 14(33.5/144)(150) = 1000$ lb/lin ft (14.6 kN/m); $V_{DL} = 27$ kips (120.1 kN); $M_{DL} = (\frac{1}{8})(1)(54)^2 = 364.5$ ft·kips (494 kN·m). By AASHTO, IF $= 50/(54 + 125) = 0.28$; $V_{total} = 27 + 1.28(30.69) = 66.28$ kips (294.8 kN); $M_{total} = 12(364.5 + 1.28 \times 372.8) = 10,100$ in·kips (1141 N·m).

In establishing the effective depth of the girder, assume that No. 4 stirrups will be supplied and that the main reinforcement will consist of three rows of No. 11 bars. AASHTO requires 1½-in (38.1-mm) insulation for the stirrups and a clear distance of 1 in (25.4 mm) between rows of bars. However, 2 in (50.8 mm) of insulation will be provided in this instance, and the center-to-center spacing of rows will be taken as 2.5 times the bar diameter. Then, $d = 5.75 + 33.5 - 2 - 0.5 - 1.375(0.5 + 2.5) = 32.62$ in (828.548 mm); $v = V/b'jd = 66,280/(14 \times 0.875 \times 32.62) = 166 < 225$ lb/in² (1144.6 < 1551.4 kPa). This is acceptable.

Compute the moment capacity of the girder at balanced design. Since the concrete is poured monolithically, the girder and slab function as a T beam. Refer to Fig. 120 and its calculation procedure.

Thus, $k_b d = 0.375(32.62) = 12.23$ in (310.642 mm); $12.23 - 5.75 = 6.48$ in (164.592 mm). At balanced design, $f_{c1} = 1200(6.48/12.23) = 636$ lb/in² (4835.2 kPa). The effective flange width of the T beam as governed by AASHTO is 64 in (1625.6 mm); and $C_b = 5.75(64)(\frac{1}{2})(1.200 + 0.636) = 338$ kips (1503 kN); $jd = 32.62 - (5.75/3)(1200 + 2 \times 636)/(1200 + 636) = 30.04$ in (763.016 mm); $M_b = 338(30.04) = 10,150$ in·kips (1146 kN·m). The concrete section is therefore slightly excessive, and the steel is stressed to capacity, or $A_s = 10,100/20(30.04) = 16.8$ in² (108.4 cm²). Use 11 no. 11 bars, arranged in three rows.

AASHTO requires that the girders be tied together by diaphragms to obtain lateral rigidity of the structure.

COMPOSITE STEEL-AND-CONCRETE BRIDGE

The bridge shown in cross section in Fig. 165 is to carry an HS20-44 loading on an effective span of 74 ft 6 in (22.7 m). The structure will be unshored during construction. The concrete strength

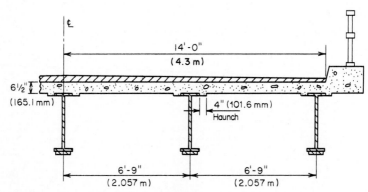

FIG. 165 Transverse section of composite bridge.

is 3000 lb/in² (20,685 kPa), and the entire slab is considered structurally effective; the allowable bending stress in the steel is 18,000 lb/in² (124.1 MPa). The dead load carried by the composite section is 250 lb/lin ft (3648 N/m). Preliminary design calculations indicate that the interior girder is to consist of 36WF150 (W36 × 150) and a cover plate 10 × 1½ in (254 × 38.1 mm) welded to the bottom flange. Determine whether the trial section is adequate and complete the design.

Calculation Procedure:

1. Record the relevant properties of the 36WF150

The design of a composite bridge consisting of a concrete slab and steel girders is governed by specific articles in the AASHTO *Specification*.

Composite behavior of the steel and concrete is achieved by adequately bonding the materials to function as a flexural unit. Loads that are present before the concrete has hardened are supported by the steel member alone; loads that are applied after hardening are supported by the composite member. Thus, the steel alone supports the concrete slab, and the steel and concrete jointly support the wearing surface.

Plastic flow of the concrete under sustained load generates a transfer of compressive stress from the concrete to the steel. Consequently, the stresses in the composite member caused by dead load are analyzed by using a modular ratio three times the value that applies for transient loads.

If a wide-flange shape is used without a cover plate, the neutral axis of the composite section is substantially above the center of the steel, and the stress in the top steel fiber is therefore far below that in the bottom fiber. Use of a cover plate depresses the neutral axis, reduces the disparity between these stresses, and thereby results in a more economical section. Let y' = distance from neutral axis of member to given point, in absolute value; $\bar{y}$ = distance from centroidal axis of WF shape to neutral axis of member. The subscripts b, ts, and tc refer to the bottom of member, top of steel, and top of concrete, respectively. The superscripts c and n refer to the composite and noncomposite member, respectively.

The relevant properties of the 36WF150 are A = 44.16 in^2 (284.920 cm^2); I = 9012 in^4 (37.511 dm^4); d = 35.84 in (910.336 mm); S = 503 in^3 (8244.2 cm^3); flange thickness = 1 in (25.4 mm), approximately.

2. Compute the section moduli of the noncomposite section where the cover plate is present

To do this, compute the static moment and moment of inertia of the section with respect to the center of the WF shape; record the results in Table 11. Refer to Fig. 166: $\bar{y}$ = $-280/59.16$ = -4.73 in (-120.142 mm); y'_b = $19.42 - 4.73$ = 14.69 in (373.126 mm); y'_{ts} = $17.92 + 4.73$ = 22.65 in (575.31 mm). By the moment-of-inertia equation, I = $5228 + 9012 - 59.16(4.73)^2$ = 12,916 in^4 (53.76 dm^4); S_b = 879 in^3 (14,406.8 cm^3); S_{ts} = 570 in^3 (9342.3 cm^3).

3. Transform the composite section, with cover plate included, to an equivalent homogeneous section of steel; compute the section moduli

In accordance with AASHTO, the effective flange width is 12(6.5) = 78 in (1981.2 mm). Using the method of an earlier calculation procedure, we see that when n = 30, $\bar{y}$ = 78/76.06 = 1.03

TABLE 11 Calculations for Girder with Cover Plate

	A	y	Ay	Ay^2	I_o
Noncomposite:					
36WF150	44.16	0	0	0	9,012
Cover plate	15.00	-18.67	-280	5,228	0
Total	59.16	......	-280	5,228	9,012
Composite, n = 30:					
Steel (total)	59.16	......	-280	5,228	9,012
Slab	16.90	21.17	358	7,574	60
Total	76.06	......	78	12,802	9,072
Composite, n = 10:					
Steel (total)	59.16	......	-280	5,228	9,012
Slab	50.70	21.17	1,073	22,722	179
Total	109.86	......	793	27,950	9,191

in (26.162 mm); $y_b' = 19.42 + 1.03 = 20.45$ in (519.43 mm); $y_{ts}' = 17.92 - 1.03 = 16.89$ in (429.006 mm); $y_{tc}' = 16.89 + 6.50 = 23.39$ in (594.106 mm); $I = 12,802 + 9072 - 76.06(1.03)^2 = 21,793$ in^4 (90.709 dm^4); $S_b = 1066$ in^3 (17,471.7 cm^3); $S_{ts} = 1,290$ in^3 (21,143.1 cm^3); $S_{tc} = 932$ in^3 (15,275.5 cm^3).

When $n = 10$: $y = 7.22$ in (183.388 mm); $y_b' = 26.64$ in (676.66 mm); $y_{ts}' = 10.70$ in (271.78 mm); $y_{tc}' = 17.20$ in (436.88 mm); $I = 27,950 + 9191 - 109.86(7.22)^2 = 31,414$ in^4 (130.7545 dm^4); $S_b = 1179$ in^3 (19,320.3 cm^3); $S_{ts} = 2936$ in^3 (48,121.0 cm^3); $S_{tc} = 1826$ in^3 (29,928.1 cm^3).

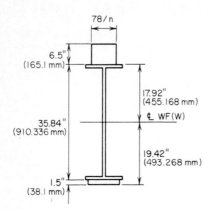

FIG. 166 Transformed section.

4. Transform the composite section, exclusive of the cover plate, to an equivalent homogeneous section of steel, and compute the values shown below

Thus, when $n = 30$, $y_b' = 23.78$ in (604.012 mm); $y_{ts}' = 12.06$ in (306.324 mm); $I = 14,549$ in^4 (60.557 dm^4); $S_b = 612$ in^3 (10,030.7 cm^3). When $n = 10$, $y_b' = 29.23$ in (742.442 mm); $y_{ts}' = 6.61$ in (167.894 mm); $I = 19,779$ in^4 (82.326 dm^4); $S_b = 677$ in^3 (11,096.0 cm^3).

5. Compute the dead load carried by the noncomposite member

Thus,

	lb/lin ft	N/m
Beam	150	2189.1
Cover plate	51	744.3
Slab: 0.54(6.75)(150)	547	7982.8
Haunch: 0.67(0.083)(150)	8	116.8
Diaphragms (approximate)	12	175.1
Shear connectors (approximate)	6	87.6
Total	774, say 780	11,383.2

6. Compute the maximum dead-load moments

Thus, $M_{DL}^c = (\frac{1}{8})(0.250)(74.5)^2(12) = 2080$ in·kips (235.00 kN·m); $M_{DL}^n = (\frac{1}{8})(0.780)(74.5)^2(12) = 6490$ in·kips (733.24 kN·m).

7. Compute the maximum live-load moment, with impact included

In accordance with the AASHTO, the distribution factor is DF $= 6.75/5.5 = 1.23$; IF $= 50/(74.5 + 125) = 0.251$, and $16(1.23)(1.251) = 24.62$ kips (109.510 kN); $4(1.23)(1.251) = 6.15$ kips (270.355 kN); $P_{LL+I} = 2(24.62) + 6.15 = 55.39$ kips (246.375 kN). Refer to Fig. 164a as a guide. Then, $M_{LL+I} = 12[(55.39 \times 39.58 \times 39.58/74.5) - 24.62(14)] = 9840$ in·kips (1111.7 kN·m).

For convenience, the foregoing results are summarized here:

	M, in·kips (kN·m)	S_b, in^3 (cm^3)	S_{ts}, in^3 (cm^3)	S_{tc}, in^3 (cm^3)
Noncomposite	6,490 (733.2)	879 (14,406.8)	570 (9,342.3)	
Composite, dead loads	2,080 (235.0)	1,066 (17,471.7)	1,290 (21,143.1)	932 (15,275.5)
Composite, moving loads	9,840 (1,111.7)	1,179 (19,323.8)	2,936 (48,121.0)	1,826 (29,928.1)

8. Compute the critical stresses in the member

To simplify the calculations, consider the sections of maximum live-load and dead-load stresses to be coincident. Then $f_b = 6490/879 + 2080/1066 + 9840/1179 = 17.68$ kips/in^2 (121.9 MPa); $f_{ts} = 6490/570 + 2080/1290 + 9840/2936 = 16.35$ kips/in^2 (112.7 MPa); $f_{tc} = 2080/(30 \times 932) + 9840/(10 \times 1826) = 0.61$ kips/in^2 (4.21 MPa). The section is therefore satisfactory.

9. Determine the theoretical length of cover plate

Let K denote the theoretical cutoff point at the left end. Let L_c = length of cover plate exclusive of the development length; b = distance from left support to K; $m = L_c/L$; d = distance from heavier exterior load to action line of resultant, as shown in Fig. 163; $r = 2d/L$.
From these definitions, $b = (L - L_c)/2 = L(1 - m)/2$; $m = 1 - b/(0.5L)$. The maximum moment at K due to live load and impact is

$$M_{LL+I} = \frac{(P_{LL+I}L)(1 - r + rm - m^2)}{4} \tag{72}$$

The diagram of dead-load moment is a parabola having its summit at midspan.
To locate K, equate the bottom-fiber stress immediately to the left of K, where the cover plate is inoperative, to its allowable value. Or, $(P_{LL+I}L)/4 = 55.39(74.5)(12)/4 = 12,380$ in·kips (1398.7 kN·m); $d = 9.33$ ft (2.844 m); $r = 18.67/74.5 = 0.251$; $6490(1 - m^2)/503 + 2080(1 - m^2)/612 + 12,380(0.749 + 0.251m - m^2)/677 = 18$ kips/in^2 (124.1 MPa); $m = 0.659$; $L_c = 0.659(74.5) = 49.10$ ft (14.97 m).
The plate must be extended toward each support and welded to the WF shape to develop its strength.

10. Verify the result obtained in step 9

Thus, $b = \frac{1}{2}(74.5 - 49.10) = 12.70$ ft (3.871 m). At K: $M_{DL}^n = 12(\frac{1}{2} \times 74.5 \times 0.780 \times 12.70 - \frac{1}{2} \times 0.780 \times 12.70^2) = 3672$ in·kips (414.86 kN·m); $M_{DL}^c = 3672(250/780) = 1177$ in·kips (132.98 kN·m). The maximum moment at K due to the moving-load system occurs when the heavier exterior load lies directly at this section. Also $M_{LL+I} = 55.39(74.5 - 12.70 - 9.33)(12.70)(12)/74.5 = 5945$ in·kips (671.7 kN·m); $f_b = 3672/503 + 1177/612 + 5945/677 = 18.0$ kips/in^2 (124.11 MPa). This is acceptable.

11. Compute V_{DL} and V_{LL+I} at the support and at K

At the support $V_{DL}^c = \frac{1}{2}(0.250 \times 74.5) = 9.31$ kips (41.411 kN); IF = 0.251.
Consider that the load at the support is not subject to distribution. By applying the necessary correction, the following is obtained: $V_{LL+I} = 55.39(74.5 - 9.33)/74.5 - 16(1.251)(0.23) = 43.85$ kips (195.045 kN). At K: $V_{DL}^c = 9.31 - 12.70(0.250) = 6.13$ kips (27.266 kN); IF = $50/(61.8 + 125) = 0.268$; $P_{LL+I} = 36(1.268)(1.23) = 56.15$ kips (249.755 kN); $V_{LL+I} = 56.15(74.5 - 12.70 - 9.33)/74.5 = 39.55$ kips (175.918 kN).

12. Select the shear connectors, and determine the allowable pitch p at the support and immediately to the right of K

Assume use of $\frac{3}{4}$-in (19.1-mm) studs, 4 in (101.6 mm) high, with four studs in each transverse row, as shown in Fig. 167. The capacity of a connector as established by AASHTO is $110d^2(f_c')^{0.5} = 110 \times 0.75^2(3000)^{0.5} = 3390$ lb (15,078.7 N). The capacity of a row of connectors = $4(3390) = 13,560$ lb (60,314.9 N).

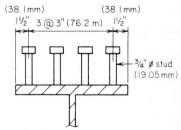

(38.1 mm)   (38.1 mm)
$1\frac{1}{2}$" 3 @ 3" (76.2 m) $1\frac{1}{2}$"

$\frac{3}{4}$" ⌀ stud (19.05 mm)

FIG. 167 Shear connectors.

The shear flow at the bottom of the slab is found by applying $q = VQ/I$, or $q_{DL}^c = 9310(16.90)(12.06 + 3.25)/14,549 = 166$ lb/lin in (29,071.0 N/m); $q_{LL+I} = 43,850(50.70)(6.61 + 3.25)/19,779 = 1108$ lb/in^2 (7639.7 kPa); $p = 13,560/(166 + 1108) = 10.6$ in (269.24 mm).
Directly to the right of K: $q_{DL}^c = 6130(16.90)(16.89 + 3.25)/21,793 = 96$ lb/lin in (16,812.2 N/m); $q_{LL+I} = 39,550(50.70)(10.70 + 3.25)/31,414 = 890$ lb/lin in (155,862.9 N/m); $p = 13,560/(96 + 890) = 13.8$ in (350.52 mm).

It is necessary to determine the allowable pitch at other sections and to devise a suitable spacing of connectors for the entire span.

13. Design the weld connecting the cover plate to the WF shape

The calculations for shear flow are similar to those in step 12. The live-load deflection of an unshored girder is generally far below the limit imposed by AASHTO. However, where an investigation is warranted, the deflection at midspan may be calculated by assuming, for simplicity, that the position of loads for maximum deflection coincides with the position for maximum moment. The theorem of reciprocal deflections, presented in an earlier calculation procedure, may conveniently be applied in calculating this deflection. The girders are usually tied together by diaphragms at the ends and at third points to obtain lateral rigidity of the structure.

Fluid Mechanics

Hydrostatics

The notational system used in hydrostatics is as follows: W = weight of floating body, lb (N); V = volume of displaced liquid, ft^3 (m^3); w = specific weight of liquid, lb/ft^3 (N/m^3); for water w = 62.4 lb/ft^3 (9802 N/m^3), unless another value is specified.

BUOYANCY AND FLOTATION

A timber member 12 ft (3.65 m) long with a cross-sectional area of 90 in^2 (580.7 cm^2) will be used as a buoy in saltwater. What volume of concrete must be fastened to one end so that 2 ft (60.96 cm) of the member will be above the surface? Use these specific weights: timber = 38 lb/ft^3 (5969 N/m^3); saltwater = 64 lb/ft^3 (10,053 N/m^3); concrete = 145 lb/ft^3 (22,777 N/m^3).

Calculation Procedure:

1. Express the weight of the body and the volume of the displaced liquid in terms of the volume of concrete required

Archimedes' principle states that a body immersed in a liquid is subjected to a vertical buoyant force equal to the weight of the displaced liquid. In accordance with the equations of equilibrium, the buoyant force on a floating body equals the weight of the body. Therefore,

$$W = Vw \tag{73}$$

Let x denote the volume of concrete. Then $W = (90/144)(12)(38) + 145x = 285 + 145x$; $V = (90/144)(12 - 2) + x = 6.25 + x$.

2. Substitute in Eq. 73 and solve for x

Thus, $285 + 145x = (6.25 + x)64$; $x = 1.42$ ft^3 (0.0402 m^3).

HYDROSTATIC FORCE ON A PLANE SURFACE

In Fig. 168, AB is the side of a vessel containing water, and CDE is a gate located in this plane. Find the magnitude and location of the resultant thrust of the water on the gate when the liquid surface is 2 ft (60.96 cm) above the apex.

Calculation Procedure:

1. State the equations for the resultant magnitude and position

In Fig. 168, FH denotes the centroidal axis of area CDE that is parallel to the liquid surface, and G denotes the point of application of the resultant force. Point G is termed the *pressure center*.

Let A = area of given surface, ft^2 (cm^2); P = hydrostatic force on given surface, lb (N); p_m = mean pressure on surface, lb/ft^2 (kPa); y_{CA} and y_{PC} = vertical distance from centroidal axis and pressure center, respectively, to liquid surface, ft (m); z_{CA} and z_{PC} = distance along plane of given surface from the centroidal axis and pressure center, respectively, to line of intersection of this plane and the liquid surface, ft (m); I_{CA} = moment of inertia of area with respect to its centroidal axis, ft⁴ (m⁴).

Consider an elemental surface of area dA at a vertical distance y below the liquid surface. The hydrostatic force dP on this element is normal to the surface and has the magnitude

$$dP = wy \, dA \tag{74}$$

By applying Eq. 74, develop the following equations for the magnitude and position of the resultant force on the entire surface:

$$P = wy_{CA}A \tag{75}$$

$$z_{PC} = \frac{I_{CA}}{Az_{CA}} + z_{CA} \tag{76}$$

2. Compute the required values, and solve the equations in step 1

Thus $A = \frac{1}{2}(5)(6) = 15$ ft² (1.39 m²); $y_{CA} = 2 + 4 \sin 60° = 5.464$ ft (166.543 cm); $z_{CA} = 2$ csc $60° + 4 = 6.309$ ft (192.3 cm); $I_{CA}/A = (bd^3/36)/(bd/2) = d^2/18 = 2$ ft² (0.186 m²); $P = 62.4(5.464)(15) = 5114$ lb (22,747 N); $z_{PC} = 2/6.309 + 6.309 = 6.626$ ft (201.960 cm); $y_{PC} = 6.626 \sin 60° = 5.738$ ft (174.894 cm). By symmetry, the pressure center lies on the centroidal axis through C.

An alternative equation for P is

$$P = p_m A \tag{77}$$

Equation 75 shows that the mean pressure occurs at the centroid of the area. The above two steps constitute method 1 for solving this problem. The next three steps constitute method 2.

3. Now construct the pressure "prism" associated with the area

In Fig. 169, construct the pressure prism associated with area CDE. The pressures are as follows: at apex, $p = 2w$; at base, $p = (2 + 6 \sin 60°)w = 7.196w$.

The force P equals the volume of this prism, and its action line lies on the centroidal plane parallel to the base. For convenience, resolve this prism into a triangular prism and rectangular pyramid, as shown.

4. Determine P by computing the volume of the pressure prism

Thus, $P = Aw[2 + \frac{2}{3}(5.196)] = Aw(2 + 3.464) = 15(62.4)(5.464) = 5114$ lb (22,747 N).

5. Find the location of the resultant thrust

Compute the distance h from the top line to the centroidal plane. Then find y_{PC}. Or, $h = [2(\frac{1}{2})(6) + 3.464(\frac{2}{3})(6)]/5.464 = 4.317$ ft (131.582 cm); $y_{PC} = 2 + 4.317 \sin 60° = 5.738$ ft (174.894 cm).

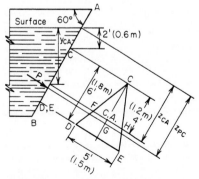

FIG. 168 Hydrostatic thrust on plane surface.

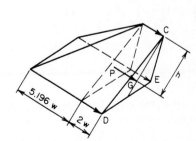

FIG. 169 Pressure prism.

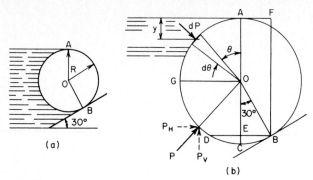

FIG. 170

HYDROSTATIC FORCE ON A CURVED SURFACE

The cylinder in Fig. 170a rests on an inclined plane and is immersed in liquid up to its top, as shown. Find the hydrostatic force on a 1-ft (30.48-cm) length of cylinder in terms of w and the radius R; locate the pressure center.

Calculation Procedure:

1. Evaluate the horizontal and vertical component of the force dP on an elemental surface having a central angle $d\theta$

Refer to Fig. 170b. Adopt this sign convention: A horizontal force is positive if directed to the right; a vertical force is positive if directed upward. The first three steps constitute method 1.

Evaluating dP yields $dP_H = wR^2(\sin\theta - \sin\theta\cos\theta)\,d\theta$; $dP_V = wR^2(-\cos\theta + \cos^2\theta)\,d\theta$.

2. Integrate these equations to obtain the resultant forces P_H and P_V; then find P

Here, $P_H = wR^2(-\cos\theta + \tfrac{1}{2}\cos^2\theta)]_0^{7\pi/6} = wR^2[-(-0.866-1) + \tfrac{1}{2}(0.75-1)] = 1.741wR^2$, to right; $P_V = wR^2(-\sin\theta + \tfrac{1}{2}\theta + \tfrac{1}{4}\sin 2\theta)]_0^{7\pi/6} = wR^2(0.5 + 1.833 + 0.217) = 2.550wR^2$, upward; $P = wR^2(1.741^2 + 2.550^2)^{0.5} = 3.087wR^2$.

3. Determine the value of θ at the pressure center

Since each elemental force dP passes through the center of the cylinder, the resultant force P also passes through the center. Thus, $\tan(180° - \theta_{PC}) = P_H/P_V = 1.741/2.550$; $\theta_{PC} = 145°41'$.

4. Evaluate P_H and P_V

Apply these principles: $P_H =$ force on an imaginary surface obtained by projecting the wetted surface on a vertical plane; $P_V = \pm$ weight of real or imaginary liquid lying between the wetted surface and the liquid surface. Use the plus sign if the *real* liquid lies below the wetted surface and the minus sign if it lies above this surface.

Then $P_H =$ force, to right, on $AC +$ force, to left, on $EC =$ force, to right, on AE; $AE = 1.866R$; $p_m = 0.933\,wR$; $P_H = 0.933\,wR(1.866R) = 1.741\,wR^2$; $P_V =$ weight of imaginary liquid above $GCB -$ weight of real liquid above $GA =$ weight of imaginary liquid in cylindrical sector $AOBG$ and in prismoid $AOBF$. Volume of sector $AOBG = [(7\pi/6)/(2\pi)](\pi R^2) = 1.833R^2$; volume of prismoid $AOBF = \tfrac{1}{2}(0.5R)(R + 1.866R) = 0.717R^2$; $P_V = wR^2(1.833 + 0.717) = 2.550wR^2$.

STABILITY OF A VESSEL

The boat in Fig. 171 is initially floating upright in freshwater. The total weight of the boat and cargo is 182 long tons (1813 kN); the center of gravity lies on the longitudinal (i.e., the fore-and-aft) axis of the boat and 8.6 ft (262.13 cm) above the bottom. A wind causes the boat to list through an angle of 6° while the cargo remains stationary relative to the boat. Compute the righting or upsetting moment (a) without applying any set equation; (b) by applying the equation for metacentric height.

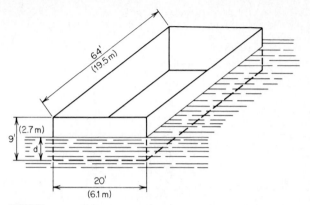

FIG. 171

Calculation Procedure:

1. Compute the displacement volume and draft when the boat is upright

The buoyant force passes through the center of gravity of the displaced liquid; this point is termed the *center of buoyancy*. Figure 172 shows the cross section of a boat rotated through an angle ϕ. The center of buoyancy for the upright position is B; B' is the center of buoyancy for the position shown, and G is the center of gravity of the boat and cargo.

In the position indicated in Fig. 172, the weight W and buoyant force R constitute a couple that tends to restore the boat to its upright position when the disturbing force is removed; their moment is therefore termed *righting*. When these forces constitute a couple that increases the rotation, their moment is said to be *upsetting*. The wedges OAC and $OA'C'$ are termed the *wedge of emersion* and *wedge of immersion*, respectively. Let h = horizontal displacement of center of buoyancy caused by rotation; h' = horizontal distance between centroids of wedge of emersion and wedge of immersion; V' = volume of wedge of emersion (or immersion). Then

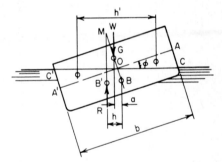

FIG. 172   Location of resultant forces on inclined vessel.

$$h = \frac{V'h'}{V} \qquad (78)$$

The displacement volume and the draft when the boat is upright are $W = 182(2240) = 407,700$ lb (1813 N); $V = W/w = 407,700/62.4 = 6530$ ft^3 (184.93 m^3); $d = 6530/[64(20)] = 5.10$ ft (155.448 cm).

2. Find h, using Eq. 78

Since ϕ is relatively small, apply this approximation: $h' = 2b/3 = 2(20)/3 = 13.33$ ft (406.298 cm), $h = \frac{1}{2}(10)(10 \tan 6°)(13.33)/[5.10(20)] = 0.687$ ft (20.940 cm).

3. Compute the horizontal distance a (Fig. 172)

Thus, $BG = 8.6 - \frac{1}{2}(5.10) = 6.05$ ft (184.404 cm); $a = 6.05 \sin 6° = 0.632$ ft (19.263 cm).

4. Compute the moment of the vertical forces

Thus, $M = W(h - a) = 407,700(0.055) = 22,400$ ft·lb (30,374.4 N·m). Since $h > a$, the moment is righting. This constitutes the solution to part a. The remainder of this procedure is concerned with part b.

In Fig. 172, let M denote the point of intersection of the vertical line through B' and the line

BG prolonged. Then *M* is termed the *metacenter* associated with this position, and the distance *GM* is called the *metacentric height*. Also *BG* is positive if *G* is above *B*, and *GM* is positive if *M* is above *G*. Thus, the moment of vertical forces is righting or upsetting depending on whether the metacentric height is positive or negative, respectively.

5. Find the lever arm of the vertical forces

Use the relation for metacentric height:

$$GM = \frac{I_{WL}}{V \cos \phi} - BG \tag{79}$$

where I_{WL} = moment of inertia of original waterline section about axis through *O*. Or, I_{WL} = $(\frac{1}{12})(64)(20)^3$ = 42,670 ft^4 (368.3 m^4); GM = 42,670/6530 cos 6° − 6.05 = 0.52 ft (15.850 cm); $h - a$ = 0.52 sin 6° = 0.054 ft (1.646 cm), which agrees closely with the previous result.

Mechanics of Incompressible Fluids

The notational system is a = acceleration; A = area of stream cross section; C = discharge coefficient; D = diameter of pipe or depth of liquid in open channel; F = force; g = gravitational acceleration; H = total head, or total specific energy; h_F = loss of head between two sections caused by friction; h_L = total loss of head between two sections; h_V = difference in velocity heads at two sections if no losses occur; L = length of stream between two sections; M = mass of body; N_R = Reynolds number; p = pressure; Q = volumetric rate of flow, or discharge; s = hydraulic gradient = $-dH/dL$; T = torque; V = velocity; w = specific weight; z = elevation above datum plane; ρ = density (mass per unit volume); μ = dynamic (or absolute) viscosity; ν = kinematic viscosity = μ/ρ; τ = shearing stress. The units used for each symbol are given in the calculation procedure where the symbol is used.

If the discharge of a flowing stream of liquid remains constant, the flow is termed *steady*. Let subscripts 1 and 2 refer to cross sections of the stream, 1 being the upstream section. From the definition of steady flow,

$$Q = A_1 V_1 = A_2 V_2 = \text{constant} \tag{80}$$

This is termed the *equation of continuity*. Where no statement is made to the contrary, it is understood that the flow is steady.

Conditions at two sections may be compared by applying the following equation, which is a mathematical statement of Bernoulli's theorem:

$$\frac{V_1^2}{2g} + \frac{p_1}{w} + z_1 = \frac{V_2^2}{2g} + \frac{p_2}{w} + z_2 + h_L \tag{81}$$

The terms on each side of this equation represent, in their order of appearance, the *velocity head, pressure head,* and *potential head* of the liquid. Alternatively, they may be considered to represent forms of specific energy, namely, kinetic, pressure, and potential energy.

The force causing a change in velocity is evaluated by applying the basic equation

$$F = Ma \tag{82}$$

Consider that liquid flows from section 1 to section 2 in a time interval *t*. At any instant, the volume of liquid bounded by these sections is *Qt*. The force required to change the velocity of this body of liquid from V_1 to V_2 is found from: $M = Qwt/g$; $a = (V_2 - V_1)/t$. Substituting in Eq. (82) gives $F = Qw(V_2 - V_1)/g$, or

$$F = \frac{A_1 V_1 w (V_2 - V_1)}{g} = \frac{A_2 V_2 w (V_2 - V_1)}{g} \tag{83}$$

VISCOSITY OF FLUID

Two horizontal circular plates 9 in (228.6 mm) in diameter are separated by an oil film 0.08 in (2.032 mm) thick. A torque of 0.25 ft·lb (0.339 N·m) applied to the upper plate causes that plate to rotate at a constant angular velocity of 4 revolutions per second (r/s) relative to the lower plate. Compute the dynamic viscosity of the oil.

Calculation Procedure:

1. Develop equations for the force and torque

Consider that the fluid film in Fig. 173a is in motion and that a fluid particle at boundary A has a velocity dV relative to a particle at B. The shearing stress in the fluid is

$$\tau = \mu \frac{dV}{dx} \tag{84}$$

Figure 173b shows a cross section of the oil film, the shaded portion being an elemental surface. Let m = thickness of film; R = radius of plates; ω = angular velocity of one plate relative

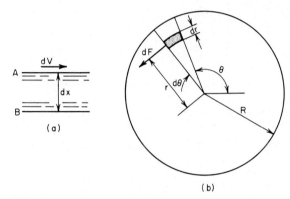

(a)

(b)

FIG. 173

to the other; dA = area of elemental surface; dF = shearing force on elemental surface; dT = torque of dF with respect to the axis through the center of the plate.

Applying Eq. 84, develop these equations: $dF = 2\pi\omega\mu r^2\, dr\, d\theta/m$; $dT = r\, dF = 2\pi\omega\mu r^2\, dr\, d\theta/m$.

2. Integrate the foregoing equation to obtain the resulting torque; solve for μ

Thus,

$$\mu = \frac{Tm}{\pi^2 \omega R^4} \tag{85}$$

$T = 0.25$ ft·lb (0.339 N·m); $m = 0.08$ in (2.032 mm); $\omega = 4$ r/s; $R = 4.5$ in (114.3 mm); $\mu = 0.25(0.08)(12)^3/[\pi^2(4)(4.5)^4] = 0.00214$ lb·s/ft² (0.1025 N·s/m²).

APPLICATION OF BERNOULLI'S THEOREM

A steel pipe is discharging 10 ft³/s (283.1 L/s) of water. At section 1, the pipe diameter is 12 in (304.8 mm), the pressure is 18 lb/in² (124.11 kPa), and the elevation is 140 ft (42.67 m). At section 2, farther downstream, the pipe diameter is 8 in (203.2 mm), and the elevation is 106 ft (32.31 m). If there is a head loss of 9 ft (2.74 m) between these sections due to pipe friction, what is the pressure at section 2?

Calculation Procedure:

1. Tabulate the given data

Thus D_1 = 12 in (304.8 mm); D_2 = 8 in (203.2 mm); p_1 = 18 lb/in^2 (124.11 kPa); p_2 = ?; z_1 = 140 ft (42.67 m); z_2 = 106 ft (32.31 m).

2. Compute the velocity at each section

Applying Eq. 80 gives V_1 = 10/0.785 = 12.7 ft/s (387.10 cm/s); V_2 = 10/0.349 = 28.7 ft/s (874.78 cm/s).

3. Compute p_2 by applying Eq. 81

Thus, $(p_2 - p_1)/w = (V_1^2 - V_2^2)/(2g) + z_1 - z_2 - h_F = (12.7^2 - 28.7^2)/64.4 + 140 - 106 - 9 = 14.7$ ft (448.06 cm); p_2 = 14.7(62.4)/144 + 18 = 24.4 lb/in^2 (168.24 kPa).

FLOW THROUGH A VENTURI METER

A venturi meter of 3-in (76.2-mm) throat diameter is inserted in a 6-in (152.4-mm) diameter pipe conveying fuel oil having a specific gravity of 0.94. The pressure at the throat is 10 lb/in^2 (68.95 kPa), and that at an upstream section 6 in (152.4 mm) higher than the throat is 14.2 lb/in^2 (97.91 kPa). If the discharge coefficient of the meter is 0.97, compute the flow rate in gallons per minute (liters per second).

Calculation Procedure:

1. Record the given data, assigning the subscript 1 to the upstream section and 2 to the throat

The loss of head between two sections can be taken into account by introducing a *discharge coefficient* C. This coefficient represents the ratio between the actual discharge Q and the discharge Q_i that would occur in the absence of any losses. Then $Q = CQ_i$, or $(V_2^2 - V_1^2)/(2g) = C^2 h_V$.

Record the given data: D_1 = 6 in (152.4 mm); p_1 = 14.2 lb/in^2 (97.91 kPa); z_1 = 6 in (152.4 mm); D_2 = 3 in (76.2 mm); p_2 = 10 lb/in^2 (68.95 kPa); z_2 = 0; C = 0.97.

2. Express V_1 in terms of V_2 and develop velocity and flow relations

Thus,

$$V_2 = C \left[\frac{2gh_V}{1 - (A_2/A_1)^2} \right]^{0.5} \tag{86a}$$

Also

$$Q = CA_2 \left[\frac{2gh_V}{1 - (A_2/A_1)^2} \right]^{0.5} \tag{86b}$$

If V_1 is negligible, these relations reduce to

$$V_2 = C(2gh_V)^{0.5} \tag{87a}$$

and

$$Q = CA_2(2gh_V)^{0.5} \tag{87b}$$

3. Compute h_v by applying Eq. 81

Thus, $h_V = (p_1 - p_2)/w + z_1 - z_2 = 4.2(144)/[0.94(62.4)] + 0.5 = 10.8$ ft (3.29 m).

4. Compute Q by applying Eq. 86b

Thus, $(A_2/A_1)^2 = (D_2/D_1)^4 = \frac{1}{16}$; A_2 = 0.0491 ft^2 (0.00456 m^2); and $Q = 0.97(0.0491)[64.4 \times 10.8/(1 - \frac{1}{16})]^{0.5} = 1.30$ ft^3/s or, by using the conversion factor of 1 ft^3/s = 449 gal/min (28.32 L/s), the flow rate is 1.30(449) = 584 gal/min (36.84 L/s).

FLOW THROUGH AN ORIFICE

Compute the discharge through a 3-in (76.2-mm) diameter square-edged orifice if the water on the upstream side stands 4 ft 8 in (1.422 m) above the center of the orifice.

Calculation Procedure:

1. Determine the discharge coefficient

For simplicity, the flow through a square-edged orifice discharging to the atmosphere is generally computed by equating the area of the stream to the area of the opening and then setting the discharge coefficient $C = 0.60$ to allow for contraction of the issuing stream. (The area of the issuing stream is about 0.62 times that of the opening.)

2. Compute the flow rate

Since the velocity of approach is negligible, use Eq. 87*b*. Or, $Q = 0.60(0.0491)(64.4 \times 4.67)^{0.5}$ $= 0.511 \text{ ft}^3/\text{s}$ (14.4675 L/s).

FLOW THROUGH THE SUCTION PIPE OF A DRAINAGE PUMP

Water is being evacuated from a sump through the suction pipe shown in Fig. 174. The entrance-end diameter of the pipe is 3 ft (91.44 cm); the exit-end diameter, 1.75 ft (53.34 cm). The exit pressure is 12.9 in (32.77 cm) of mercury vacuum. The head loss at the entry is one-fifteenth of the velocity head at that point, and the head loss in the pipe due to friction is one-tenth of the velocity head at the exit. Compute the discharge flow rate.

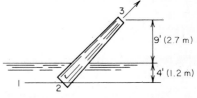

FIG. 174

Calculation Procedure:

1. Convert the pressure head to feet of water

The discharge may be found by comparing the conditions at an upstream point 1, where the velocity is negligible, with the conditions at point 3 (Fig. 174). Select the elevation of point 1 as the datum.

Converting the pressure head at point 3 to feet of water and using the specific gravity of mercury as 13.6, we have $p_3/w = -(12.9/12)13.6 = -14.6 \text{ ft}$ (-4.45 m).

2. Express the velocity head at 2 in terms of that at 3

By the equation of continuity, $V_2 = A_3V_3/A_2 = (1.75/3)^2 V_3 = 0.34V_3$.

3. Evaluate V_3 by applying Eq. 81; then determine Q

Thus, $V_1^2/(2g) + p_1/w + z_1 = V_3^2/(2g) + p_3/w + z_3 + (\frac{1}{15})V_2^2/(2g) + (\frac{1}{10})V_3^2/(2g)$, or $0 + 4 + 0 = V_3^2/(2g) - 14.6 + 13 + [V_3^2/(2g)](\frac{1}{15} \times 0.34^2 + \frac{1}{10})$; $V_3 = 18.0 \text{ ft/s}$ (548.64 cm/s); then $Q_3 = A_3V_3 = 0.785(1.75)^2(18.0) = 43.3 \text{ ft}^3/\text{s}$ (1225.92 L/s).

POWER OF A FLOWING LIQUID

A pump is discharging 8 ft³/s (226.5 L/s) of water. Gages attached immediately upstream and downstream of the pump indicate a pressure differential of 36 lb/in² (248.2 kPa). If the pump efficiency is 85 percent, what is the horsepower output and input?

Calculation Procedure:

1. Evaluate the increase in head of the liquid

Power is the rate of performing work, or the amount of work performed in a unit time. If the fluid flows with a specific energy H, the total energy of the fluid discharged in a unit time is

QwH. This expression thus represents the work that the flowing fluid can perform in a unit time and therefore the power associated with this discharge. Since 1 hp = 550 ft·lb/s,

$$1 \text{ hp} = \frac{QwH}{550} \tag{88}$$

In this situation, the power developed by the pump is desired. Therefore, H must be equated to the specific energy added by the pump.

To evaluate the increase in head, consider the differences of the two sections being considered. Since both sections have the same velocity and elevation, only their pressure heads differ. Thus, $p_2/w - p_1/w = 36(144)/62.4 = 83.1$ ft (2532.89 cm).

2. Compute the horsepower output and input

Thus, $\text{hp}_{\text{out}} = 8(62.4)(83.1)/550 = 75.4$ hp; $\text{hp}_{\text{in}} = 75.4/0.85 = 88.7$ hp.

DISCHARGE OVER A SHARP-EDGED WEIR

Compute the discharge over a sharp-edged rectangular weir 4 ft (121.9 cm) high and 10 ft (304.8 cm) long, with two end contractions, if the water in the canal behind the weir is 4 ft 9 in (144.78 cm) high. Disregard the velocity of approach.

Calculation Procedure:

1. Adopt a standard relation for this weir

The discharge over a sharp-edged rectangular weir without end contractions in which the velocity of approach is negligible is given by the Francis formula as

$$Q = 3.33bh^{1.5} \tag{89a}$$

where b = length of crest and h = head on weir, i.e., the difference between the elevation of the crest and that of the water surface upstream of the weir.

2. Modify the Francis equation for end contractions

With two end contractions, the discharge of the weir is

$$Q = 3.33(b - 0.2h)h^{1.5} \tag{89b}$$

Substituting the given values yields $Q = 3.33(10 - 0.2 \times 0.75)0.75^{1.5} = 21.3$ ft^3/s (603.05 L/s).

LAMINAR FLOW IN A PIPE

A tank containing crude oil discharges 340 gal/min (21.4 L/s) through a steel pipe 220 ft (67.1 m) long and 8 in (203.2 mm) in diameter. The kinematic viscosity of the oil is 0.002 ft^2/s (1.858 cm^2/s). Compute the difference in elevation between the liquid surface in the tank and the pipe outlet.

Calculation Procedure:

1. Identify the type of flow in the pipe

To investigate the discharge in a pipe, it is necessary to distinguish between two types of fluid flow—*laminar* and *turbulent*. Laminar (or *viscous*) flow is characterized by the telescopic sliding of one circular layer of fluid past the adjacent layer, each fluid particle traversing a straight line. The velocity of the fluid flow varies parabolically from zero at the pipe wall to its maximum value at the pipe center, where it equals twice the mean velocity.

Turbulent flow is characterized by the formation of eddy currents, with each fluid particle traversing a sinuous path.

In any pipe the type of flow is ascertained by applying a dimensionless index termed the *Reynolds number*, defined as

$$N_R = \frac{DV}{\nu} \qquad (90)$$

Flow is considered laminar if $N_R < 2100$ and turbulent if $N_R > 3000$.

In laminar flow the head loss due to friction is

$$h_F = \frac{32L\nu V}{gD^2} \qquad (91a)$$

or

$$h_F = \left(\frac{64}{N_R}\right)\left(\frac{L}{D}\right)\left(\frac{V^2}{2g}\right) \qquad (91b)$$

Let 1 denote a point on the liquid surface and 2 a point at the pipe outlet. The elevation of 2 will be taken as datum.

To identify the type of flow, compute N_R. Thus, $D = 8$ in (203.2 mm); $L = 220$ ft (6705.6 cm); $\nu = 0.002$ ft^2/s (1.858 cm^2/s); $Q = 340/449 = 0.757$, converting from gallons per minute to cubic feet per second. Then $V = Q/A = 0.757/0.349 = 2.17$ ft/s (66.142 cm/s). And $N_R = 0.667(2.17)/0.002 = 724$. Therefore, the flow is laminar because N_R is less than 2100.

2. Express all losses in terms of the velocity head

By Eq. 91*b*, $h_F = (64/724)(220/0.667)V^2/(2g) = 29.2V^2/(2g)$. Where $L/D > 500$, the following may be regarded as negligible in comparison with the loss due to friction: loss at pipe entrance, losses at elbows, velocity head at the discharge, etc. In this instance, include the secondary items. The loss at the pipe entrance is $h_E = 0.5V^2/(2g)$. The total loss is $h_L = 29.7V^2/(2g)$.

3. Find the elevation of 1 by applying Eq. 81

Thus, $z_1 = V_2^2/(2g) + h_L = 30.7V_2^2/(2g) = 30.7(2.17)^2/64.4 = 2.24$ ft (68.275 cm).

TURBULENT FLOW IN PIPE—APPLICATION OF DARCY-WEISBACH FORMULA

Water is pumped at the rate of 3 ft^3/s (85.0 L/s) through an 8-in (203.2-mm) fairly smooth pipe 2600 ft (792.48 m) long to a reservoir where the water surface is 180 ft (50.86 m) higher than the pump. Determine the gage pressure at the pump discharge.

Calculation Procedure:

1. Compute h_F

Turbulent flow in a pipe flowing full may be investigated by applying the Darcy-Weisbach formula for friction head

$$h_F = \frac{fLV^2}{2gD} \qquad (92)$$

where f is a friction factor. However, since the friction head does not vary precisely in the manner implied by this equation, f is dependent on D and V, as well as the degree of roughness of the pipe. Values of f associated with a given set of values of the independent quantities may be obtained from Fig. 175.

Accurate equations for h_F are the following:

Extremely smooth pipes:

$$h_F = \frac{0.30LV^{1.75}}{1000D^{1.25}} \qquad (93a)$$

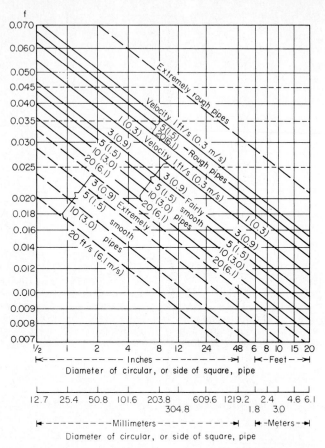

FIG. 175 Flow of water in pipes. *(From E. W. Schoder and F. M. Dawson, Hydraulics, McGraw-Hill Book Company, New York, 1934. By permission of the publishers.)*

Fairly smooth pipes:

$$h_F = \frac{0.38 L V^{1.86}}{1000 D^{1.25}} \tag{93b}$$

Rough pipes:

$$h_F = \frac{0.50 L V^{1.95}}{1000 D^{1.25}} \tag{93c}$$

Extremely rough pipes:

$$h_F = \frac{0.69 L V^2}{1000 D^{1.25}} \tag{93d}$$

Using Eq. 93b gives $V = Q/A = 3/0.349 = 8.60$ ft/s (262.128 cm/s); $h_F = 0.38(2.6)(8.60)^{1.86}/0.667^{1.25} = 89.7$ ft (27.34 m).

2. Alternatively, determine h_F using Eq. 92

First obtain the appropriate f value from Fig. 175, or $f = 0.020$ for this pipe. Then $h_F = 0.020(2,600/0.667)(8.60^2/64.4) = 89.6$ ft (27.31 m).

3. Compute the pressure at the pump discharge

Use Eq. 81. Since $L/D > 500$, ignore the secondary items. Then $p_1/w = z_2 + h_F = 180 + 89.6 = 269.6$ ft (82.17 m); $p_1 = 269.6(62.4)/144 = 117$ lb/in² (806.7 kPa).

DETERMINATION OF FLOW IN A PIPE

Two reservoirs are connected by a 7000-ft (2133.6-m) fairly smooth cast-iron pipe 10 in (254.0 mm) in diameter. The difference in elevation of the water surfaces is 90 ft (27.4 m). Compute the discharge to the lower reservoir.

Calculation Procedure:

1. Determine the fluid velocity and flow rate

Since the secondary items are negligible, the entire head loss of 90 ft (27.4 m) results from friction. Using Eq. 93b and solving for V, we have $90 = 0.38(7)V^{1.86}/0.833^{1.25}$; $V = 5.87$ ft/s (178.918 cm/s). Then $Q = VA = 5.87(0.545) = 3.20$ ft³/s (90.599 L/s).

2. Alternatively, assume a value of f and compute V

Referring to Fig. 175, select a value for f. Then compute V by applying Eq. 92. Next, compare the value of f corresponding to this result with the assumed value of f. If the two values differ appreciably, assume a new value of f and repeat the computation. Continue this process until the assumed and actual values of f agree closely.

PIPE-SIZE SELECTION BY THE MANNING FORMULA

A cast-iron pipe is to convey water at 3.3 ft³/s (93.430 L/s) on a grade of 0.001. Applying the Manning formula with $n = 0.013$, determine the required size of pipe.

Calculation Procedure:

Compute the pipe diameter

The Manning formula, which is suitable for both open and closed conduits, is

$$V = \frac{1.486R^{2/3}s^{1/2}}{n} \qquad (94)$$

where n = roughness coefficient; R = hydraulic radius = ratio of cross-sectional area of pipe to the wetted perimeter of the pipe; s = hydraulic gradient = dH/dL. If the flow is uniform, i.e., the area and therefore the velocity are constant along the stream, then the loss of head equals the drop in elevation, and the grade of the conduit is s.

For a circular pipe flowing full, Eq. 94 becomes

$$D = \left(\frac{2.159Qn}{s^{1/2}}\right)^{3/8} \qquad (94a)$$

Substituting numerical values gives $D = (2.159 \times 3.3 \times 0.013/0.001^{1/2})^{3/8} = 1.50$ ft (45.72 cm). Therefore, use an 18-in (457.2-mm) diameter pipe.

LOSS OF HEAD CAUSED BY SUDDEN ENLARGEMENT OF PIPE

Water flows through a pipe at 4 ft³/s (113.249 L/s). Compute the loss of head resulting from a change in pipe size if (a) the pipe diameter increases abruptly from 6 to 10 in (152.4 to 254.0 mm); (b) the pipe diameter increases abruptly from 6 to 8 in (152.4 to 203.2 mm) at one section and then from 8 to 10 in (203.2 to 254.0 mm) at a section farther downstream.

Calculation Procedure:

1. Evaluate the pressure-head differential required to decelerate the liquid

Where there is an abrupt increase in pipe size, the liquid must be decelerated upon entering the larger pipe, since the fluid velocity varies inversely with area. Let subscript 1 refer to a section immediately downstream of the enlargement, where the higher velocity prevails, and let subscript 2 refer to a section farther downstream, where deceleration has been completed. Disregard the frictional loss.

Using Eq. 83 we see $p_2/w = p_1/w + (V_1 V_2 - V_2^2)/g$.

2. Combine the result of step 1 with Eq. 81

The result is Borda's formula for the head loss h_E caused by sudden enlargement of the pipe cross section:

$$h_E = \frac{(V_1 - V_2)^2}{2g} \tag{95}$$

As this investigation shows, only part of the drop in velocity head is accounted for by a gain in pressure head. The remaining head h_E is dissipated through the formation of eddy currents at the entrance to the larger pipe.

3. Compute the velocity in each pipe

Thus

Pipe diam, in (mm)	Pipe area, ft² (m²)	Fluid velocity, ft/s (cm/s)
6 (152.4)	0.196 (0.0182)	20.4 (621.79)
8 (203.2)	0.349 (0.0324)	11.5 (350.52)
10 (254.0)	0.545 (0.0506)	7.3 (222.50)

4. Find the head loss for part a

Thus, $h_E = (20.4 - 7.3)^2/64.4 = 2.66$ ft (81.077 cm).

5. Find the head loss for part b

Thus, $h_E = [(20.4 - 11.5)^2 + (11.5 - 7.3)^2]/64.4 = 1.50$ ft (45.72 cm). Comparison of these results indicates that the eddy-current loss is attenuated if the increase in pipe size occurs in steps.

DISCHARGE OF LOOPING PIPES

A pipe carrying 12.5 ft³/s (353.90 L/s) of water branches into three pipes of the following diameters and lengths; $D_1 = 6$ in (152.4 mm); $L_1 = 1000$ ft (304.8 m); $D_2 = 8$ in (203.2 mm); $L_2 = 1300$ ft (396.2 m); $D_3 = 10$ in (254.0 mm); $L_3 = 1200$ ft (365.8 m). These pipes rejoin at their downstream ends. Compute the discharge in the three pipes, considering each as fairly smooth.

Calculation Procedure:

1. Express Q as a function of D and L

Since all fluid particles have the same energy at the juncture point, irrespective of the loops they traversed, the head losses in the three loops are equal. The flow thus divides itself in a manner that produces equal values of h_F in the loops.

Transforming Eq. 93b,

$$Q = \frac{kD^{2.67}}{L^{0.538}} \tag{96}$$

where k is a constant.

2. Establish the relative values of the discharges; then determine the actual values

Thus, $Q_2/Q_1 = (8/6)^{2.67}/1.3^{0.538} = 1.87$; $Q_3/Q_1 = (10/6)^{2.67}/1.2^{0.538} = 3.55$. Then $Q_1 + Q_2 + Q_3 = Q_1(1 + 1.87 + 3.55) = 12.5$ ft^3/s (353.90 L/s). Solving gives $Q_1 = 1.95$ ft^3/s (55.209 L/s); $Q_2 = 3.64$ ft^3/s (103.056 L/s); $Q_3 = 6.91$ ft^3/s (195.637 L/s).

FLUID FLOW IN BRANCHING PIPES

The pipes AM, MB, and MC in Fig. 176 have the diameters and lengths indicated. Compute the water flow in each pipe if the pipes are considered rough.

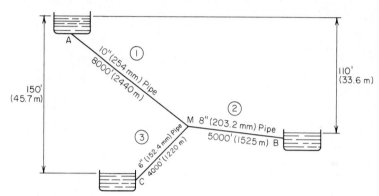

FIG. 176 Branching pipes.

Calculation Procedure:

1. Write the basic equations governing the discharges

Let subscripts 1, 2 and 3 refer to AM, MB, and MC, respectively. Then $h_{F1} + h_{F2} = 110$; $h_{F1} + h_{F3} = 150$, Eq. a; $Q_1 = Q_2 + Q_3$, Eq. b.

2. Transform Eq. 93c

The transformed equation is

$$Q = 38.7D^{2.64}\left(\frac{h_F}{L}\right)^{0.513} \tag{97}$$

3. Assume a trial value for h_{F1} and find the discharge; test the result

Use Eqs. a and 97 to find the discharges. Test the results for compliance with Eq. b. If we assume $h_{F1} = 70$ ft (21.3 m), then $h_{F2} = 40$ ft (12.2 m) and $h_{F3} = 80$ ft (24.4 m); $Q_1 = 38.7(0.833)^{2.64}(70/8000)^{0.513} = 2.10$ ft^3/s (59.455 L/s). Similarly, $Q_2 = 1.12$ ft^3/s (31.710 L/s) and $Q_3 = 0.83$ ft^3/s (23.499 L/s); $Q_2 + Q_3 = 1.95 < Q_1$. The assumed value of h_{F1} is excessive.

4. Make another assumption for h_{F1} and the corresponding revisions

Assume $h_{F1} = 66$ ft (20.1 m). Then $Q_1 = 2.10(66/70)^{0.513} = 2.04$ ft^3/s (57.757 L/s). Similarly, $Q_2 = 1.18$ ft^3/s (33.408 L/s); $Q_3 = 0.85$ ft^3/s (24.065 L/s). $Q_2 + Q_3 = 2.03$ ft^3/s (57.736 L/s). These results may be accepted as sufficiently precise.

UNIFORM FLOW IN OPEN CHANNEL—DETERMINATION OF SLOPE

It is necessary to convey 1200 ft^3/s (33,974.6 L/s) of water from a dam to a power plant in a canal of rectangular cross section, 24 ft (7.3 m) wide and 10 ft (3.0 m) deep, having a roughness coefficient of 0.016. The canal is to flow full. Compute the required slope of the canal in feet per mile (meters per kilometer).

Calculation Procedure:

1. Apply Eq. 94

Thus, $A = 24(10) = 240 \text{ ft}^2 (22.3 \text{ m}^2)$; wetted perimeter $= \text{WP} = 24 + 2(10) = 44 \text{ ft} (13.4 \text{ m})$; $R = 240/44 = 5.45 \text{ ft} (1.661 \text{ m})$; $V = 1200/240 = 5 \text{ ft/s} (152.4 \text{ cm/s})$; $s = [nV/(1.486R^{2/3})]^2$ $= [0.016 \times 5/(1.486 \times 5.45^{2/3})]^2 = 0.000302$; slope $= 0.000302(5280 \text{ ft/mi}) = 1.59 \text{ ft/mi} (0.302 \text{ m/km})$.

REQUIRED DEPTH OF CANAL FOR SPECIFIED FLUID FLOW RATE

A trapezoidal canal is to carry water at $800 \text{ ft}^3/\text{s}$ (22,649.7 L/s). The grade of the canal is 0.0004; the bottom width is 25 ft (7.6 m); the slope of the sides is 1½ horizontal to 1 vertical; the roughness coefficient is 0.014. Compute the required depth of the canal, to the nearest tenth of a foot.

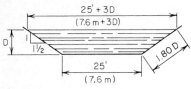

FIG. 177

Calculation Procedure:

1. Transform Eq. 94 and compute $AR^{2/3}$

Thus, $AR^{2/3} = nQ/(1.486s^{1/2})$, Eq. 94b. Or, $AR^{2/3} = 0.014(800)/[1.486(0.0004)^{1/2}] = 377$.

2. Express the area and wetted perimeter in terms of D (Fig. 177)

Side of canal $= D(1^2 + 1.5^2)^{0.5} = 1.80D$. $A = D(25 + 1.5D)$; $\text{WP} = 25 + 360D$.

3. Assume the trial values of D until Eq. 94b is satisfied

Thus, assume $D = 5 \text{ ft} (152.4 \text{ cm})$; $A = 162.5 \text{ ft}^2 (15.10 \text{ m}^2)$; $\text{WP} = 43 \text{ ft} (1310.6 \text{ cm})$; $R = 3.78$ ft (115.2 cm); $AR^{2/3} = 394$. The assumed value of D is therefore excessive because the computed $AR^{2/3}$ is greater than the value computed in step 1.

Next, assume a lower value for D, or $D = 4.9 \text{ ft} (149.35 \text{ cm})$; $A = 158.5 \text{ ft}^2 (14.72 \text{ m}^2)$; $\text{WP} = 42.64 \text{ ft} (1299.7 \text{ cm})$; $R = 3.72 \text{ ft} (113.386 \text{ cm})$; $AR^{2/3} = 381$. This is acceptable. Therefore, $D = 4.9 \text{ ft} (149.35 \text{ cm})$.

ALTERNATE STAGES OF FLOW; CRITICAL DEPTH

A rectangular channel 20 ft (609.6 cm) wide is to discharge $500 \text{ ft}^3/\text{s}$ (14,156.1 L/s) of water having a specific energy of 4.5 ft·lb/lb (1.37 J/N). (a) Using $n = 0.013$, compute the required slope of the channel. (b) Compute the maximum potential discharge associated with the specific energy of 4.5 ft·lb/lb (1.37 J/N). (c) Compute the minimum specific energy required to maintain a flow of $500 \text{ ft}^3/\text{s}$ (14,156.1 L/s).

Calculation Procedure:

1. Evaluate the specific energy of an elemental mass of liquid at a distance z above the channel bottom

To analyze the discharge conditions at a given section in a channel, it is advantageous to evaluate the specific energy (or head) by taking the elevation of the bottom of the channel *at the given section* as datum. Assume a uniform velocity across the section, and let $D = $ depth of flow, ft (cm); $H_e = $ specific energy as computed in the prescribed manner; $Q_u = $ discharge through a unit width of channel, $\text{ft}^3/(\text{s·ft})$ [L/(s·cm)].

Evaluating the specific energy of an elemental mass of liquid at a given distance z above the channel bottom, we get

$$H_e = \frac{Q_u^2}{2gD^2} + D \tag{98}$$

Thus, H_e is constant across the entire section. Moreover, if the flow is uniform, as it is here, H_e is constant along the entire stream.

2. Apply the given values and solve for D

Thus, H_e = 4.5 ft·lb/lb (1.37 J/N); Q_u = 500/20 = 25 ft³/(s·ft) [2323 L/(s·m)]. Rearrange Eq. 98 to obtain

$$D^2(H_e - D) = \frac{Q_u^2}{2g} \tag{98a}$$

Or, $D^2(4.5 - D)$ = $25^2/64.4$ = 9.705. This cubic equation has two positive roots, D = 1.95 ft (59.436 cm) and D = 3.84 ft (117.043 cm). There are therefore two stages of flow that accommodate the required discharge with the given energy. [The third root of the equation is D = -1.29 ft (-39.319 cm), an impossible condition.]

3. Compute the slope associated with the computed depths

Using Eq. 94, at the lower stage we have D = 1.95 ft (59.436 cm); A = 20(1.95) = 39.0 ft² (36,231.0 cm²); WP = 20 + 2(1.95) = 23.9 ft (728.47 cm); R = 39.0/23.9 = 1.63 ft (49.682 cm); V = 25/1.95 = 12.8 ft/s (390.14 cm/s); s = $[nV/(1.486R^{2/3})]^2$ = (0.013 × 12.8/1.486 × $1.63^{2/3})^2$ = 0.00654.

At the upper stage D = 3.84 ft (117.043 cm); A = 20(3.84) = 76.8 ft² (71,347.2 cm²); WP = 20 + 2(3.84) = 27.68 ft (843.686 cm); R = 76.8/27.68 = 2.77 ft (84.430 cm); V = 25/3.84 = 6.51 ft/s (198.4 cm/s); s = $[0.013 × 6.51/(1.486 × 2.77^{2/3})]^2$ = 0.000834. This constitutes the solution to part a.

4. Plot the D-Q_u curve

For part b, consider H_e as remaining constant at 4.5 ft·lb/lb (1.37 J/N) while Q_u varies. Plot the D-Q_u curve as shown in Fig. 178a. The depth that provides the maximum potential discharge is called the *critical depth* with respect to the given specific energy.

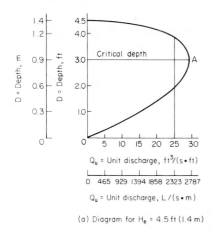

(a) Diagram for H_e = 4.5 ft (1.4 m)

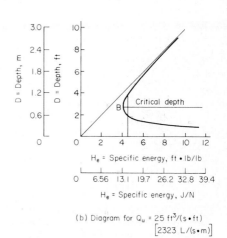

(b) Diagram for Q_u = 25 ft³/(s·ft)
[2323 L/(s·m)]

FIG. 178

5. Differentiate Eq. 98 to find the critical depth; then evaluate $Q_{u,max}$

Differentiating Eq. 98 and setting dQ_u/dD = 0 yield

$$\text{Critical depth } D_c = \tfrac{2}{3}H_e \tag{99}$$

Or, D_c = $\tfrac{2}{3}(4.5)$ = 3.0 ft (91.44 cm); $Q_{u,max}$ = $[64.4(4.5 × 3.0^2 - 3.0^2)]^{0.5}$ = 29.5 ft³/(s·ft) [2741 L/(s·m)]; Q_{max} = 29.5(20) = 590 ft³/s (16,704.2 L/s). This constitutes the solution to part b.

6. Plot the D-H_e curve

For part c, consider Q_u as remaining constant at 25 ft³/(s·ft) [2323 L/(s·m)] while H_e varies. Plot the D-H_e curve as shown in Fig. 178b. (This curve is asymptotic with the straight lines $D = H_e$ and $D = 0$.) The depth at which the specific energy is minimum is called the *critical depth* with respect to the given unit discharge.

7. Differentiate Eq. 98 to find the critical depth; then evaluate H_{e,min}

Differentiating gives

$$D_c = \left(\frac{Q_u^2}{g}\right)^{1/3} \tag{100}$$

Then $D_c = (25^2/32.2)^{1/3} = 2.69$ ft (81.991 cm). Then $H_{e,min} = 25^2/[64.4(2.69)^2] + 2.69 = 4.03$ ft·lb/lb (1.229 J/N).

The values of D as computed in part *a* coincide with those obtained by referring to the two graphs in Fig. 178. The equations derived in this procedure are valid solely for rectangular channels, but analogous equations pertaining to other channel profiles may be derived in a similar manner.

DETERMINATION OF HYDRAULIC JUMP

Water flows over a 100-ft (30.5-m) long dam at 7500 ft³/s (212,400 L/s). The depth of tailwater on the level apron is 9 ft (2.7 m). Determine the depth of flow immediately upstream of the hydraulic jump.

Calculation Procedure:

1. Find the difference in hydrostatic forces per unit width of channel required to decelerate the liquid

Refer to Fig. 179. *Hydraulic jump* designates an abrupt transition from lower-stage to upper-stage flow caused by a sharp decrease in slope, sudden increase in roughness, encroachment of

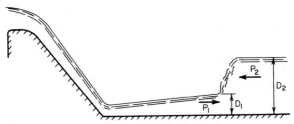

FIG. 179 Hydraulic jump on apron of dam.

backwater, or some other factor. The deceleration of liquid requires an increase in hydrostatic pressure, but only part of the drop in velocity head is accounted for by a gain in pressure head. The excess head is dissipated in the formation of a turbulent standing wave. Thus, the phenomenon of hydraulic jump resembles the behavior of liquid in a pipe at a sudden enlargement, as analyzed in an earlier calculation procedure.

Let D_1 and D_2 denote the depth of flow immediately upstream and downstream of the jump, respectively. Then $D_1 < D_c < D_2$. Refer to Fig. 178b. Since the hydraulic jump is accompanied by a considerable drop in energy, the point on the D-H_e diagram that represents D_2 lies both above and to the left of that representing D_1. Therefore, the upstream depth is less than the depth that would exist in the absence of any loss.

Using literal values, apply Eq. 83 to find the difference in hydrostatic forces per unit width of channel that is required to decelerate the liquid. Solve the resulting equation for D_1:

$$D_1 = -\frac{D_2}{2} + \left(\frac{2V_2^2 D_2}{g} + \frac{D_2^2}{4}\right)^{0.5} \tag{101}$$

215

2. Substitute numerical values in Eq. 101

Thus, $Q_u = 7500/100 = 75$ ft^3/(s·ft) [6969 L/(s·m)]; $V_2 = 75/9 = 8.33$ ft/s (2.538 m/s); $D_1 = -\frac{3}{2} + (2 \times 8.33^2 \times 9/32.2 + 9^2/4)^{0.5} = 3.18$ ft (0.969 m).

RATE OF CHANGE OF DEPTH IN NONUNIFORM FLOW

The unit discharge in a rectangular channel is 28 ft^3/(s·ft) [2602 L/(s·m)]. The energy gradient is 0.0004, and the grade of the channel bed is 0.0010. Determine the rate at which the depth of flow is changing in the downstream direction (i.e., the grade of the liquid surface with respect to the channel bed) at a section where the depth is 3.2 ft (0.97 m).

Calculation Procedure:

1. Express H as a function of D

Let H = total specific energy at a given section as evaluated by selecting a fixed horizontal reference plane; L = distance measured in downstream direction; z = elevation of given section with respect to datum plane; s_b = grade of channel bed = $-dz/dL$; s_e = energy gradient = $-dH/dL$.

Express H as a function of D by annexing the potential-energy term to Eq. 98. Thus,

$$H = \frac{Q_u^2}{2gD^2} + D + z \qquad (102)$$

2. Differentiate this equation with respect to L to obtain the rate of change of D; substitute numerical values

Differentiating gives

$$\frac{dD}{dL} = \frac{s_b - s_e}{1 - Q_u^2/(gD^3)} \qquad (103a)$$

or in accordance with Eq. 100,

$$\frac{dD}{dL} = \frac{s_b - s_e}{1 - D_c^3/D^3} \qquad (103b)$$

Substituting yields $Q_u^2/(gD^3) = 28^2/(32.2 \times 3.2^3) = 0.743$; $dD/dL = (0.0010 - 0.0004)/(1 - 0.743) = 0.00233$ ft/ft (0.00233 m/m). The depth is increasing in the downstream direction, and the water is therefore being decelerated.

As Eq. 103b reveals, the relationship between the actual depth at a given section and the critical depth serves as a criterion in ascertaining whether the depth is increasing or decreasing.

DISCHARGE BETWEEN COMMUNICATING VESSELS

In Fig. 180, liquid is flowing from tank A to tank B through an orifice near the bottom. The area of the liquid surface is 200 ft^2 (18.58 m^2) in A and 150 ft^2 (13.93 m^2) in B. Initially, the difference in water levels is 14 ft (4.3 m), and the discharge is 2 ft^3/s (56.6 L/s). Assuming that the discharge coefficient remains constant, compute the time required for the water level in tank A to drop 1.8 ft (0.54 m).

Calculation Procedure:

1. By expressing the change in h during an elemental time interval, develop the time-interval equation

Let A_a and A_b denote the area of the liquid surface in tanks A and B, respectively; let subscripts 1 and 2 refer to the beginning and end, respectively, of a time interval t. Then

$$t = \frac{2A_aA_b(h_1 - [h_1h_2]^{0.5})}{Q_1(A_a + A_b)} \qquad (104)$$

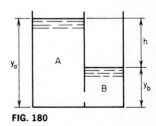

FIG. 180

2. **Find the value of h when y_a diminishes by 1.8 ft (0.54 m)**

Thus, $\Delta y_b = (-A_a/A_b)(\Delta y_a) = -(200/150)(-1.8) = 2.4$ ft (0.73 m); $\Delta h = \Delta y_a - \Delta y_b = -1.8 - 2.4 = -4.2$ ft $(-1.28$ m); $h_1 = 14$ ft (4.3 m); $h_2 = 14 - 4.2 = 9.8$ ft (2.99 m).

3. **Substitute numerical values in Eq. 104 and solve for t**

Thus, $t = 2(200)(150)[14 - (14 \times 9.8)^{0.5}]/[2(200 + 150)] = 196$ s $= 3.27$ min.

VARIATION IN HEAD ON A WEIR WITHOUT INFLOW TO THE RESERVOIR

Water flows over a weir of 60-ft (18.3-m) length from a reservoir having a surface area of 50 acres (202,350 m^2). If the inflow to the reservoir ceases when the head on the weir is 2 ft (0.6 m), what will the head be at the expiration of 1 h? Consider that the instantaneous discharge is given by Eq. 89a.

Calculation Procedure:

1. **Develop the time-interval equation**

Let A = surface area of reservoir and C = numerical constant in discharge equation; and subscripts 1 and 2 refer to the beginning and end, respectively, of a time interval t. By expressing the change in head during an elemental time interval,

$$t = \frac{2A}{Cb(1/h_2^{0.5} - 1/h_1^{0.5})} \qquad (105)$$

2. **Substitute numerical values in Eq. 105; solve for h_2**

Thus, $A = 50(43,560) = 2,178,000$ ft^2 (202,336.2 m^2); $t = 3600$ s; solving gives $h_2 = 1.32$ ft (0.402 m).

VARIATION IN HEAD ON A WEIR WITH INFLOW TO THE RESERVOIR

Water flows over an 80-ft (24.4-m) long weir from a reservoir having a surface area of 6,000,000 ft^2 (557,400.0 m^2) while the rate of inflow to the reservoir remains constant at 2175 ft^3/s (61,578.9 L/s). How long will it take for the head on the weir to increase from zero to 95 percent of its maximum value? Consider that the instantaneous rate of flow over the weir is $3.4bh^{1.5}$.

Calculation Procedure:

1. **Compute the maximum head on the weir by equating outflow to inflow**

The water in the reservoir reaches its maximum height when equilibrium is achieved, i.e., when the rate of outflow equals the rate of inflow. Let Q_i = rate of inflow; Q_o = rate of outflow at a given instant; t = time elapsed since the start of the outflow.

Equating outflow to inflow yields $3.4(80h_{max}^{1.5}) = 2175$; $h_{max} = 4.0$ ft (1.2 m); $0.95h_{max} = 3.8$ ft (1.16 m).

2. **Using literal values, determine the time interval dt during which the water level rises a distance dh**

Thus, with C = numerical constant in the discharge equation,

$$dt = \frac{A}{Q_i - Cbh^{1.5}} dh \qquad (106)$$

The right side of this equation is not amenable to direct integration. Consequently, the only feasible way of computing the time is to perform an approximate integration.

3. **Obtain the approximate value of the required time**

Select suitable increments of h, calculate the corresponding increments of t, and total the latter to obtain an approximate value of the required time. In calculating Q_o, apply the mean value of h associated with each increment.

The precision inherent in the result thus obtained depends on the judgment used in selecting the increments of h, and a clear visualization of the relationship between h and t is essential. Let $m = dt/dh = A/(Q_i - Cbh^{1.5})$. The m-h curve is shown in Fig. 181a. Then, $t = \int m \, dh =$ area between the m-h curve and h axis.

This area is approximated by summing the areas of the rectangles as indicated in Fig. 181, the length of each rectangle being equal to the value of m at the center of the interval. Note that as h increases, the increments Δh should be made progressively smaller to minimize the error introduced in the procedure.

Select the increments shown in Table 12, and perform the indicated calculations. The symbols h_b and h_m denote the values of h at the beginning and center, respectively, of an interval. The following calculations for the third interval illustrate the method: $h_m = \frac{1}{2}(2.0 + 2.8) = 2.4$ ft (0.73 m); $m = 6,000,000/(2175 - 3.4 \times 80 \times 2.4^{1.5}) = 5160$ s/ft (16,929.1 s/m); $\Delta t = m \, \Delta h = 5160(0.8) = 4130$ s. From Table 12, the required time is $t = 24,090$ s $= 6$ h 41.5 min.

The t-h curve is shown in Fig. 181b. The time required for the water to reach its maximum height is difficult to evaluate with precision because m becomes infinitely large as h approaches h_{max}; that is, the water level rises at an imperceptible rate as it nears its limiting position.

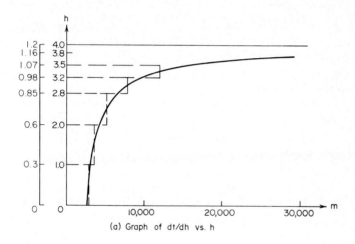

(a) Graph of dt/dh vs. h

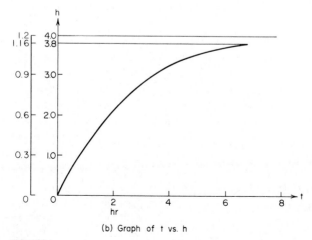

(b) Graph of t vs. h

FIG. 181

TABLE 12 Approximate Integration of Eq. 106

Δh, ft (m)	h_b, ft (m)	h_m, ft (m)	m, s/ft (s/m)	Δt, s
1.0 (0.30)	0 (0.00)	0.5 (0.15)	2,890 (9,633.3)	2,890
1.0 (0.30)	1.0 (0.30)	1.5 (0.46)	3,580 (11,933.3)	3,580
0.8 (0.24)	2.0 (0.61)	2.4 (0.73)	5,160 (17,308.3)	4,130
0.4 (0.12)	2.8 (0.85)	3.0 (0.91)	7,870 (26,250.0)	3,150
0.3 (0.09)	3.2 (0.98)	3.35 (1.02)	11,830 (39,444.4)	3,550
0.2 (0.06)	3.5 (1.07)	3.6 (1.10)	18,930 (63,166.7)	3,790
0.1 (0.03)	3.7 (1.13)	3.75 (1.14)	30,000 (100,000)	3,000
Total				24,090

DIMENSIONAL ANALYSIS METHODS

The velocity of a raindrop in still air is known or assumed to be a function of these quantities: gravitational acceleration, drop diameter, dynamic viscosity of the air, and the density of both the water and the air. Develop the dimensionless parameters associated with this phenomenon.

Calculation Procedure:

1. Using a generalized notational system, record the units in which the six quantities of this situation are expressed

Dimensional analysis is an important tool both in theoretical investigations and in experimental work because it clairfies the relationships intrinsic in a given situation.

A quantity that appears in every dimensionless parameter is termed repeating; a quantity that appears in only one parameter is termed *nonrepeating*. Since the engineer is usually more accustomed to dealing with units of force rather than of mass, the force-length-time system of units is applied here. Let F, L, and T denote units of force, length, and time, respectively.

By using this generalized notational system, it is convenient to write the appropriate USCS units and then replace these with the general units. For example, with respect to acceleration: USCS units, ft/s^2; general units, L/T^2 or LT^{-2}. Similarly, with respect to density (w/g): USCS units, (lb/ft^3)/(ft/s^2); general units, FL^{-3}/LT^{-2} or $FL^{-4}T^2$.

The results are shown in the following table.

Quantity	Units
V = velocity of raindrop	LT^{-1}
g = gravitational acceleration	LT^{-2}
D = diameter of drop	L
μ_a = air viscosity	$FL^{-2}T$
ρ_w = water density	$FL^{-4}T^2$
ρ_a = air density	$FL^{-4}T^2$

2. Compute the number of dimensionless parameters present

This phenomenon contains six physical quantities and three units. Therefore, as a consequence of Buckingham's pi theorem, the number of dimensionless parameters is $6 - 3 = 3$.

3. Select the repeating quantities

The number of repeating quantities must equal the number of units (three here). These quantities should be independent, and they should collectively contain all the units present. The quantities g, D, and μ_a satisfy both requirements and therefore are selected as the repeating quantities.

4. Select the dependent variable V as the first nonrepeating quantity

Then write $\pi_1 = g^x D^y \mu_a^z V$, Eq. a, where π_1 is a dimensionless parameter and x, y, and z are unknown exponents that may be evaluated by experiment.

5. Transform Eq. a to a dimensional equation

Do this by replacing each quantity with the units in which it is expressed. Then perform the necessary expansions and multiplications. Or, $F^0 L^0 T^0 = (LT^{-2})^x L^y \times (FL^{-2}T)^z LT^{-1}$, $F^0 L^0 T^0 = F^z L^{x+y-2z+1} T^{-2x+z-1}$, Eq. b.

Every equation must be dimensionally homogeneous; i.e., the units on one side of the equation must be consistent with those on the other side. Therefore, the exponent of a unit on one side of Eq. b must equal the exponent of that unit on the other side.

6. Evaluate the exponents x, y, and z

Do this by applying the principle of dimensional homogeneity to Eq. b. Thus, $0 = z$; $0 = x + y - 2z + 1$; $0 = -2x + z - 1$. Solving these simultaneous equations yields $x = -\frac{1}{2}$; $y = -\frac{1}{2}$; $z = 0$.

7. Substitute these values in Eq. a

Thus, $\pi_1 = g^{-1/2} D^{-1/2} V$, or $\pi_1 = V/(gD)^{1/2}$.

8. Follow the same procedure for the remaining nonrepeating quantities

Select ρ_w and ρ_a in turn as the nonrepeating quantities. Follow the same procedure as before to obtain the following dimensionless parameters: $\pi_2 = \rho_w(gD^3)^{1/2}/\mu_a$, and $\pi_3 = \rho_a(gD^3)^{1/2}/\mu_a = (gD^3)^{1/2}/\nu_a$, where ν_a = kinematic viscosity of air.

HYDRAULIC SIMILARITY AND CONSTRUCTION OF MODELS

A dam discharges 36,000 ft³/s (1,019,236.7 L/s) of water, and a hydraulic jump occurs on the apron. The power loss resulting from this jump is to be determined by constructing a geometrically similar model having a scale of 1:12. (a) Determine the required discharge in the model. (b) Determine the power loss on the dam if the power loss on the model is found to be 0.18 hp (0.134 kW).

Calculation Procedure:

1. Determine the value of Q_m

Two systems are termed similar if their corresponding variables have a constant ratio. A hydraulic model and its prototype must possess three forms of similarity: geometric, or similarity of shape; kinematic, or similarity of motion; and dynamic, or similarity of forces.

In the present instance, the ratio associated with the geometric similarity is given, i.e., the ratio of a linear dimension in the model to the corresponding linear dimension in the prototype. Let r_g denote this ratio, and let subscripts m and p refer to the model and prototype, respectively.

Apply Eq. 89a to evaluate Q_m. Or $Q = C_1 b h^{1.5}$, where C_1 is a constant. Then $Q_m/Q_p = (b_m/b_p)(h_m/h_p)^{1.5}$. But $b_m/b_p = h_m/h_p = r_g$; therefore, $Q_m/Q_p = r_g^{2.5} = (\frac{1}{12})^{2.5} = 1/499$; $Q_m = 36,000/499 = 72$ ft³/s (2038.5 L/s).

2. Evaluate the power loss on the dam

Apply Eq. 88 to evaluate the power loss on the dam. Thus, hp = $C_2 Q h$, where C_2 is a constant. Then $hp_p/hp_m = (Q_p/Q_m)(h_p/h_m)$. But $Q_p/Q_m = (1/r_g)^{2.5}$, and $h_p/h_m = 1/r_g$; therefore, $hp_p/hp_m = (1/r_g)^{3.5} = 12^{3.5} = 5990$. Hence, $hp_p = 5990(0.18) = 1078$ hp (803.86 kW).

Surveying and Route Design

PLOTTING A CLOSED TRAVERSE

Complete the following table for a closed traverse.

Course	Bearing	Length, ft (m)
a	N32°27′E	110.8 (33.77)
b		83.6 (25.48)
c	S8°51′W	126.9 (38.68)
d	S73°31′W	
e	N18°44′W	90.2 (27.49)

Calculation Procedure:

1. Draw the known courses; then form a closed traverse

Refer to Fig. 182a. A line PQ is described by expressing its length L and its bearing α with respect to a reference meridian NS. For a closed traverse, such as $abcde$ in Fig. 182b, the algebraic sum

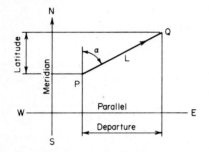

(a) Latitude and departure   (b) Closure of traverse

FIG. 182

of the latitudes and the algebraic sum of the departures must equal zero. A positive latitude corresponds to a northerly bearing, and a positive departure corresponds to an easterly bearing.

In Fig. 183, draw the known courses a, c, and e. Then introduce the hypothetical course f to form a closed traverse.

2. Calculate the latitude and departure of the courses

Use these relations:

$$\text{Latitude} = L \cos \alpha \tag{107}$$

$$\text{Departure} = L \sin \alpha \tag{108}$$

Computing the results for courses a, c, e, and f, we have the values shown in the following table.

Course	Latitude, ft (m)	Departure, ft (m)
a	+93.5	+59.5
c	−125.4	−19.5
e	+85.4	−29.0
Total	+53.5 (+16.306)	+11.0 (+3.35)
f	−53.5 (−16.306)	−11.0 (−3.35)

3. Find the length and bearing of f

Thus, $\tan \alpha_f = 11.0/53.5$; therefore, the bearing of f = S11°37′W; length of f = 53.5/cos α_f = 54.6 ft (16.64 m).

4. *Complete the layout*

Complete Fig. 183 by drawing line d through the upper end of f with the specified bearing and by drawing a circular arc centered at the lower end of f having a radius equal to the length of b.

5. *Find the length of d and the bearing of b*

Solve the triangle fdb to find the length of d and the bearing of b. Thus, $B = 73°31' - 11°37' = 61°54'$. By the law of sines, sin $F = f$ sin $B/b = 54.6$ sin $61°54'/83.6$; $F = 35°11'$; $D = 180° - (61°54' + 35°11') = 82°55'$; $d = b$ sin $D/$sin $B = 83.6$ sin $82°55'$ $/$sin $61°54' = 94.0$ ft (28.65 m); $\alpha_b = 180° - (73°31' + 35°11') = 71°18'$. The bearing of b is S71°18'E.

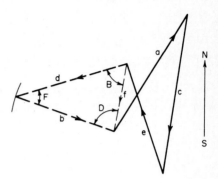

AREA OF TRACT WITH RECTILINEAR BOUNDARIES

The balanced latitudes and departures of a closed transit-and-tape traverse are recorded in the table below. Compute the area of the tract by the DMD method.

FIG. 183

Course	Latitude, ft (m)	Departure, ft (m)
AB	−132.3 (−40.33)	−135.6 (−41.33)
BC	+9.6 (2.93)	−77.5 (−23.62)
CD	+97.9 (29.84)	−198.5 (−60.50)
DE	+161.9 (49.35)	+143.6 (43.77)
EF	−35.3 (−10.76)	+246.7 (75.19)
FA	−101.8 (−31.03)	+21.3 (6.49)

Calculation Procedure:

1. *Plot the tract*

Refer to Fig. 184. The sum of m_1 and m_2 is termed the *double meridian distance* (DMD) of course AB. Let D denote the departure of a course. Then

$$\text{DMD}_n = \text{DMD}_{n-1} + D_{n-1} + D_n \tag{109}$$

where the subscripts refer to two successive courses.

The area of trapezoid $ABba$, which will be termed the *projection area* of AB, equals half the product of the DMD and latitude of the course. A projection area may be either positive or negative.

Plot the tract in Fig. 185. Since D is the most westerly point, pass the reference meridian through D, thus causing all DMDs to be positive.

2. *Establish the DMD of each course by successive applications of Eq. 109*

Thus, $\text{DMD}_{DE} = 143.6$ ft (43.77 m); $\text{DMD}_{EF} = 143.6 + 143.6 + 246.7 = 533.9$ ft (162.73 m); $\text{DMD}_{FA} = 533.9 + 246.7 + 21.3 = 801.9$ ft (244.42 m); $\text{DMD}_{AB} = 801.9 + 21.3 - 135.6 = 687.6$ ft (209.58 m); $\text{DMD}_{BC} = 687.6 - 135.6 - 77.5 = 474.5$ ft (144.62 m); $\text{DMD}_{CD} = 474.5 - 77.5 - 198.5 = 198.5$ ft (60.50 m). This is acceptable.

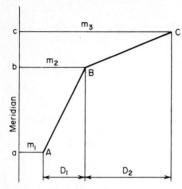

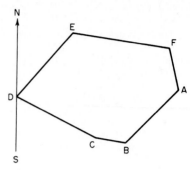

FIG. 184 Double meridian distance.   **FIG. 185**

3. Calculate the area of the tract

Use the following theorem: The area of a tract is numerically equal to the aggregate of the projection areas of its courses. The results of this calculation are

Course	Latitude	×	DMD	=	2 × Projection area
AB	−132.3		687.6		−90,970
BC	+9.6		474.5		+4,555
CD	+97.9		198.5		+19,433
DE	+161.9		143.6		+23,249
EF	−35.3		533.9		−18,847
FA	−101.8		801.9		−81,634
Total					−144,214

$$\text{Area} = \tfrac{1}{2}(144{,}214) = 72{,}107 \text{ ft}^2 \ (6698.74 \text{ m}^2)$$

PARTITION OF A TRACT

The tract in the previous calculation procedure is to be divided into two parts by a line through E, the part to the west of this line having an area of 30,700 ft² (2852.03 m²). Locate the dividing line.

Calculation Procedure:

1. Ascertain the location of the dividing line EG

This procedure requires the solution of an oblique triangle. Refer to Fig. 186. It will be necessary to apply the following equations, which may be readily developed by drawing the altitude *BD*:

$$\text{Area} = \tfrac{1}{2} \, bc \sin A \tag{110}$$

$$\tan C = \frac{c \sin A}{b - c \cos A} \tag{111}$$

In Fig. 187, let *EG* represent the dividing line of this tract. By scaling the dimensions and making preliminary calculations or by using a planimeter, ascertain that *G* lies between *B* and *C*.

2. Establish the properties of the hypothetical course EC

By balancing the latitudes and departures of *DEC*, latitude of $EC = -(+161.9 + 97.9) = -259.8$ ft $(-79.18$ m$)$; departure of $EC = -(+143.6 - 198.5) = +54.9$ ft $(+16.73$ m$)$; length

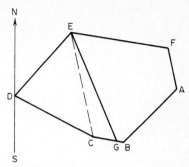

FIG. 187 Partition of tract.

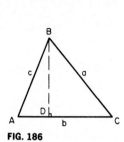

FIG. 186

of $EC = (259.8^2 + 54.9^2)^{0.5} = 265.5$ ft (80.92 m). Then $\text{DMD}_{DE} = 143.6$ ft (43.77 m); $\text{DMD}_{EC} = 143.6 + 143.6 + 54.9 = 342.1$ ft (104.27 m); $\text{DMD}_{CD} = 342.1 + 54.9 - 198.5 = 198.5$ ft (60.50 m). This is acceptable.

3. Determine angle GCE by finding the bearings of courses EC and BC

Thus $\tan \alpha_{EC} = 54.9/259.8$; bearing of $EC = S11°55.9'E$; $\tan \alpha_{BC} = 77.5/9.6$; bearing of $BC = N82°56.3'W$; angle $GCE = 180° - (82°56.3' - 11°55.9') = 108°59.6'$.

4. Determine the area of triangle GCE

Calculate the area of triangle DEC; then find the area of triangle GCE by subtraction. Thus

Course	Latitude	× DMD	= 2 × Projection area
CD	+97.9	198.5	+19,433
DE	+161.9	143.6	+23,249
EC	−259.8	342.1	−88,878
Total			−46,196

So the area of $DEC = \frac{1}{2}(46,196) = 23,098$ ft^2 (2145.8 m^2); area of $GCE = 30,700 - 23,098 = 7602$ ft^2 (706.22 m^2).

5. Solve triangle GCE completely

Apply Eqs. 110 and 111. To ensure correct substitution, identify the corresponding elements, making A the known angle GCE and c the known side EC. Thus

Fig. 186	Fig. 187	Known values	Calculated values
A	GCE	108°59.6'	
B	CEG		11°21.6'
C	EGC		59°38.8'
a	EG		291.0 ft (88.70 m)
b	GC		60.6 ft (18.47 m)
c	EC	265.5 ft (80.92 m)	

By Eq. 110, $7602 = \frac{1}{2}GC(265.5 \sin 108°59.6')$; solving gives $GC = 60.6$ ft (18.47 m). By Eq. 111, $\tan EGC = 265.5 \sin 108°59.6'/(60.6 - 265.5 \cos 108°59.6')$; $EGC = 59°38.8'$. By the law of sines, $EG/\sin GCE = EC/\sin EGC$; $EG = 291.0$ ft (88.70 m); $CEG = 180° - (108°59.6' + 59°38.8') = 11°21.6'$.

6. Find the bearing of course EG

Thus, $\alpha_{EG} = \alpha_{EC} + CEG = 11°55.9' + 11°21.6' = 23°17.5'$; bearing of $EG = S23°17.5'E$.

The surveyor requires the length and bearing of EG to establish this line of demarcation. She or he is able to check the accuracy of both the fieldwork and the office calculations by ascertaining that the point G established in the field falls on BC and that the measured length of GC agrees with the computed value.

AREA OF TRACT WITH MEANDERING BOUNDARY: OFFSETS AT IRREGULAR INTERVALS

The offsets below were taken from stations on a traverse line to a meandering stream, all data being in feet. What is the encompassed area?

Station	0 + 00	0 + 25	0 + 60	0 + 75	1 + 10
Offset	29.8	64.6	93.2	58.1	28.5

Calculation Procedure:

1. Assume a rectilinear boundary between successive offsets; develop area equations

Refer to Fig. 188. When a tract has a meandering boundary, this boundary is approximated by measuring the perpendicular offsets of the boundary from a straight line AB. Let d_r denote the distance along the traverse line between the first and the rth offset, and let $h_1, h_2, \ldots, h_n$ denote the offsets.

Developing the area equations yields

$$\text{Area} = \tfrac{1}{2}[d_2(h_1 - h_3) + d_3(h_2 - h_4) + \cdots + d_{n-1}(h_{n-2} - h_n) + d_n(h_{n-1} + h_n)] \quad (112)$$

Or,

$$\text{Area} = \tfrac{1}{2}[h_1 d_2 + h_2 d_3 + h_3(d_4 - d_2) + h_4(d_5 - d_3) + \cdots + h_n(d_n - d_{n-1})] \quad (113)$$

2. Determine the area, using Eq. 112

Thus, area $= \tfrac{1}{2}[25(29.8 - 93.2) + 60(64.6 - 58.1) + 75(93.2 - 28.5) + 110(58.1 + 28.5)] = 6590$ ft^2 (612.2 m^2).

3. Determine the area, using Eq. 113

Thus, area $= \tfrac{1}{2}[29.8 \times 25 + 64.6 \times 60 + 93.2(75 - 25) + 58.1(110 - 60) + 28.5(110 - 75)] = 6590$ ft^2 (612.2 m^2). Hence, both equations yield the same result. The second equation has a distinct advantage over the first because it has only positive terms.

DIFFERENTIAL LEVELING PROCEDURE

Complete the following level notes, and show an arithmetic check.

Point	BS, ft (m)	HI	FS ft (m)	Elevation, ft (m)
BM42	2.076 (0.63)	...	......	180.482 (55.01)
TP1	3.408 (1.04)	...	8.723 (2.66)	
TP2	1.987 (0.61)	...	9.826 (2.99)	
TP3	2.538 (0.77)	...	10.466 (3.19)	
TP4	2.754 (0.84)	...	8.270 (2.52)	
BM43	...	...	11.070 (3.37)	

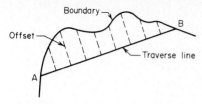

FIG. 188   Tract with irregular boundary.

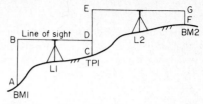

FIG. 189   Differential leveling.

Calculation Procedure:

1.   *Obtain the elevation for each point*

Differential leveling is used to ascertain the difference in elevation between two successive bench-marks by finding the elevations of several convenient intermediate points, called *turning points* (TP). In Fig. 189, consider that the instrument is set up at $L1$ and C is selected as a turning point. The rod reading AB represents the backsight (BS) of BM1, and rod reading CD represents the foresight (FS) of TP1. The elevation of BD represents the height of instrument (HI). The instrument is then set up at $L2$, and rod readings CE and FG are taken. Let a and b designate two successive turning points. Then

$$\text{Elevation}_a + \text{BS}_a = \text{HI} \tag{114}$$

$$\text{HI} - \text{FS}_b = \text{elevation}_b \tag{115}$$

Therefore,

$$\text{Elevation BM2} - \text{elevation BM1} = \Sigma\text{BS} - \Sigma\text{FS} \tag{116}$$

Apply Eqs. 114 and 115 successively to obtain the elevations recorded in the accompanying table.

Point	BS, ft (m)	HI, ft (m)	FS, ft (m)	Elevation, ft (m)
BM42	2.076 (0.63)	182.558 (55.64)	. . . . . .	180.482 (55.01)
TP1	3.408 (1.04)	177.243 (54.02)	8.723 (2.66)	173.835 (52.98)
TP2	1.987 (0.61)	169.404 (51.63)	9.826 (2.99)	167.417 (51.03)
TP3	2.538 (0.77)	161.476 (49.22)	10.466 (3.19)	158.938 (48.44)
TP4	2.754 (0.84)	155.960 (47.54)	8.270 (2.52)	153.206 (46.70)
BM43	. . . . . .	. . . . . .	11.070 (3.37)	144.890 (44.16)
Total	12.763 (3.89)	. . . . . .	48.355 (14.73)	

2.   *Verify the result by summing the backsights and foresights*

Substitute the results in Eq. 116: $144.890 - 180.482 = 12.763 - 48.355 = -35.592$.

STADIA SURVEYING

The following stadia readings were taken with the instrument at a station of elevation 483.2 ft (147.28 m), the height of instrument being 5 ft (1.5 m). The stadia interval factor is 100, and the value of C is negligible. Compute the horizontal distances and elevations.

Point	Rod intercept, ft (m)	Vertical angle
1	5.46 (1.664)	$+2°40'$ on 8 ft (2.4 m)
2	6.24 (1.902)	$+3°12'$ on 3 ft (0.9 m)
3	4.83 (1.472)	$-1°52'$ on 4 ft (1.2 m)

Calculation Procedure:

1. State the equations used in stadia surveying

Refer to Fig. 190 for the notational system pertaining to stadia surveying. The transit is set up over a reference point O, the rod is held at a control point N, and the telescope is sighted at a point Q on the rod; P and R represent the apparent locations of the stadia hairs on the rod.

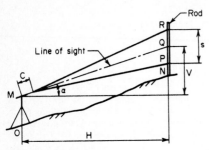

The first column in these notes presents the rod intercepts s, and the second column presents the vertical angle α and the distance NQ. Then

$$H = Ks \cos^2\alpha + C \cos \alpha \qquad (117)$$

$$V = \tfrac{1}{2} Ks \sin 2\alpha + C \sin \alpha \qquad (118)$$

Elevation of N = elevation of O

$$+ OM + V - NQ \qquad (119)$$

FIG. 190 Stadia surveying.

where K = stadia interval factor; C = distance from center of instrument to principal focus.

2. Substitute numerical values in the above equations

The results obtained are shown:

Point	H, ft (m)	V, ft (m)	Elevation, ft (m)
1	544.8 (166.06)	25.4 (7.74)	505.6 (154.11)
2	622.0 (189.59)	34.8 (10.61)	520.0 (158.50)
3	482.5 (147.07)	−15.7 (−4.79)	468.5 (142.80)

VOLUME OF EARTHWORK

Figure 191a and b represent two highway cross sections 100 ft (30.5 m) apart. Compute the volume of earthwork to be excavated, in cubic yards (cubic meters). Apply both the average-end-area method and the prismoidal method.

Calculation Procedure:

1. Resolve each section into an isosceles trapezoid and a triangle; record the relevant dimensions

Let A_1 and A_2 denote the areas of the end sections, L the intervening distance, and V the volume of earthwork to be excavated or filled.

Method 1: The average-end-area method equates the average area to the mean of the two end areas. Then

$$V = \frac{L(A_1 + A_2)}{2}. \qquad (120)$$

Figure 191c shows the first section resolved into an isosceles trapezoid and a triangle, along with the relevant dimensions.

2. Compute the end areas, and apply Eq. 120

Thus: $A_1 = [24(40 + 64) + (32 - 24)64]/2 = 1504$ ft^2 (139.72 m^2); $A_2 = [36(40 + 76) + (40 - 36)76]/2 = 2240$ ft^2 (208.10 m^2); $V = 100(1504 + 2240)/[2(27)] = 6933$ yd^3 (5301.0 m^3).

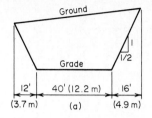

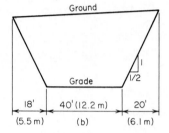

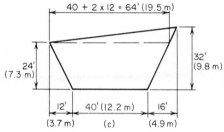

FIG. 191

3. Apply the prismoidal method

Method 2: The prismoidal method postulates that the earthwork between the stations is a prismoid (a polyhedron having its vertices in two parallel planes). The volume of a prismoid is

$$V = \frac{L(A_1 + 4A_m + A_2)}{6} \tag{121}$$

where A_m = area of center section.

Compute A_m. Note that each coordinate of the center section of a prismoid is the arithmetic mean of the corresponding coordinates of the end sections. Thus, $A_m = [30(40 + 70) + (36 - 30)70]/2 = 1860$ ft^2 (172.8 m^2).

4. Compute the volume of earthwork

Using Eq. 121 gives $V = 100(1504 + 4 \times 1860 + 2240)/[6(27)] = 6904$ yd^3 (5278.8 m^3).

DETERMINATION OF AZIMUTH OF A STAR BY FIELD ASTRONOMY

An observation of the sun was made at a latitude of 41°20′N. The altitude of the center of the sun, after correction for refraction and parallax, was 46°48′. By consulting a solar ephemeris, it was found that the declination of the sun at the instant of observation was 7°58′N. What was the azimuth of the sun?

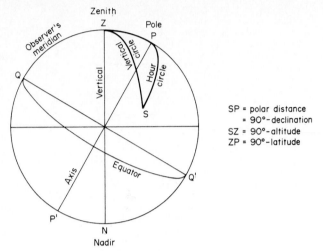

FIG. 192   The celestial sphere.

Calculation Procedures:

1.   *Calculate the azimuth of the body*

Refer to Fig. 192. The *celestial sphere* is an imaginary sphere on the surface of which the celestial bodies are assumed to be located; this sphere is of infinite radius and has the earth as its center. The *celestial equator*, or *equinoctial*, is the great circle along which the earth's equatorial plane intersects the celestial sphere. The *celestial axis* is the prolongation of the earth's axis of rotation. The *celestial poles* are the points at which the celestial axis pierces the celestial sphere. An *hour circle*, or a *meridian*, is a great circle that passes through the celestial poles.

The *zenith* and *nadir* of an observer are the points at which the vertical (plumb) line at the observer's site pierces the celestial sphere, the former being visible and the latter invisible to the observer. A *vertical circle* is a great circle that passes through the observer's zenith and nadir. The *observer's meridian* is the meridian that passes through the observer's zenith and nadir; it is both a meridian and a vertical circle.

In Fig. 192, P is the celestial pole, S is the apparent position of a star on the celestial sphere, and Z is the observer's zenith.

The coordinates of a celestial body *relative to the observer* are the *azimuth*, which is the angular distance from the observer's meridian to the vertical circle through the body as measured along the observer's horizon in a clockwise direction; and the *altitude*, which is the angular distance of the body from the observer's horizon as measured along a vertical circle.

The *absolute* coordinates of a celestial body are the *right ascension*, which is the angular distance between the vernal equinox and the hour circle through the body as measured along the celestial equator; and the *declination*, which is the angular distance of the body from the celestial equator as measured along an hour circle.

The relative coordinates of a body at a given instant are obtained by observation; the absolute coordinates are obtained by consulting an almanac of astronomical data. The latitude of the observer's site equals the angular distance of the observer's zenith from the celestial equator as measured along the meridian. In the astronomical triangle PZS in Fig. 192, the arcs represent the indicated coordinates, and angle Z represents the azimuth of the body as measured from the north.

Calculating the azimuth of the body yields

$$\tan^2 \tfrac{1}{2} Z = \frac{\sin (S - L) \sin (S - h)}{\cos S \cos (S - p)} \tag{122}$$

where L = latitude of site; h = altitude of star; p = polar distance = $90°$ − declination; S = $\frac{1}{2}(L + h + p)$; L = $41°20'$; h = $46°48'$; p = $90°$ − $7°58'$ = $82°02'$; S = $\frac{1}{2}(L + h + p)$ = $85°05'$; $S - L$ = $43°45'$; $S - h$ = $38°17'$; $S - p$ = $3°03'$.

Then

log sin 43°45′ =		9.839800
log sin 38°17′ =		9.792077
		9.631877
log cos 85°05′ =	8.933015	
log cos 3°03′ =	9.999384	8.932399
2 log tan ½Z =		0.699478
log tan ½Z =		0.349739
½Z = 65°55′03.5″	Z = 131°50′07″	

2. Verify the solution by calculating Z in an alternative manner

To do this, introduce an auxiliary angle M, defined by

$$\cos^2 M = \frac{\cos p}{\sin h \sin L} \tag{123}$$

Then

$$\cos (180° - Z) = \tan h \tan L \sin^2 M \tag{124}$$

Then

log cos 82°02′ =		9.141754
log sin 46°48′ =	9.862709	
log sin 41°20′ =	9.819832	9.682541
2 log cos M =		9.459213
log cos M =		9.729607
log sin M =		9.926276
2 log sin M =		9.852552
log tan 46°48′ =		0.027305
log tan 41°20′ =		9.944262
log cos (180° − Z) =		9.824119
Z = 131°50′07″, as before		

TIME OF CULMINATION OF A STAR

Determine the Eastern Standard Time (75th meridian time) of the upper culmination of Polaris at a site having a longitude 81°W of Greenwich. Reference to an almanac shows that the Greenwich Civil Time (GCT) of upper culmination for the date of observation is $3^h20^m05^s$.

Calculation Procedure:

1. Convert the longitudes to the hour-minute-second system

The rotation of the earth causes a star to appear to describe a circle on the celestial sphere centered at the celestial axis. The star is said to be *at culmination* or *transit* when it appears to cross the observer's meridian.

In Fig. 193, P and M represent the position of Polaris and the mean sun, respectively, when Polaris is at the Greenwich meridian, and P' and M' represent the position of these bodies when Polaris is at the observer's meridian. The distances h and h' represent, respectively, the time of culmination of Polaris at Greenwich and at the observer's site, measured from local noon. Since the apparent velocity of the mean sun is less than that of the stars, h' is less than h, the difference being approximately 10 s/h of longitude.

By converting the longitudes, 360° corresponds to 24 h; therefore, 15° corresponds to 1 h. Longitude of site = 81° = 5.4^h = $5^h24^m00^s$; standard longitude = 75° = 5^h.

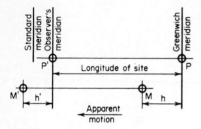

FIG. 193 Culmination of Polaris.

2. Calculate the time of upper culmination at the site

Correct this result to Eastern Standard Time. Since the standard meridian is east of the observer's meridian, the standard time is greater. Thus

GCT of upper culmination at Greenwich	$3^h20^m05^s$
Correction for longitude, 5.4 × 10 s	54^s
Local civil time of upper culmination at site	$3^h19^m11^s$
Correction to standard meridian	24^m00^s
EST of upper culmination at site	$3^h43^m11^s$a.m.

PLOTTING A CIRCULAR CURVE

A horizontal circular curve having an intersection angle of 28° is to have a radius of 1200 ft (365.7 m). The point of curve is at station 82 + 30. (a) Determine the tangent distance, long chord, middle ordinate, and external distance. (b) Determine all the data necessary to stake the curve if the *chord* distance between successive stations is to be 100 ft (30.5 m). (c) Calculate all the data necessary to stake the curve if the *arc* distance between successive stations is to be 100 ft (30.5 m).

Calculation Procedure:

1. Determine the geometric properties of the curve

Part a. Refer to Fig. 194: A is termed the *point of curve* (PC), B is the *point of tangent* (PT), and V the *point of intersection* (PI), or vertex. The notational system is Δ = intersection angle = angle between back and forward tangents = central angle AOB; R = radius of curve; T = tangent distance = AV = VB; C = long chord = AB; M = middle ordinate = DC; E = external distance = CV.

From the geometric relationships,

$$T = R \tan \tfrac{1}{2} \Delta$$
$$T = 1200(0.2493) = 299.2 \text{ ft (91.20 m)} \tag{125}$$

Also

$$C = 2R \sin \tfrac{1}{2} \Delta$$
$$C = 2(1200)(0.2419) = 580.6 \text{ ft (176.97 m)} \tag{126}$$

And,

$$M = R(1 - \cos \tfrac{1}{2} \Delta)$$
$$M = 1200(1 - 0.9703) = 35.6 \text{ ft (10.85 m)} \tag{127}$$

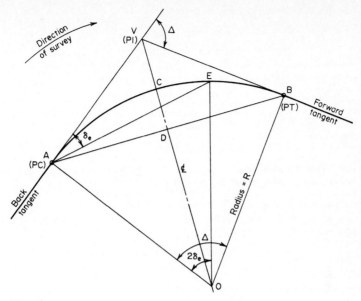

FIG. 194 Circular curve.

Lastly,

$$E = R \tan \tfrac{1}{2} \Delta \tan \tfrac{1}{4} \Delta \tag{128}$$
$$E = 1200(0.2493)(0.1228) = 36.7 \text{ ft } (11.19 \text{ m})$$

2. *Verify the results in step 1*

Use the pythagorean theorem on triangle *ADV*. Or, $AD = \tfrac{1}{2}(580.6) = 290.3$ ft (88.48 m); $DV = 35.6 + 36.7 = 72.3$ ft (22.04 m); then $290.3^2 + 72.3^2 = 89,500$ ft^2 (8314.6 m^2), to the nearest hundred; $299.2^2 = 89,500$ ft^2 (8314.6 m^2); this is acceptable.

3. *Calculate the degree of curve D*

Part b. In Fig. 194, let *E* represent a station along the curve. Angle *VAE* is termed the *deflection angle* δ_e of this station; it is equal to one-half the central angle *AOE*. In the field, the curve is staked by setting up the transit at the *PC* and then locating each station by means of its deflection angle and its chord distance from the preceding station.

Calculate the *degree of curve D*. This is the central angle formed by the radii to two successive stations or, what is the same in this instance, the central angle subtended by a *chord* of 100 ft (30.5 m). Then

$$\sin \tfrac{1}{2} D = \frac{50}{R} \tag{129}$$

So $\tfrac{1}{2}D = \arcsin 50/1200 = \arcsin 0.04167$; $\tfrac{1}{2}D = 2°23.3'$; $D = 4°46.6'$.

4. *Determine the station at the PT*

Number of stations on the curve $= 28°/4°46.6' = 5.862$; station of PT $= (82 + 30) + (5 + 86.2) = 88 + 16.2$.

5. *Calculate the deflection angle of station 83 and the difference between the deflection angles of station 88 and the PT*

For simplicity, assume that central angles are directly proportional to their corresponding chord lengths; the resulting error is negligible. Then $\delta_{83} = 0.70(2°23.3') = 1°40.3'$; $\delta_{PT} - \delta_{88} = 0.162(2°23.3') = 0°23.2'$.

6. Calculate the deflection angle of each station

Do this by adding ½D to that of the preceding station. Record the results thus:

Station	Deflection angle
82 + 30	0
83	1°40.3′
84	4°03.6′
85	6°26.9′
86	8°50.2′
87	11°13.5′
88	13°36.8′
88 + 16.2	14°00′

7. Calculate the degree of curve in the present instance

Part c. Since the subtended central angle is directly proportional to its arc length, $D/100 = 360/(2\pi R)$; therefore,

$$D = 18{,}000/\pi R = 5729.58/R \text{ degrees} \qquad (130)$$

Then, $D = 5729.58/1200 = 4.7747° = 4°46.5′$.

8. Repeat the calculations in steps 4, 5, and 6

INTERSECTION OF CIRCULAR CURVE AND STRAIGHT LINE

In Fig. 195, MN represents a straight railroad spur that intersects the curved higway route AB. Distances on the route are measured along the arc. Applying the recorded data, determine the station of the intersection point P.

Calculation Procedure:

1. Apply trigonometric relationships to determine three elements in triangle ONP

Draw line OP. The problem resolves itself into the calculation of the central angle AOP, and this may be readily found by solving the oblique triangle ONP. Applying trigonometric relationships gives $AV = T = 800 \tan 54° = 1101.1$ ft (335.62 m); $AM = 1101.1 - 220 = 881.1$ ft (268.56 m); $AN = AM \tan 28° = 468.5$ ft (142.80 m); $ON = 800 - 468.5 = 331.5$ ft (101.04 m); $OP = 800$ ft (243.84 m); $ONP = 90° + 28° = 118°$.

2. Establish the station of P

Solve triangle ONP to find the central angle; then calculate arc AP and establish the station of P. By the law of sines, $\sin OPN = \sin ONP(ON)/OP$; therefore, $OPN = 21°27.7′$; $AOP = 180° - (118° + 21°27.7′) = 40°32.3′$; arc $AP = 2\pi(800)(40°32.3′)/360° = 566.0$ ft (172.52 m); station of $P = (22 + 00) + (5 + 66) = 27 + 66$.

REALIGNMENT OF CIRCULAR CURVE BY DISPLACEMENT OF FORWARD TANGENT

In Fig. 196, the horizontal circular curve AB has a radius of 720 ft (219.5 m) and an intersection angle of 126°. The curve is to be realigned by rotating the forward tangent through an angle of 22° to the new position $V'B$ while maintaining the PT at B. Compute the radius, and locate the PC of the new curve.

Calculation Procedure:

1. Find the tangent distance of the new curve

Solve triangle $BV'V$ to find the tangent distance of the new curve and the location of V'. Thus, $\Delta' = 126° - 22° = 104°$; $VB = 720 \tan 63° = 1413.1$ ft (430.71 m). By the law of sines, $V'B$

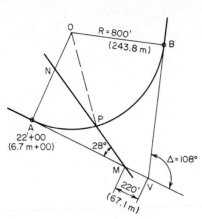

FIG. 195 Intersection of curve and straight line.   **FIG. 196** Displacement of forward tangent.

$= 1413.1 \sin 126°/\sin 104° = 1178.2$ ft (359.12 m); $V'V = 1413.1 \sin 22°/\sin 104° = 545.6$ ft (166.30 m).

2. Compute the radius R'

By Eq. 125, $R' = 1178.2 \cot 52° = 920.5$ ft (280.57 m).

3. Determine the station of A'

Thus, $AV = VB = 1413.1$ ft (403.71 m); $A'V' = V'B = 1178.2$ ft (359.12 m); $A'A = A'V' + V'V - AV = 310.7$ ft (94.70 m); station of new PC $= (34 + 41) - (3 + 10.7) = 31 + 30.3$.

4. Verify the foregoing results

Draw the long chords AB and $A'B$. Then apply the computed value of R' to solve triangle $BA'A$ and thereby find $A'A$. By Eq. 126, $A'B = 2R' \sin \frac{1}{2}\Delta' = 1450.7$ ft (442.17 m); $AA'B = \frac{1}{2}\Delta' = 52°$; $A'AB = 180° - \frac{1}{2}\Delta = 117°$; $ABA' = 180° - (52° + 117°) = 11°$. By the law of sines, $A'A = 1450.7 \sin 11°/\sin 117° = 310.7$ ft (94.70 m). This is acceptable.

CHARACTERISTICS OF A COMPOUND CURVE

The tangents to a horizontal curve intersect at an angle of 68°22′. To fit the curve to the terrain, it is necessary to use a compound curve having tangent lengths of 955 ft (291.1 m) and 800 ft (243.8 m), as shown in Fig. 197. The minimum allowable radius is 1000 ft (304.8 m). Compute the larger radius and the two central angles.

Calculation Procedure:

1. Calculate the latitudes and departures of the known sides

A *compound curve* is a curve that comprises two successive circular arcs of unequal radii that are tangent at their point of intersection, the centers of the arcs lying on the same side of their common tangent. (Where the centers lie on opposite sides of this tangent, the curve is termed a *reversed curve*). In Fig. 197, C is the point of intersection of the arcs, and DE is the common tangent.

This situation is analyzed without applying any set equation to illustrate the general method of solution for compound and reversed curves. There are two unknown quantities: the radius R_1 and a central angle. (Since $\Delta_1 + \Delta_2 = \Delta$, either central angle may be considered the unknown.)

If the polygon $AVBO_2O_1$ is visualized as a closed traverse, the latitudes and departures of its sides are calculated, and the sum of the latitudes and sum of the departures are equated to zero,

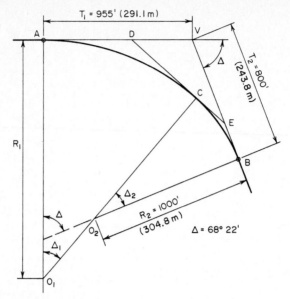

FIG. 197 Compound curve.

two simultaneous equations containing these two unknowns are obtained. For convenience, select O_1A as the reference meridian. Then

Side	Length, ft (m)	Bearing	Latitude	Departure
AV	955 (291.1)	90°	0	+955.00
VB	800 (243.8)	21°38′	−743.65	+294.93
BO_2	1000 (304.8)	68°22′	−368.67	−929.56
Total	...	...	−1112.32	+320.37

2. Express the latitudes and departures of the unknown sides in terms of R_1 and Δ_1

Thus, for side O_2O_1: length $= R_1 - 1000$; bearing $= \Delta_1$; latitude $= -(R_1 - 1000)\cos\Delta_1$; departure $= -(R_1 - 1000)\sin\Delta_1$.

Also, for side O_1A: length $= R_1$; bearing $= 0$; latitude $= R_1$; departure $= 0$.

3. Equate the sum of the latitudes and sum of the departures to zero; express Δ_1 as a function of R_1

Thus, $\Sigma\text{lat} = R_1 - (R_1 - 1000)\cos\Delta_1 - 1112.32 = 0$; $\cos\Delta_1 = (R_1 - 1112.32)/(R_1 - 1000)$, or $1 - \cos\Delta_1 = 112.32/(R_1 - 1000)$, Eq. *a*. Also, $\Sigma\text{dep} = -(R_1 - 1000)\sin\Delta_1 + 320.37 = 0$; $\sin\Delta_1 = 320.37/(R_1 - 1000)$, Eq. *b*.

4. Divide Eq. a by Eq. b, and determine the central angles

Thus, $(1 - \cos\Delta_1)/\sin\Delta_1 = \tan\frac{1}{2}\Delta_1 = 112.32/320.37$; $\frac{1}{2}\Delta_1 = 19°19′13″$; $\Delta_1 = 38°38′26″$; $\Delta_2 = 68°22′ - \Delta_1 = 29°43′34″$.

5. Substitute the value of Δ_1 in Eq. b to find R_1

Thus, $R_1 = 1513.06$ ft (461.181 m).

6. Verify the foregoing results by analyzing triangle DEV

Thus, $AD = R_1 \tan \frac{1}{2}\Delta_1 = 530.46$ ft (161.684 m); $DV = 955 - 530.46 = 424.54$ ft (129.400 m); $EB = R_2 \tan \frac{1}{2}\Delta_2 = 265.40$ ft (80.894 m); $VE = 800 - 265.40 = 534.60$ ft (162.946 m); $DE = 530.46 + 265.40 = 795.86$ ft (242.578 m). By the law of cosines, $\cos \Delta = -(DV^2 + VE^2 - DE^2)/[2(DV)(VE)]$; $\Delta = 68°22'$. This is correct.

ANALYSIS OF A HIGHWAY TRANSITION SPIRAL

A horizontal circular curve for a highway is to be designed with transition spirals. The PI is at station 34 + 93.81, and the intersection angle is 52°48'. In accordance with the governing design criteria, the spirals are to be 350 ft (106.7 m) long and the degree of curve of the circular curve is to be 6° (arc definition). The approach spiral will be staked by setting the transit at the TS and locating 10 stations on the spiral by means of their deflection angles from the main tangent. Compute all data needed for staking the approach spiral. Also, compute the long tangent, short tangent, and external distance.

Calculation Procedure:

1. Calculate the basic values

In the design of a road, a spiral is interposed between a straight-line segment and a circular curve to effect a gradual transition from rectilinear to circular motion, and vice versa. The type of spiral most frequently used is the clothoid, which has the property that the curvature at a given point is directly proportional to the distance from the start of the curve to the given point, measured along the curve.

Refer to Fig. 198. The key points are identified by the following notational system: PI = point of intersection of main tangents; TS = point of intersection of main tangent and approach spiral (tangent-to-spiral point); SC = point of intersection of approach spiral and circular curve (spiral-to-curve point); CS = point of intersection of circular curve and departure spiral (curve-to-spiral point); ST = point of intersection of departure spiral and main tangent (spiral-to-tangent point); PC and PT = point at which tangents to the circular curve prolonged are parallel to the main tangents (also referred to as the *offsets*). Distances are designated in the following manner: L_s = length of spiral from TS to SC; L = length of spiral from TS to given point on spiral; R_c radius of circular curve; R = radius of curvature at given point on spiral; T_s = length of main tangent from TS to PI; E_s = external distance, i.e., distance from PI to midpoint of circular curve.

In addition, there is a long tangent (LT), short tangent (ST), and long chord (LC), as indicated with respect to the departure spiral.

Place the origin of coordinates at the TS and the x axis on the main tangent. Then x_c and y_c = coordinates of SC; k and p = abscissa and ordinate, respectively, of PC. The coordinates of the SC and PC are useful as parameters in the calculation of required distances. The distance p is termed the *throw*, or *shift*, of the curve; it represents the displacement of the circular curve from the main tangent resulting from interposition of the spiral.

The basic angles are Δ = angle between main tangents, or intersection angle; Δ_c = angle between radii at SC and CS, or central angle of circular curve; θ_s = angle between radii of spiral at TS and SC, or central angle of entire spiral; D_c = degree of curve of circular curve; D = degree of curve at given point on spiral; δ_s = deflection angle of SC from main tangent, with transit at TS; δ = deflection angle of given point on spiral from main tangent, with transit at TS.

Although extensive tables of spiral values have been compiled, this example is solved without recourse to these tables in order to illuminate the relatively simple mathematical relationships that inhere in the clothoid. Consider that a vehicle starts at the TS and traverses the approach spiral at constant speed. The degree of curve, which is zero at the TS, increases at a uniform rate to become D_c at the SC. The basic equations are

$$\theta_s = \frac{L_s D_c}{200} = \frac{L_s}{2R_c} \tag{131}$$

$$x_c = L_s\left(1 - \frac{\theta_s^2}{10}\right) \tag{132}$$

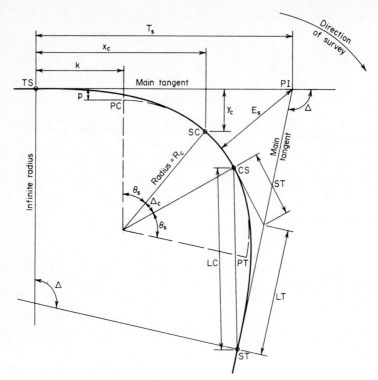

FIG. 198 Notational system for transition spirals.

$$y_c = L_s \left(\frac{\theta_s}{3} - \frac{\theta_s^3}{42} \right) \tag{133}$$

$$k = x_c - R_c \sin \theta_s \tag{134}$$

$$p = y_c - R_c(1 - \cos \theta_s) \tag{135}$$

$$\delta_s = \frac{L_s}{6R_c} = \frac{\theta_s}{3} \tag{136}$$

$$y = \left(\frac{L}{L_s} \right)^3 y_c \tag{137}$$

$$\delta = \left(\frac{L}{L_s} \right)^2 \delta_s \tag{138}$$

$$T_s = (R_c + p) \tan \tfrac{1}{2}\Delta + k \tag{139}$$

$$E_s = (R_c + p)(\sec \tfrac{1}{2}\Delta - 1) + p \tag{140}$$

$$LT = x_c - y_c \cot \theta_s \tag{141}$$

$$ST = y_c \csc \theta_s \tag{142}$$

Even though several of the foregoing equations are actually approximations, their use is valid when the value of D_c is relatively small.

Calculating the basic values yields Δ = 52°48'; L_s = 350 ft (106.7 m); D_c = 6°; $\theta_s = L_s D_c/$ 200 = 350(6)/200 = 10.5° = 10°30', or θ_s = 10.5(0.017453) = 0.18326 rad; Δ_c = 52°48' − 2(10°30') = 31°48'; D_c = 6(0.017453) = 0.10472 rad; R_c = 100/D_c = 954.93 ft (291.063 m); x_c = 350(1 − 0.18326²/10) = 348.83 ft (106.323 m); y_c = 350(0.18326/3 − 0.18326²/42) = 21.33 ft (6.501 m); k = 348.83 − 954.93 sin 10°30' = 174.80 ft (53.279 m); p = 21.33 − 954.93(1 − cos 10°30') = 5.34 ft (1.628 m).

2.   Locate the TS and SC

Thus, T_s = (954.93 + 5.34) tan 26°24' + 174.80 = 651.47; station of TS = (34 + 93.81) − (6 + 51.47) = 28 + 42.34; station of SC = (28 + 42.34) + (3 + 50.00) = 31 + 92.34.

3.   Calculate the deflection angles

Thus, δ_s = 10°30'/3 = 3°30' = 3.5°. Apply Eq. 138 to find the deflection angles at the intermediate stations. For example, for point 7, $\delta = 0.7^2(3.5°)$ = 1.715° = 1°42.9'.

TABLE 13   Deflection Angles on Approach Spiral

Point	Station	Deflection angle
TS	28 + 42.34	0
1	77.34	0°02.1'
2	29 + 12.34	0°08.4'
3	47.34	0°18.9'
4	82.34	0°33.6'
5	30 + 17.34	0°52.5'
6	52.34	1°15.6'
7	87.34	1°42.9'
8	31 + 22.34	2°14.4'
9	57.34	2°50.1'
SC	92.34	3°30.0'

Record the results in Table 13. The chord lengths between successive stations differ from the corresponding arc lengths by negligible amounts, and therefore each chord length may be taken as 35.00 ft (1066.8 cm).

4.   Compute the LT, ST, and E_s

Thus, LT = 348.83 − 21.33 cot 10°30' = 233.75 ft (7124.7 cm); ST = 21.33 csc 10°30' = 117.04 ft (3567.4 cm); E_s = (954.93 + 5.34)(sec 26°24' − 1) + 5.34 = 117.14 ft (3570.4 cm).

5.   Verify the last three calculations by substituting in the following test equation

Thus

$$\frac{ST + R_c \tan \frac{1}{2}\Delta_c}{\cos \frac{1}{2}\Delta} = \frac{T_s - LT}{\cos \frac{1}{2}\Delta_c} = \frac{E_s - R_c (\sec \frac{1}{2}\Delta_c - 1)}{\sin \theta_s} \qquad (143)$$

TRANSITION SPIRAL: TRANSIT AT INTERMEDIATE STATION

Referring to the transition spiral in the previous calculation procedure, assume that lack of visibility from the TS makes these setups necessary: Points 4, 5, 6, and 7 will be located with the transit at point 3; points 8 and 9 and the SC will be located with the transit at point 7. Compute the orientation and deflection angles.

Calculation Procedure:

1.   Consider that the transit is set up at point 3 and a backsight is taken to the TS; find the orientation angle

In Fig. 199, assume that the spiral has been staked up to P with the transit set up at the TS and that the remainder of the spiral is to be staked with the transit set up at P. Deflection angles are measured from the tangent through P (the *local* tangent). The instrument is oriented by backsighting to a preceding point B and then turning the angle δ_b. The orientation angle to B and deflection angle to a subsequent point F are

$$\delta_b = (2L_p + L_b)(L_p - L_b)\frac{\theta_s}{3L_s^2} \qquad (144)$$

$$\delta_f = (2L_p + L_f)(L_f - L_p)\frac{\theta_s}{3L_s^2} \qquad (145)$$

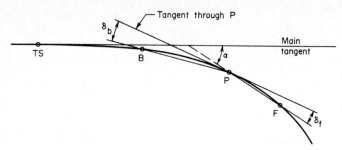

FIG. 199 Deflection angles from local tangent to spiral.

If B, P, and F are points obtained by dividing the spiral into an integral number of arcs, these equations may be converted to these more suitable forms:

$$\delta_b = (2n_p + n_b)(n_p - n_b)\frac{\theta_s}{3n_s^2} \tag{144a}$$

$$\delta_f = (2n_p + n_f)(n_f - n_p)\frac{\theta_s}{3n_s^2} \tag{145a}$$

where n denotes the number of arcs to the designated point.

Applying Eq. 144a to find the orientation angle and using data from the previous calculation procedure, we find $\theta_s = 10.5°$, $\theta_s/(3n_s^2) = 10.5°/[3(10^2)] = 0.035° = 2.1'$; $n_b = 0$; $n_p = 3$; $\delta_b = 6(3)(2.1') = 0°37.8'$.

2. Find the deflection angles from the tangent through point 3

Thus, by Eq. 145a: point 4, $\delta = (6 + 4)(2.1') = 0°21'$; point 5, $\delta = (6 + 5)(2)(2.1') = 0°46.2'$; point 6, $\delta = (6 + 6)(3)(2.1') = 1°15.6'$; point 7, $\delta = (6 + 7)(4)(2.1') = 1°49.2'$.

3. Consider that the transit is set up at point 7 and a backsight is taken to point 3; compute the orientation angle

Thus $n_b = 3$; $n_p = 7$; $\delta_b = (14 + 3)(4)(2.1') = 2°22.8'$.

4. Compute the deflection angles from the tangent through point 7

Thus point 8, $\delta = (14 + 8)(2.1') = 0°46.2'$; point 9, $\delta = (14 + 9)(2)(2.1') = 1°36.6'$; SC, $\delta = (14 + 10)(3)(2.1') = 2°31.2'$.

5. Test the results obtained

In Fig. 199, extend chord PF to its intersection with the main tangent, and let α denote the angle between these lines. Then

$$\alpha = (n_f^2 + n_f n_p + n_p^2)\frac{\theta_s}{3n_s^2} \tag{146}$$

This result should equal the sum of the angles applied in staking the curve from the TS to F. This procedure will be shown with respect to point 9.

For point 9, let P and F refer to points 7 and 9, respectively. Then $\alpha = (9^2 + 9 \times 7 + 7^2)(2.1') = 6°45.3'$. Summing the angles leading from the TS to point 9, we get

Deflection angle from main tangent to point 3	0°18.9'
Orientation angle at point 3	0°37.8'
Deflection angle from local tangent to point 7	1°49.2'
Orientation angle at point 7	2°22.8'
Deflection angle from local tangent to point 9	1°36.6'
Total	6°45.3'

This test may be applied to each deflection angle beyond point 3.

PLOTTING A PARABOLIC ARC

A grade of −4.6 percent is followed by a grade of +1.8 percent, the grades intersecting at station 54 + 20 of elevation 296.30 ft (90.312 m). The change in grade is restricted to 2 percent in 100 ft (30.5 m). Compute the elevation of every 50-ft (15.24-m) station on the parabolic curve, and locate the sag (lowest point of the curve). Apply both the average-grade method and the tangent-offset method.

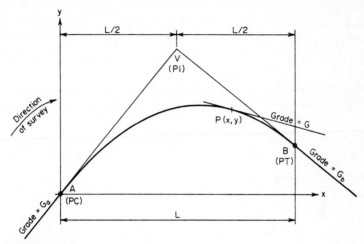

FIG. 200 Parabolic arc.

Calculation Procedure:

1. Compute the required length of curve

Using the notation in Figs. 200 and 201, we have G_a = −4.6 percent; G_b = +1.8 percent; r = rate of change in grade = 0.02 percent per foot; $L = (G_b - G_a)/r = [1.8 - (-4.6)]/0.02 = 320$ ft (97.5 m).

2. Locate the PC and PT

The station of the PC = station of the PI − $L/2$ = (54 + 20) − (1 + 60) = 52 + 60; station of PT = (54 + 20) + (1 + 60) = 55 + 80; elevation of PC = elevation of PI − $G_aL/2$ = 296.30 + 0.046(160) = 303.66 ft (92.556 m); elevation of PT = 296.30 + 0.018(160) = 299.18 ft (91.190 m).

3. Use the average-grade method to find the elevation of each station

Calculate the grade at the given station; calculate the average grade between the PI and that station, and multiply the average grade by the horizontal distance to find the ordinate. Equations used in analyzing a parabolic arc are

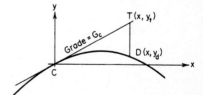

FIG. 201 Tangent offset.

$$y = \frac{rx^2}{2} + G_a x \tag{147}$$

$$G = rx + G_a \tag{148}$$

$$y = (G_a + G)\frac{x}{2} \tag{149}$$

$$DT = -\frac{rx^2}{2} \tag{150a}$$

$$DT = (G_c - G_d)\frac{x}{2} \tag{150b}$$

where DT = distance in Fig. 201.

Applying Eq. 148 with respect to station $53 + 00$ yields $x = 40$ ft (12.2 m); $G = 0.0002(40) - 0.046 = -0.038$; $G_{av} = (-0.046 - 0.038)/2 = -0.042$; $y = -0.042(40) = -1.68$ ft (-51.206 cm); elevation $= 303.66 - 1.68 = 301.98$ ft (9204.350 cm). Perform these calculations for each station, and record the results in tabular form as shown:

Station	x, ft (m)	G	G_{av}	y, ft (m)	Elevation, ft (m)
52 + 60	0 (0)	−0.046	−0.046	0 (0)	303.66 (92.56)
53 + 00	40 (12.2)	−0.038	−0.042	−1.68 (−0.51)	301.98 (92.04)
53 + 50	90 (27.4)	−0.028	−0.037	−3.33 (−1.01)	300.33 (91.54)
54 + 00	140 (42.7)	−0.018	−0.032	−4.48 (−1.37)	299.18 (91.19)
54 + 50	190 (57.9)	−0.008	−0.027	−5.13 (−1.56)	298.53 (90.99)
55 + 00	240 (73.2)	+0.002	−0.022	−5.28 (−1.61)	298.38 (90.95)
55 + 50	290 (88.4)	+0.012	−0.017	−4.93 (−1.50)	298.73 (91.05)
55 + 80	320 (97.5)	+0.018	−0.014	−4.48 (−1.37)	299.18 (91.19)

4. Verify the foregoing results

Apply the principle that for a uniform horizontal spacing the "second differences" between the ordinate are equal. The results are shown:

Calculation of Differences

Elevations, ft (m)	First differences, ft (m)	Second differences, ft (m)
301.98 (92.04)		
	1.65 (0.5029)	
300.33 (91.54)		0.50 (0.1524)
	1.15 (0.3505)	
299.18 (91.19)		0.50 (0.1525)
	0.65 (0.1981)	
298.53 (90.99)		0.50 (0.1524)
	0.15 (0.0457)	
298.38 (90.95)		0.50 (0.1524)
	−0.35 (0.10668)	
298.73 (91.05)		

5. Apply the tangent-offset method to find the elevation of each station

Since this method is based on Eq. 147, substitute directly in that equation. For the present case, the equation becomes $y = rx^2/2 + G_a x = 0.0001x^2 - 0.046x$. Record the calculations for y in tabular form. The results, as shown, agree with those obtained by the average-grade method.

Tangent-Offset Method

Station	x, ft (m)	$0.0001x^2$, ft (m)	$0.046x$, ft (m)	y, ft (m)
52 + 60	0 (0)	0 (0)	0 (0)	0 (0)
53 + 00	40 (12.19)	0.16 (0.05)	1.84 (0.56)	−1.68 (−0.51)
53 + 50	90 (27.43)	0.81 (0.25)	4.14 (1.26)	−3.33 (−1.01)
54 + 00	140 (42.67)	1.96 (0.60)	6.44 (1.96)	−4.48 (−1.37)
54 + 50	190 (57.91)	3.61 (1.10)	8.74 (2.66)	−5.13 (−1.56)
55 + 00	240 (73.15)	5.76 (1.76)	11.04 (3.36)	−5.28 (−1.61)
55 + 50	290 (88.39)	8.41 (2.56)	13.34 (4.07)	−4.93 (−1.50)
55 + 80	320 (97.54)	10.24 (3.12)	14.72 (4.49)	−4.48 (−1.37)

6. Locate the sag S

Since the grade is zero at this point, Eq. 148 yields $G_s = rx_s + G_a = 0$; therefore $x_s = -G_a/r$ = $-(-0.046/0.0002) = 230$ ft (70.1 m); station of sag = $(52 + 60) + (2 + 30) = 54 + 90$; $G_{av} = \frac{1}{2} G_a = -0.023$; $y_s = -0.023(230) = -5.29$ ft (1.61 m); elevation of sag $= 303.66 - 5.29 = 298.37$ ft (90.943 m).

7. Verify the location of the sag

Do this by ascertaining that the offsets of the PC and PT from the tangent through S, which is horizontal, satisfy the tangent-offset principle. From the preceding results, tangent offset of PC = 5.29 ft (1.612 m); tangent offset of PT = $5.29 - 4.48 = 0.81$ ft (0.247 m); distance to PC = 230 ft (70.1 m); distance to PT = $320 - 230 = 90$ ft (27.4 m); $5.29/0.81 = 6.53$; $230^2/90^2 = 6.53$. Therefore, the results are verified.

LOCATION OF A SINGLE STATION ON A PARABOLIC ARC

The PC of a vertical parabolic curve is at station $22 + 00$ of elevation 165.30, and the grade at the PC is + 3.2 percent. The elevation of the station $24 + 00$ is 168.90 ft (51.481 m). What is the elevation of station $25 + 50$?

Calculation Procedure:

1. Compute the offset of P_1 from the tangent through the PC

Refer to Fig. 202. The tangent-offset principle offers the simplest method of solution. Thus $x_1 = 200$ ft (61.0 m); $y_1 = 168.90 - 165.30 = 3.60$ ft (1.097 m); $Q_1T_1 = 200(0.032) = 6.40$ ft (1.951 m); $P_1T_1 = 6.40 - 3.60 = 2.80$ ft (0.853 m).

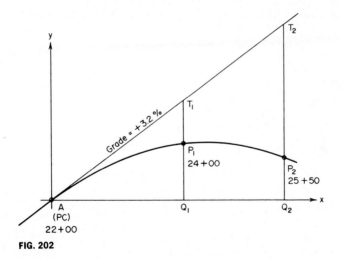

FIG. 202

2. Compute the offset of P_2 from the tangent through the PC; find the elevation of P_2

Thus $x_2 = 350$ ft (106.7 m); $P_2T_2/(P_1T_1) = x_2^2/x_1^2$; $P_2T_2 = 2.80(^{350}\!/_{200})^2 = 8.575$ ft (2.6137 m); $Q_2T_2 = 350(0.032) = 11.2$ ft (3.41 m); $Q_2P_2 = 11.2 - 8.575 = 2.625$ ft (0.8001 m); elevation of $P_2 = 165.30 + 2.625 = 167.925$ ft (51.184 m).

LOCATION OF A SUMMIT

An approach grade of +1.5 percent intersects a grade of − 2.5 percent at station $29 + 00$ of elevation 226.30 ft (68.976 m). The connecting parabolic curve is to be 800 ft (243.8 m) long. Locate the summit.

Calculation Procedure:

1. Locate the PC

Draw a freehand sketch of the curve, and record all values in the sketch as they are obtained. Thus, station of PC = station of PI − $L/2$ = 25 + 00; elevation of PC = 226.30 − 400(0.015) = 220.30 ft (67.147 m).

2. Calculate the rate of change in grade; locate the summit

Apply the average-grade method to locate the summit. Thus, r = (−2.5 − 1.5)/800 = −0.005 percent per foot.

Place the origin of coordinates at the PC. By Eq. 148, x_s = −G_a/r = 1.5/0.005 = 300 ft (91.44 m). From the PC to the summit, G_{av} = ½G_a = 0.75 percent. Then y_s = 300(0.0075) = 2.25 ft (0.686 m); station of summit = (25 + 00) + (3 + 00) = 28 + 00; elevation of summit = 220.30 + 2.25 = 222.55 ft (67.833 m). The summit can also be located by the tangent-offset method.

PARABOLIC CURVE TO CONTAIN A GIVEN POINT

A grade of −1.6 percent is followed by a grade of +3.8 percent, the grades intersecting at station 42 + 00 of elevation 210.00 ft (64.008 m). The parabolic curve connecting these grades is to pass through station 42 + 60 of elevation 213.70 ft (65.136 m). Compute the required length of curve.

Calculation Procedure:

1. Compute the tangent offsets; establish the horizontal location of P in terms of L

Refer to Fig. 203, where P denotes the specified point. The given data enable computation of the tangent offsets CP and DP, thus giving a relationship between the horizontal distances from A to P and from P to B. Since the distance from the centerline of curve to P is known, the length of curve may readily be found.

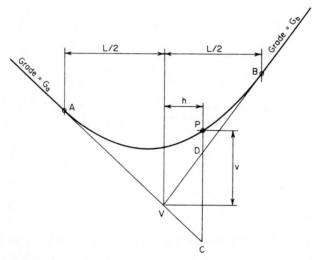

FIG. 203

Computing the tangent offsets gives $CP = v − G_a h$ and $DP = v − G_b h$; but CP/DP = $(L/2 + h)^2/(L/2 − h)^2 = (L + 2h)^2/(L − 2h)^2$; therefore,

$$\frac{L + 2h}{L - 2h} = \left(\frac{v - G_a h}{v - G_b h}\right)^{1/2} \tag{151}$$

2. Substitute numerical values and solve for L

Thus, $G_a = -1.6$ percent; $G_b = +3.8$ percent; $h = 60$ ft (18.3 m); $v = 3.70$ ft (1.128 m); then $(L + 120)/(L - 120) = [(3.7 \times 0.016 \times 60)/(3.7 - 0.038 \times 60)]^{1/2} = 1.81$; so $L = 416$ ft (126.8 m).

3. Verify the solution

There are many ways of verifying the solution. The simplest way is to compare the offsets of P and B from a tangent through A. By Eq. 150b, offset of B from tangent through A = 208(0.016 + 0.038) = 11.232 ft (3.4235 m). From the preceding calculations, offset of P from tangent through A = 4.66 ft (1.4203 m); 4.66/11.232 = 0.415; $(208 + 60)^2/(416)^2 = 0.415$. This is acceptable.

SIGHT DISTANCE ON A VERTICAL CURVE

A vertical summit curve has tangent grades of +2.6 and −1.5 percent. Determine the minimum length of curve that is needed to provide a sight distance of 450 ft (137.2 m) to an object 4 in (101.6 mm) in height. Assume that the eye of the motorist is 4.5 ft (1.37 m) above the roadway.

Calculation Procedure:

1. State the equation for minimum length when S < L

The vertical curvature of a road must be limited to ensure adequate visibility across the summit. Consequently, the distance across which a given change in grade may be effected is subject to a lower limit imposed by the criterion of sight distance.

Let S denote the required sight distance and L the minimum length of curve. In Fig. 204, let E denote the position of the motorist's eye and P the top of an object. Assume that the curve has the maximum allowable curvature, so that the distance from E to P equals S.

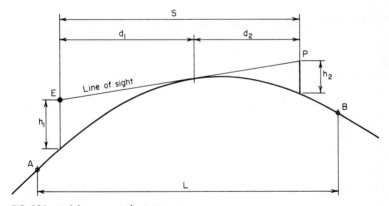

FIG. 204   Visibility on vertical summit curve.

Applying Eq. 150a gives

$$L = \frac{AS^2}{100[(2h_1)^{1/2} + (2h_2)^{1/2}]^2} \tag{152}$$

whee $A = G_a - G_b$, in percent.

2. State the equation for L when S > L

Thus,

$$L = 2S - \frac{200(h_1^{1/2} + h_2^{1/2})^2}{A} \tag{153}$$

3. Assuming, tentatively, that S < L, compute L

Thus $h_1 = 4.5$ ft (1.37 m); $h_2 = 4$ in $= 0.33$ ft (0.1 m); $A = 2.6 + 1.5 = 4.1$ percent; $L = 4.1(450)^2/[100(9^{1/2} + 0.67^{1/2})^2] = 570$ ft (173.7 m). Therefore, the assumption that $S < L$ is valid because $450 < 570$.

MINE SURVEYING: GRADE OF DRIFT

A vein of ore has a strike of S38°20′E and a northeasterly dip of 33°14′. What is the grade of a drift having a bearing of S43°10′E?

Calculation Procedure:

1. Express β as a function of α and θ

A vein of ore is generally assumed to have plane faces. The *strike*, or *trend*, of the vein is the bearing of any horizontal line in a face, and the *dip* is the angle of inclination of its face. A *drift* is a slightly sloping passage that follows the vein. Any line in a plane perpendicular to the horizontal is a *dip line*. The dip line is the steepest line in a plane, and the dip of the plane equals the angle of inclination of this line. With reference to the inclined plane $ABCD$ in Fig. 205a, let α

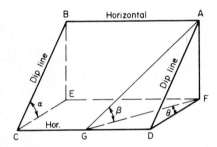

(a) Isometric view of inclined plane   (b) Strike-and-dip diagram (plan)

FIG. 205

$=$ dip of plane; $\beta =$ angle of inclination of arbitrary line AG; $\theta =$ angle between horizontal projections of AG and dip line.

By expressing β as a function of α and θ: $\tan \beta = AF/GF$; $\tan \alpha = AF/DF$; $\tan \beta / \tan \alpha = DF/GF = \cos \theta$.

$$\tan \beta = \tan \alpha \cos \theta \qquad (154)$$

2. Find the grade of the drift

Apply Eq. 154. In Fig. 205b, OA is a horizontal line in the vein, OB is the horizontal projection of the drift, and the arrow indicates the direction of dip. Then angle $COD = 43°10′ - 38°20′ = 4°50′$; $\theta =$ angle $CDO = 90° - 4°50′ = 85°10′$; $\tan \beta = \tan 33°14′ \cos 85°10′ = 0.0552$; grade of drift $= 5.52$ percent.

3. Alternatively, solve without the use of Eq. 154

In Fig. 205b, set $OD = 100$ ft (30.5 m); let D' denote the point on the face of the vein vertically below D. Then $CD = 100 \sin 4°50′ = 8.426$ ft (2.5682 m); drop in elevation from O to $D' =$ drop in elevation from C to $D' = 8.426 \tan 33°14′ = 5.52$ ft (1.682 m). Therefore, grade $= 5.52$ percent.

DETERMINING STRIKE AND DIP FROM TWO APPARENT DIPS

Three points on the hanging wall (upper face) of a vein of ore have been located by vertical boreholes. These points, designated P, Q, and R, have these relative positions: P is 142 ft (43.3 m)

245

above Q and 130 ft (39.6 m) above R; horizontal projection of PQ, length = 180 ft (54.9 m) and bearing = S55°32'W; horizontal projection of PR, length = 220 ft (67.1 m) and bearing = N19°26'W. Determine the strike and dip of the vein by both graphical construction and trigonometric calculations.

Calculation Procedure:

1. Plot the given data for the graphical procedure

In Fig. 206a, draw lines PQ and PR in plan in accordance with the given data for their horizontal projections. The angle of inclination of any line other than a dip line is an *apparent dip* of the vein.

2. Draw the elevations

In Fig. 206b and c, draw elevations normal to PQ and PR, respectively, locating the points in accordance with the given differences in elevation. Find the points S and T lying on PQ and PR, respectively, at an arbitrary distance v below P.

3. Draw the representation of the strike of the vein

Locate points S and T in Fig. 206a, and connect them with a straight line. This line is horizontal, and its bearing ϕ, therefore, represents the strike of the vein.

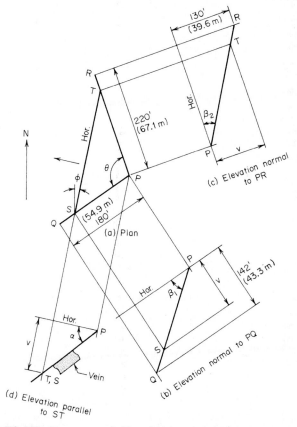

FIG. 206 Determination of strike and dip by orthographic projections.

4. Draw an edge view of the vein

In Fig. 206d, draw an elevation parallel to ST. Since this is an edge view of one line in the face, it is an edge view of the vein itself; it therefore represents the dip α of the vein in its true magnitude.

5. Determine the strike and dip

Scale angles ϕ and α, respectively. In Fig. 206a, the direction of dip is represented by the arrow perpendicular to ST.

6. Draw the dip line for the trigonometric solution

In Fig. 207, draw an isometric view of triangle PST, and draw the dip line PW. Its angle of inclination α equals the dip of the vein. Let O denote the point on a vertical line through P at the same elevation as S and T. Let β_1 and β_2 denote the angle of inclination of PS and PT, respectively; and let θ = angle SOT, θ_1 = angle SOW, θ_2 = angle TOW, m = tan β_2/tan β_1.

7. Express θ_1 in terms of the known angles β_1, β_2, and θ

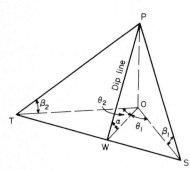

FIG. 207

Then substitute numerical values to find the strike ϕ of the vein. Thus

$$\tan \theta_1 = \frac{m - \cos \theta}{\sin \theta} \qquad (155)$$

For this vein, m = tan β_2/tan β_1 = 130(180)/ [220(142)] = 0.749040; θ = 180° − (55°32′ + 19°26′) = 105°02′. Substituting gives tan θ_1 = (0.749040 + 0.259381)/0.965775; θ_1 = 46°14′ 15″; ϕ = 55°32′ + 46°14′15″ − 90° = 11°46′ 15″; strike of vein = N11°46′15″E.

8. Compute the dip of the vein

Use Eq. 154, considering PS as the line of known inclination. Thus, tan α = tan β_1/cos θ_1; α = 48°45′25″.

9. Verify these results

Apply Eq. 154, considering PT as the line of known inclination. Thus θ_2 = θ − θ_1 = 105°02′ − 46°14′15″ = 58°47′45″; tan α = tan β_2/cos θ_2; α = 48°45′25″. This value agrees with the earlier computed value.

DETERMINATION OF STRIKE, DIP, AND THICKNESS FROM TWO SKEW BOREHOLES

In Fig. 208a, A and B represent points on the earth's surface through which skew boreholes were sunk to penetrate a vein of ore. Point B is 110 ft (33.5 m) due south of A. The data for these boreholes are as follows. *Borehole through A:* surface elevation = 870 ft (265.2 m); inclination = 49°; bearing of horizontal projection = N58°30′E. The hanging wall and footwall (lower face of vein) were struck at distances of 55 ft (16.8 m) and 205 ft (62.5 m), respectively, measured along the borehole. *Borehole through B:* surface elevation = 842 ft (256.6 m); inclination = 73°; bearing of horizontal projection = S44°50′E. The hanging wall and footwall were struck at distances of 98 ft (29.9 m) and 182 ft (55.5 m), respectively, measured along the borehole. Determine the strike, dip, and thickness of the vein by both graphical construction and trigonometric calculations.

Calculation Procedure:

1. Draw horizontal projections of the boreholes

Since the vein is assumed to have uniform thickness, the hanging wall and footwall are parallel. Two straight lines determine a plane. In the present instance, two points on the hanging wall and

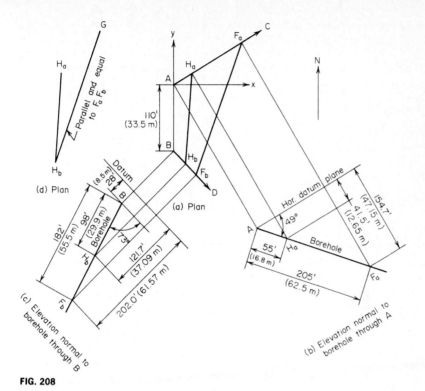

FIG. 208

two points on the footwall are given, enabling one line to be drawn in each of two parallel planes. These planes may be located by using these principles:

a. Consider a plane P and line L parallel to each other. If through any point on P a line is drawn parallel to L, this line lies in plane P.

b. Lines that are parallel and equal in length appear to be parallel and equal in length in all orthographic views.

These principles afford a means of locating a second line in the hanging wall or footwall.

Applying the specified bearings, draw the horizontal projections AC and BD of the boreholes in Fig. 208a.

2. Locate the points of intersection with the hanging wall and footwall in elevation

In Fig. 208b, draw an elevation normal to the borehole through A; locate the points of intersection H_a and F_a with the hanging wall and footwall, respectively. Select the horizontal plane through A as datum.

3. Repeat the foregoing construction with respect to borehole through B

This construction is shown in Fig. 208c.

4. Locate the points of intersection in plan

In Fig. 208a, locate the points of intersection. Draw lines H_aH_b and F_aF_b. The former lies in the hanging wall and the latter in the footwall. To avoid crowding, reproduce the plan of line H_aH_b in Fig. 208d.

5. Draw the plan of a line H_bG that is parallel and equal in length to F_aF_b

Do this by applying the second principle given above. In accordance with principle a, H_bG lies in the hanging wall, and this plane is therefore determined. The ensuing construction parallels that in the previous calculation procedure.

6. Establish a system of rectangular coordinate axes

Use A as the origin (Fig. 208a). Make x the east-west axis, y the north-south axis, and z the vertical axis.

7. Apply the given data to obtain the coordinates of the intersection points and point G

For example, with respect to F_a, $y = 205 \cos 49° \cos 58°30'$. The coordinates of G are obtained by adding to the coordinates of H_b the differences between the coordinates of F_a and F_b. The results are shown:

Point	x, ft (m)	y, ft (m)	z, ft (m)
H_a	30.8 (9.39)	18.9 (5.76)	−41.5 (−12.65)
H_b	20.2 (6.16)	−130.3 (−39.72)	−121.7 (−37.09)
F_a	114.7 (34.96)	70.3 (21.43)	−154.7 (−47.15)
F_b	37.5 (11.43)	−147.7 (−45.02)	−202.0 (−61.57)
G	97.4 (29.69)	87.7 (26.73)	−74.4 (−22.68)

8. For convenience, reproduce the plan of the intersection points, and G

This is shown in Fig. 209a.

9. Locate the point S at the same elevation as G

In Fig. 209b, draw an elevation normal to H_aH_b, and locate the point S on this line at the same elevation as G.

10. Establish the strike of the plane

Locate S in Fig. 209a, and draw the horizontal line SG. Since both S and G lie on the hanging wall, the strike of this plane is now established.

11. Complete the graphical solution

In Fig. 209c, draw an elevation parallel to SG. The line through H_a and H_b and that through F_a and F_b should be parallel to each other. This drawing is an edge view of the vein, and it presents the dip α and thickness t in their true magnitude. The graphical solution is now completed.

12. Reproduce the plan view

For convenience, reproduce the plan of H_a, H_b, and G in Fig. 209d. Draw the horizontal projection of the dip line, and label the angles as indicated.

13. Compute the lengths of lines H_aH_b and H_aG

Compute these lengths as projected on each coordinate axis and as projected on a horizontal plane. Use absolute values. Thus, line H_aH_b: $L_x = 30.8 - 20.2 = 10.6$ ft (3.23 m); $L_y = 18.9 - (-130.3) = 149.2$ ft (45.48 m); $L_z = -41.5 - (-121.7) = 80.2$ ft (24.44 m); $L_{hor} = (10.6^2 + 149.2^2)^{0.5} = 149.6$ ft (45.60 m). Line H_aG: $L_x = 97.4 - 30.8 = 66.6$ ft (20.30 m); $L_y = 87.7 - 18.9 = 68.8$ ft (20.97 m); $L_z = -41.5 - (-74.4) = 32.9$ ft (10.03 m); $L_{hor} = (66.6^2 + 68.8^2)^{0.5} = 95.8$ ft (29.20 m).

14. Compute the bearing and inclination of lines H_aH_b and H_aG

Let ϕ_1 = bearing of H_aH_b; ϕ_2 = bearing of H_aG; β_1 = angle of inclination of H_aH_b; β_2 = angle of inclination of H_aG. Then $\tan \phi_1 = 10.6/149.2$; $\phi_1 = S4°04'W$; $\tan \phi_2 = 66.6/68.8$; $\phi_2 = N44°04'E$; $\tan \beta_1 = 80.2/149.6$; $\tan \beta_2 = 32.9/95.8$.

15. Compute angle θ shown in Fig. 209d; determine the strike of the vein, using Eq. 155

Thus, $\theta = 180° + \phi_1 - \phi_2 = 140°00'$; $m = \tan \beta_2/\tan \beta_1 = 0.6406$; $\tan \theta_1 = (m - \cos 140°00')/\sin 140°00'$; $\theta_1 = 65°26'$; $\theta_2 = 74°34'$. The bearing of the horizontal projection of the dip line $= \theta_1 - \phi_1 = S61°22'E$; therefore, the strike of the vein $= N28°38'E$.

16. Compute the dip α of the vein

By Eq. 154, $\tan \alpha = \tan \beta_1/\cos \theta_1$; $\alpha = 52°12'$; or $\tan \alpha = \tan \beta_2/\cos \theta_2$; $\alpha = 52°14'$. This slight discrepancy between the two computed values falls within the tolerance of these calculations. Use the average value $\alpha = 52°13'$.

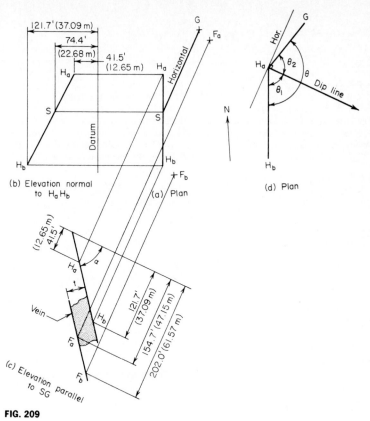

FIG. 209

17. Establish the relationship between the true thickness t of a vein and its apparent thickness t' as measured along a skew borehole

Refer to Figs. 208 and 209. Let δ = angle of inclination of borehole; γ = angle in plan between downward-sloping segments of borehole and dip line of vein. Then

$$t = t'(\cos \alpha \sin \delta - \sin \alpha \cos \delta \cos \gamma) \tag{156}$$

18. Find the true thickness, using Eq. 156

Thus, borehole through A: $\delta = 49°$; $\gamma = 180° - (58°30' + 61°22') = 60°08'$; $t' = 205 - 55 = 150$ ft (45.7 m); $t = 150(\cos 52°13' \sin 49° - \sin 52°13' \cos 49° \cos 60°08') = 30.6$ ft (9.3 m). For the borehole through B: $\delta = 73°$; $\gamma = 61°22' - 44°50' = 16°32'$; $t' = 182 - 98 = 84$ ft (25.6 m); $t = 84(\cos 52°13' \sin 73° - \sin 52°13' \cos 73° \cos 16°32') = 30.6$ ft (9.3 m). This agrees with the value previously computed.

Aerial Photogrammetry

FLYING HEIGHT REQUIRED TO YIELD A GIVEN SCALE

At what altitude above sea level must an aircraft fly to obtain vertical photography having an average scale of 1 cm = 120 m if the camera lens has a focal length of 152 mm and the average elevation of the terrain to be surveyed is 290 m?

Calculation Procedure:

1. *Write the equation for the scale of a vertical photograph*

In aerial photogrammetry, the term *photograph* generally refers to the positive photograph, and the plane of this photograph is considered to lie on the object side of the lens. A photograph is said to be *vertical* if the optical axis of the lens is in a vertical position at the instant of exposure. Since the plane of the photograph is normal to the optical axis, this plane is horizontal.

In Fig. 210*a*, point *L* is the *front nodal point* of the lens; a ray of light directed at this point leaves the lens without undergoing a change in direction. The point *o* at which the optical axis intersects the plane of the photograph is called the *principal point*. The distance from the ground to the camera may be considered infinite in relation to the dimensions of the lens, and so the distance *Lo* is equal to the focal length of the lens. The aircraft is assumed to be moving in a horizontal straight line, termed the *line of flight*, and the elevation of *L* above the horizontal

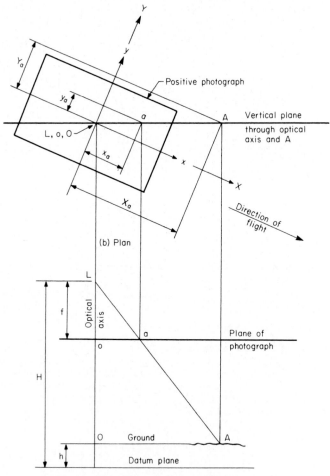

(b) Plan

(a) Elevation normal to vertical plane through optical axis and point A

FIG. 210

datum plane is called the *flying height*. The position of L in space at the instant of exposure is called the *exposure station*. Where the area to be surveyed is relatively small, the curvature of the earth may be disregarded.

Since the plane of the photograph is horizontal, Fig. 210*b* is a view normal to this plane and so presents all distances in this plane in their true magnitude. In the photograph, the origin of coordinates is placed at *o*. The *x* axis is placed parallel to the line of flight, with *x* values increasing in the direction of flight, and the *y* axis is placed normal to the *x* axis.

In Fig. 210, A is a point on the ground, a is the image of A on the photograph, and O is a point at the same elevation as A that lies on the prolongation of Lo. Thus, o is the image of O. The *scale* of a photograph, expressed as a fraction, is the ratio of a distance in the photograph to the corresponding distance along the ground. In this case, the ratio is 1 cm/120 m = 0.01 m/120 m = 1/12,000.

Let H = flying height; h = elevation of A above datum; f = focal length; S = scale of photograph, expressed as a fraction. From Fig. 210, $S = oa/OA$, and by similar triangles $S = f/(H - h)$.

2. Solve this equation for the flying height

Take sea level as datum. From the foregoing equation, with the meter as the unit of length, $H = h + f/S = 290 + 0.152/(1/12,000) = 290 + 0.152(12,000) = 2114$ m. This is the required elevation of L above sea level.

DETERMINING GROUND DISTANCE BY VERTICAL PHOTOGRAPH

Two points A and B are located on the ground at elevations of 250 and 190 m, respectively, above sea level. The images of A and B on a vertical aerial photograph are a and b, respectively. After correction for film shrinkage and lens distortion, the coordinates of a and b in the photograph are $x_a = -73.91$ mm, $y_a = +44.78$ mm, $x_b = +84.30$ mm, and $y_b = -21.65$ mm, where the subscript identifies the point. The focal length is 209.6 mm, and the flying height is 2540 m above sea level. Determine the distance between A and B as measured along the ground.

Calculation Procedure:

1. Determine the relationship between coordinates in the photograph and those in the datum plane

Refer to Fig. 210, and let X and Y denote coordinate axes that are vertically below the x and y axes, respectively, and in the datum plane. Omitting the subscript, we have $x/X = y/Y = oa/OA = S = f/(H - h)$, giving $X = x(H - h)/f$ and $Y = y(H - h)/f$.

2. Compute the coordinates of A and B in the datum plane

For A, $H - h = 2540 - 250 = 2290$ m. Substituting gives $X_A = (-0.07391)(2290)/0.2096 = -807.5$ m and $Y_A = (+0.04478)(2290)/0.2096 = +489.2$ m. For B, $H - h = 2540 - 190 = 2350$ m. Then $X_B = (+0.08430)(2350)/0.2096 = +945.2$ m, and $Y_B = (-0.02165)(2350)/0.2096 = -242.7$ m.

3. Compute the required distance

Let $\Delta X = X_A - X_B$, $\Delta Y = Y_A - Y_B$, and AB = distance between A and B as measured along the ground. Disregarding the difference in elevation of the two points, we have $(AB)^2 = (\Delta X)^2 + (\Delta Y)^2$. Then $\Delta X = -1752.7$ m, $\Delta Y = 731.9$ m, and $(AB)^2 = (1752.7)^2 + (731.9)^2$, or $AB = 1899$ m.

DETERMINING THE HEIGHT OF A STRUCTURE BY VERTICAL PHOTOGRAPH

In Fig. 211, points A and B are located at the top and bottom, respectively, and on the vertical centerline of a tower. These points have images a and b, respectively, on a vertical aerial photograph having a scale of 1:10,800 with reference to the ground, which is approximately level. In the photograph, $oa = 76.61$ mm and $ob = 71.68$ mm. The focal length is 210.1 mm. Find the height of the tower.

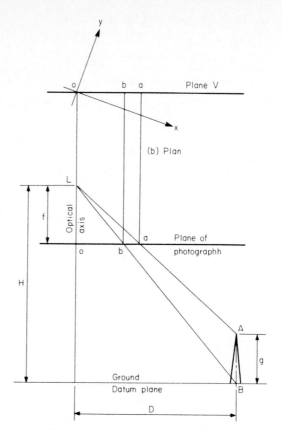

(b) Plan

(a) Elevation normal to vertical plane through
optical axis and center of tower (plane V)

FIG. 211

Calculation Procedure:

1. Compute the flying height with reference to the ground

Take the ground as datum. Then scale $S = f/H$, or $H = f/S = 0.2101/(1/10,800) = 0.2101(10,800) = 2269$ m.

2. Establish the relationship between height of tower and distances in the photograph

Let g = height of tower. In Fig. 211, $oa/D = f/(H - g)$ and $ob/D = f/H$. Thus, $oa/ob = H/(H - g)$. Solving gives $g = H(1 - ob/oa)$.

3. Compute the height of tower

Substituting in the foregoing equation yields $g = 2269(1 - 71.68/76.61) = 146$ m.

   Related Calculations: Let A denote a point at an elevation h above the datum, let B denote a point that lies vertically below A and in the datum plane, and let a and b denote the images of A and B, respectively. As Fig. 211 shows, a and b lie on a straight line that passes through o, which is called a *radial line*. The distance $d = ba$ is the displacement of the image of A resulting from its elevation above the datum, and it is termed the *relief displacement* of A. Thus, the relief displacement of a point is radially outward if that point lies above datum and radially inward if it lies below datum. From above, $ob/oa = (H - h)/H$, where H = flying height above datum. Then $d = oa - ob = (oa)h/H$.

DETERMINING GROUND DISTANCE BY TILTED PHOTOGRAPH

Two points A and B are located on the ground at elevations of 180 and 130 m, respectively, above sea level. Points A and B have images a and b, respectively, on an aerial photograph, and the coordinates of the images are $x_a = +40.63$ mm, $y_a = -73.72$ mm, $x_b = -78.74$ mm, and $y_b = +20.32$ mm. The focal length is 153.6 mm, and the flying height is 2360 m above sea level. By use of ground control points, it was established that the photograph has a tilt of 2°54′ and a swing of 162°. Determine the distance between A and B.

Calculation Procedure:

1. *Compute the transformed coordinates of the images*

Refer to Fig. 212, where L again denotes the front nodal point of the lens and o denotes the principal point. A photograph is said to be *tilted*, or *near vertical*, if by inadvertence the optical

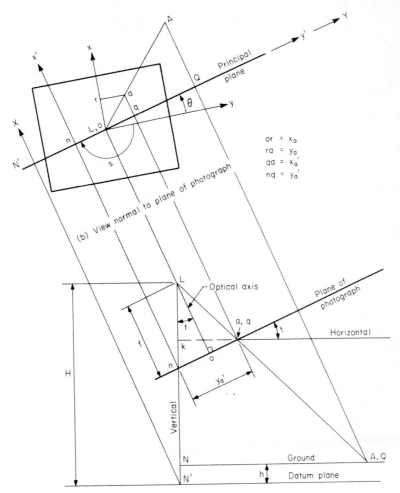

$$or = x_a$$
$$ra = y_a$$
$$qa = x_a'$$
$$nq = y_a'$$

(b) View normal to plane of photograph

(a) Elevation normal to principal plane

FIG. 212

axis of the lens is displaced slightly from the vertical at the time of exposure. The *tilt t* is the angle between the optical axis and the vertical. The *principal plane* is the vertical plane through the optical axis. Since the plane of the photograph is normal to the optical axis, it is normal to the principal plane. Therefore, Fig. 212*a* is an edge view of the plane of the photograph. Moreover, the angle between the plane of the photograph and the horizontal equals the tilt. In Fig. 212, *A* is a point on the ground and *a* is its image. Line *AQ* is normal to the principal plane, *Q* lies in that plane, and *q* is the image of *Q*.

Consider the vertical line through *L*. The points *n* and *N* at which this line intersects the plane of the photograph and the ground are called the *nadir point* and *ground nadir point*, respectively. The line of intersection of the principal plane and the plane of the photograph, which is line *no* prolonged, is termed the *principal line*. Now consider the vertical plane through *o* parallel to the line of flight. In the photograph, the *x* axis is placed on the line at which this vertical plane intersects the plane of the photograph, with *x* values increasing in the direction of flight. The *y* axis is normal to the *x* axis, and the origin lies at *o*. The *swing s* is the angle in the plane of the photograph, measured in a clockwise direction, between the positive side of the *y* axis and the radial line extending from *o* to *n*.

Transform the *x* and *y* axes in this manner: First, rotate the axes in a counterclockwise direction until the *y* axis lies on the principal line with its positive side on the upward side of the photograph; then displace the origin from *o* to *n*. Let *x′* and *y′* denote, respectively, the axes to which the *x* and *y* axes have been transformed. The *x′* axis is horizontal. Let θ denote the angle through which the axes are rotated in the first step of the transformation. From Fig. 212*b*, $\theta = 180° - s$.

The transformed coordinates of a point in the plane of the photograph are $x' = x \cos \theta + y \sin \theta$; $y' = -x \sin \theta + y \cos \theta + f \tan t$. In this case, $t = 2°54'$ and $\theta = 180° - 162° = 18°$. Then $x_a' = +40.63 \cos 18° - 73.72 \sin 18° = +15.86$ mm; $y_a' = -(+40.63) \sin 18° + (-73.72) \cos 18° + 153.6 \tan 2°54' = -74.89$ mm. Similarly, $x_b' = -78.74 \cos 18° + 20.32 \sin 18° = -68.61$ mm; $y_b' = -(-78.74) \sin 18° + 20.32 \cos 18° + 153.6 \tan 2°54' = +51.44$ mm.

2. Write the equations of the datum-plane coordinates

Let *X* and *Y* denote coordinate axes that lie in the datum plane and in the same vertical planes as the *x′* and *y′* axes, respectively, as shown in Fig. 212. Draw the horizontal line *kq* in the principal plane. Then $kq = y_a' \cos t$ and $Lk = f \sec t - y_a' \sin t$. From Fig. 212*b*, $QA/qa = LQ/Lq$. From Fig. 212*a*, $LQ/Lq = LN/Lk = (H - h)/(f \sec t - y_a' \sin t)$. Setting $QA = X_A$, we have $qa = x_a'$, and omitting subscripts gives $X = x'(H - h)/(f \sec t - y' \sin t)$, Eq. *a*. From Fig. 212*a*, $NQ/kq = LN/Lk = (H - h)/(f \sec t - y_a' \sin t)$. Setting $NQ = Y_A$ and omitting subscripts, we get $Y = y' [(H - h)/(f \sec t - y' \sin t)] \cos t$, Eq. *b*.

3. Compute the datum-plane coordinates

First compute $f \sec t = 153.6 \sec 2°54' = 153.8$ mm. For *A*, $H - h = 2360 - 180 = 2180$ m, and $f \sec t - y' \sin t = 153.8 - (-74.89) \sin 2°54' = 157.6$ mm. Then $(H - h)/(f \sec t - y' \sin t) = 2180/0.1576 = 13,830$. By Eq. *a*, $X_A = (+0.01586)(13,830) = +219.3$ m. By Eq. *b*, $Y_A = (-0.07489)(13,830) \cos 2°54' = -1034.4$ m.

Similarly, for *B*, $H - h = 2360 - 130 = 2230$ m and $f \sec t - y' \sin t = 153.8 - 51.44 \sin 2°54' = 151.2$ mm. Then $(H - h)/(f \sec t - y' \sin t) = 2230/0.1512 = 14,750$. By Eq. *a*, $X_B = (-0.06861)(14,750) = -1012.0$ m. By Eq. *b*, $Y_B = (+0.05144)(14,750) \cos 2°54' = +757.8$ m.

4. Compute the required distance

Disregarding the difference in elevation of the two points and proceeding as in the second previous calculation procedure, we have $\Delta X = +219.3 - (-1012.0) = +1231.3$ m, and $\Delta Y = -1034.4 - 757.8 = -1792.2$ m. Then $(AB)^2 = (1231.3)^2 + (1792.2)^2$, or $AB = 2174$ m.

Related Calculations: The *X* and *Y* coordinates found in step 3 can be verified by assuming that these values are correct, calculating the corresponding *x′* and *y′* coordinates, and comparing the results with the values in step 2. The procedure is as follows. In Fig. 212*a*, let v_A = angle *NLQ*. Then $\tan v_A = NQ/LN = Y_A/(H - h)$. Also, angle $oLq = v_A - t$. Now, $x_a'/X_A = Lq/LQ = f \sec (v_A - t)/[(H - h) \sec v_A]$. Rearranging and omitting subscripts, we get $x' = Xf \cos v_A/[(H - h) \cos (v_A - t)]$, Eq. *c*. Similarly, $y_a' = no + oq = f \tan t + f \tan (v_A - t)$. Omitting the subscript gives $y' = f [\tan t + \tan (v_A - t)]$, Eq. *d*.

As an illustration, consider point A in the present calculation procedure, which has the computed coordinates $X_A = +219.3$ m and $Y_A = -1034.4$ m. Then $\tan v_A = -1034.4/2180 = -0.4745$. Thus, $v_A = -25°23'$ and $v_A - t = -25°23' - 2°54' = -28°17'$. By Eq. c, $x' = (+219.3)(0.1536)(0.9035)/(2180)(0.8806) = +0.01585$ m $= +15.85$ mm. Applying Eq. d with $t = 2°54'$ gives $y' = 153.6(0.0507 - 0.5381) = -74.86$ mm. If we allow for roundoff effects, these values agree with those in step 1.

The following equation, which contains the four coordinates x', y', X, and Y, can be applied to test these values for consistency:

$$\frac{f^2 + (y' - f \tan t)^2}{x'^2} = \frac{(H - h)^2 + Y^2}{X^2}$$

DETERMINING ELEVATION OF A POINT BY OVERLAPPING VERTICAL PHOTOGRAPHS

Two overlapping vertical photographs contain point P and a control point C that lies 284 m above sea level. The air base is 768 m, and the focal length is 152.6 mm. The micrometer readings on a parallax bar are 15.41 mm for P and 11.37 mm for C. By measuring the displacement of the initial principal point and obtaining its micrometer reading, it was established that the parallax of a point equals its micrometer reading plus 76.54 mm. Find the elevation of P.

Calculation Procedure:

1. Establish the relationship between elevation and parallax

Two successive photographs are said to overlap if a certain amount of terrain appears in both. The ratio of the area that is common to the two photographs to the total area appearing in one photograph is called the *overlap*. (In practice, this value is usually about 60 percent.) The distance between two successive exposure stations is termed the *air base*. If a point on the ground appears in both photographs, its image undergoes a displacement from the first photograph to the second, and this displacement is known as the *parallax* of the point. This quantity is evaluated by using the micrometer of a *parallax bar* and then increasing or decreasing the micrometer reading by some constant.

Assume that there is no change in the direction of flight. As stated, the x axis in the photograph is parallel to the line of flight, with x values increasing in the direction of flight. Refer to Fig. 213, where photographs 1 and 2 are two successive photographs and the subscripts correspond to the photograph numbers. Let A denote a point in the overlapping terrain, and let a denote its image, with the proper subscript. Figure 213c discloses that $y_{1a} = y_{2a}$; thus, parallax occurs solely in the direction of flight. Let $p =$ parallax and $B =$ air base. Then $p = x_{1a} - x_{2a} = o_1m_1 - o_2m_2 = o_1m_1 + m_2o_2$. Thus, $m_1m_2 = B - p$. By proportion, $(B - p)/B = (H - h - f)/(H - h)$, giving $p/B = f/(H - h)$, or $p = Bf/(H - h)$, Eq. a. Thus, the parallax of a point is inversely proportional to the vertical projection of its distance from the front nodal point of the lens.

2. Determine the flying height

From the given data, $B = 768$ m and $f = 152.6$ mm. Take sea level as datum. For the control point, $h = 284$ m and $p = 11.37 + 76.54 = 87.91$ mm. From Eq. a, $H = h + Bf/p$, or $H = 284 + 768(0.1526)/0.08791 = 1617$ m.

3. Compute the elevation of P

For this point, $p = 15.41 + 76.54 = 91.95$ mm. From Eq. a, $h = H - Bf/p$, or $h = 1617 - 768(0.1526)/0.09195 = 342$ m above sea level.

DETERMINING AIR BASE OF OVERLAPPING VERTICAL PHOTOGRAPHS BY USE OF TWO CONTROL POINTS

The air base of two successive vertical photographs is to be found by using two control points, R and S, that lie in the overlapping area. The images of R and S are r and s, respectively. The following data were all obtained by measurement: The length of the straight line RS is 2073 m.

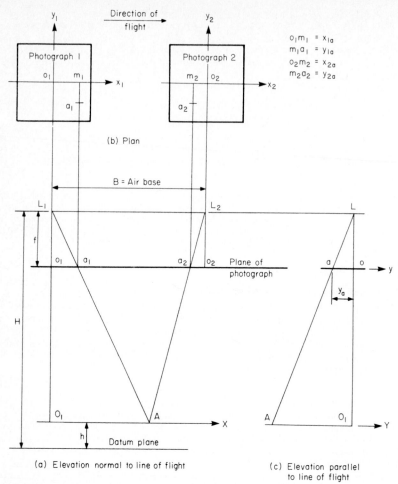

FIG. 213

The parallax of R is 92.03 mm, and that of S is 91.85 mm. The coordinates of the images in the left photograph are $x_r = +86.46$ mm, $y_r = -54.32$ mm, $x_s = +29.41$ mm, and $y_s = +56.93$ mm. Compute the air base.

Calculation Procedure:

1. Express the ground coordinates of the endpoints in terms of the air base

Refer to Fig. 213, and let X and Y denote coordinate axes that lie vertically below the x_1 and y_1 axes, respectively, and at the same elevation as A. Thus, O_1 is the origin of this system of coordinates. With reference to point A, by proportion, $X_A/x_{1a} = Y_A/y_{1a} = (H - h)/f$. Thus, O_1 is the origin of this system of coordinates. With reference to point A, by proportion, $X_A/x_{1a} = Y_A/y_{1a} = (H - h)/f$. From the previous calculation procedure, $(H - h)/f = B/p$. Omitting the subscript 1, we have $X_A = (x_a/p)B$ and $Y_A = (y_a/p)B$. Then $X_R = (+86.46/92.03)B = +0.9395B$; $Y_R = (-54.32/92.03)B = -0.5902B$; $X_S = (+29.41/91.85)B = +0.3202B$; $Y_S = (+56.93/91.85)B = +0.6198B$.

2. *Express the distance between the control points in terms of the air base; solve the resulting equation*

Disregarding the difference in elevation of the two points, we have $(RS)^2 = (X_R - X_S)^2 + (Y_R - Y_S)^2$. Now, $X_R - X_S = +0.6193B$ and $Y_R - Y_S = -1.2100B$. Then $2073^2 = [(0.6193)^2 + (1.2100)^2]B^2$, or $B = 1525$ m.

DETERMINING SCALE OF OBLIQUE PHOTOGRAPH

In a high-oblique aerial photograph, the distance between the apparent horizon and the principal point as measured along the principal line is 86.85 mm. The flying height is 2925 m above sea level, and the focal length is 152.7 mm. What is the scale of this photograph along a line that is normal to the principal line and at a distance of 20 mm above the principal point as measured along the principal line?

Calculation Procedure:

1. *Locate the true horizon in the photograph*

Refer to Fig. 214. An *oblique* aerial photograph is one that is taken with the optical axis intentionally displaced from the vertical, and a *high-oblique* photograph is one in which this displace-

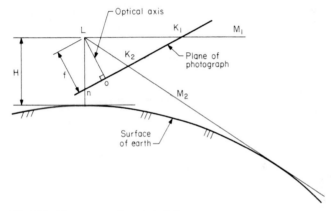

FIG. 214 Elevation normal to principal plane.

ment is sufficiently large to bring the earth's surface into view. By definition, the principal plane is the vertical plane that contains the optical axis, and the principal line is the line of intersection of this vertical plane and the plane of the photograph.

Assume that the terrain is truly level. The *apparent horizon* is the slightly curved boundary line in the photograph between earth and sky. Consider a conical surface that has its vertex at the front nodal point L and that is tangent to the spherical surface of the earth. If atmospheric refraction were absent, the apparent horizon would be the arc along which this conical surface intersected the plane of the photograph. The *true horizon* is the straight line along which the horizontal plane through L intersects the plane of the photograph; it is normal to the principal line. In Fig. 214, M_1 and M_2 are lines in the principal plane that pass through L; line M_1 is horizontal, and M_2 is tangent to the earth's surface. Points K_1 and K_2 are the points at which M_1 and M_2, respectively, intersect the plane of the photograph; these points lie on the principal line. Point K_1 lies on the true horizon; if atmospheric refraction is tentatively disregarded, K_2 lies on the apparent horizon.

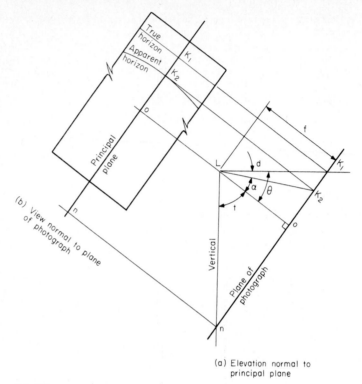

(a) Elevation normal to
principal plane

FIG. 215

Refer to Fig. 215a. The principal plane contains the *angle of dip d*, which is angle K_2LK_1; the *apparent depression angle α*, which is angle oLK_2; the (true) *depression angle θ*, which is angle oLK_1. Then $\theta = d + \alpha$. Let H = flying height above sea level in meters, and d' = angle of dip in minutes. Then $d' = 1.775 \sqrt{H}$, Eq. *a*. This relationship is based on the mean radius of the earth, and it includes allowance for atmospheric refraction. From Fig. 215a, $\tan \alpha = oK_2/f$, Eq. *b*. Then $d' = 1.775 \sqrt{2925} = 96.0'$, or $d = 1°36'$. Also, $\tan \alpha = 86.85/152.7 = 0.5688$, giving $\alpha = 29°38'$. Thus, $\theta = 1°36' + 29°38' = 31°14'$. From Fig. 215a, $oK_1 = f \tan \theta$, or $oK_1 = 152.7(0.6064) = 92.60$ mm. This dimension serves to establish the true horizon.

2. *Write the equation for the scale of a constant-scale line*

Since the optical axis is inclined, the scale S of the photograph is constant only along a line that is normal to the principal line, and so such a line is called a *constant-scale line*. As we shall find, every constant-scale line has a unique value of S.

Refer to Fig. 216, where A is a point on the ground and a is its image. Line AQ is normal to the principal plane, Q lies in that plane, and q is the image of Q. Line Rq is a horizontal line in the principal plane. If the terrain is truly level and curvature of the earth may be disregarded, the vertical projection of the distance from A to L is H. Let e = distance in photograph from true horizon to line qa. Along this line, $S = qa/QA = Lq/LQ = LR/LN$. But $LR = e \cos \theta$ and $LN = H$. Thus, $S = (e \cos \theta)/H$, Eq. *c*.

3. *Compute the scale along the specified constant-scale line*

From above, $\theta = 31°14'$ and $oK_1 = 92.60$ mm. Then $e = 92.60 - 20 = 72.60$ mm. By Eq. *c*, $S = (0.07260)(0.8551)/2925 = 1/47,120$.

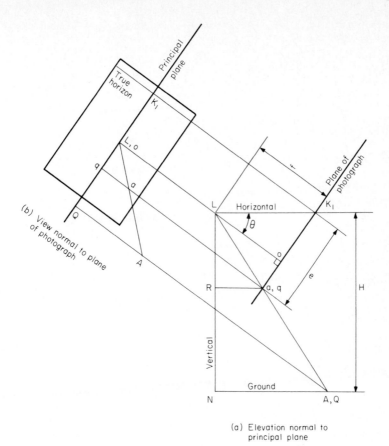

(b) View normal to plane of photograph

(a) Elevation normal to principal plane

FIG. 216

Soil Mechanics

The basic notational system used is c = unit cohesion; s = specific gravity; V = volume; W = total weight; w = specific weight; ϕ = angle of internal friction; τ = shearing stress; σ = normal stress.

COMPOSITION OF SOIL

A specimen of moist soil weighing 122 g has an apparent specific gravity of 1.82. The specific gravity of the solids is 2.53. After the specimen is oven-dried, the weight is 104 g. Compute the void ratio, porosity, moisture content, and degree of saturation of the original mass.

Calculation Procedure:

1. Compute the weight of moisture, volume of mass, and volume of each ingredient

In a three-phase soil mass, the voids, or pores, between the solid particles are occupied by moisture and air. A mass that contains moisture but not air is termed *fully saturated;* this constitutes a two-phase system. The term *apparent specific gravity* denotes the specific gravity of the mass.

Let the subscripts s, w, and a refer to the solids, moisture, and air, respectively. Where a subscript is omitted, the reference is to the entire mass. Also, let e = void ratio = $(V_w + V_a)/V_s$; n = porosity = $(V_w + V_a)/V$; MC = moisture content = W_w/W_s; S = degree of saturation = $V_w/(V_w + V_a)$.

Refer to Fig. 217. A horizontal line represents volume, a vertical line represents specific gravity, and the area of a rectangle represents the weight of the respective ingredient in grams.

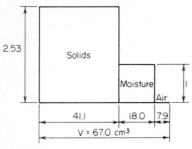

Computing weight and volume gives W = 122 g; W_s = 104 g; W_w = 122 − 104 = 18 g; V = 122/1.82 = 67.0 cm³; V_s = 104/2.53 = 41.1 cm³; V_w = 18.0 cm³; V_a = 67.0 − (41.1 + 18.0) = 7.9 cm³.

FIG. 217 Soil ingredients.

2. Compute the properties of the original mass

Thus, e = 100(18.0 + 7.9)/41.1 = 63.0 percent; n = 100(18.0 + 7.9)/67.0 = 38.7 percent; MC = 100(18)/104 = 17.3 percent; S = 100(18.0)/(18.0 + 7.9) + 69.5 percent. The factor of 100 is used to convert to percentage.

SPECIFIC WEIGHT OF SOIL MASS

A specimen of sand has a porosity of 35 percent, and the specific gravity of the solids is 2.70. Compute the specific weight of this soil in pounds per cubic foot (kilograms per cubic meter) in the saturated and in the submerged state.

Calculation Procedure:

1. Compute the weight of the mass in each state

Set V = 1 cm³. The (apparent) weight of the mass when submerged equals the true weight less the buoyant force of the water. Thus, $V_w + V_a = nV$ = 0.35 cm³; V_s = 0.65 cm³. In the saturated state, W = 2.70(0.65) + 0.35 = 2.105 g. In the submerged state, W = 2.105 − 1 = 1.105 g; or W = (2.70 − 1)0.65 = 1.105 g.

2. Find the weight of the soil

Multiply the foregoing values by 62.4 to find the specific weight of the soil in pounds per cubic foot. Thus: saturated, w = 131.4 lb/ft³ (2104.82 kg/m³); submerged, w = 69.0 lb/ft³ (1105.27 kg/m³).

ANALYSIS OF QUICKSAND CONDITIONS

Soil having a void ratio of 1.05 contains particles having a specific gravity of 2.72. Compute the hydraulic gradient that will produce a quicksand condition.

Calculation Procedure:

1. Compute the minimum gradient causing quicksand

As water percolates through soil, the head that induces flow diminishes in the direction of flow as a result of friction and viscous drag. The drop in head in a unit distance is termed the *hydraulic gradient*. A quicksand condition exists when water that is flowing upward has a sufficient momentum to float the soil particles.

Let i denote the hydraulic gradient in the vertical direction and i_c the minimum gradient that causes quicksand. Equate the buoyant force on a soil mass to the submerged weight of the mass to find i_c. Or

$$i_c = \frac{s_s - 1}{1 + e} \tag{157}$$

For this situation, i_c = (2.72 − 1)/(1 + 1.05) = 0.84.

MEASUREMENT OF PERMEABILITY BY FALLING-HEAD PERMEAMETER

A specimen of soil is placed in a falling-head permeameter. The specimen has a cross-sectional area of 66 cm² and a height of 8 cm; the standpipe has a cross-sectional area of 0.48 cm². The head on the specimen drops from 62 to 40 cm in 1 h 18 min. Determine the coefficient of permeability of the soil, in centimeters per minute.

Calculation Procedure:

1. Using literal values, equate the instantaneous discharge in the specimen to that in the standpipe

The velocity at which water flows through a soil is a function of the *coefficient of permeability*, or *hydraulic conductivity*, of the soil. By Darcy's law of laminar flow,

$$v = ki \tag{158}$$

where i = hydraulic gradient, k = coefficient of permeability, v = velocity.

In a falling-head permeameter, water is allowed to flow vertically from a standpipe through a soil specimen. Since the water is not replenished, the water level in the standpipe drops as flow continues, and the velocity is therefore variable. Let A = cross-sectional area of soil specimen; a = cross-sectional area of standpipe; h = head on specimen at given instant; h_1 and h_2 = head at beginning and end, respectively, of time interval T; L = height of soil specimen; Q = discharge at a given instant.

Using literal values, we have $Q = Aki = -a\,dh/dt$.

2. Evaluate k

Since the head h is dissipated in flow through the soil, $i = h/L$. By substituting and rearranging, $(Ak/L)dT = -a\,dh/h$; integrating gives $AkT/L = a \ln(h_1/h_2)$, where ln denotes the natural logarithm. Then

$$k = \frac{aL}{AT} \ln \frac{h_1}{h_2} \tag{159}$$

Substituting gives $k = (0.48 \times 8/66 \times 78) \ln (62/40) = 0.000326$ cm/min.

CONSTRUCTION OF FLOW NET

State the Laplace equation as applied to two-dimensional flow of moisture through a soil mass, and list three methods of constructing a flow net that are based on this equation.

Calculation Procedure:

1. Plot flow lines and equipotential lines

The path traversed by a water particle flowing through a soil mass is termed a *flow line*, *streamline*, or *path of percolation*. A line that connects points in the soil mass at which the head on the water has some assigned value is termed an *equipotential line*. A diagram consisting of flow lines and equipotential lines is called a *flow net*.

In Fig. 218a, where water flows under a dam under a head H, lines AB and CD are flow lines and EF and GH are equipotential lines.

2. Discuss the relationship of flow and equipotential lines

Since a water particle flowing from one equipotential line to another of smaller head will traverse the shortest path, it follows that flow lines and equipotential lines intersect at right angles, thus forming a system of orthogonal curves. In a flow net, the equipotential lines should be so spaced that the difference in head between successive lines is a constant, and the flow lines should be so spaced that the discharge through the space between successive lines is a constant. A flow net constructed in compliance with these rules illustrates the basic characteristics of the flow. For example, a close spacing of equipotential lines signifies a rapid loss of head in that region.

262

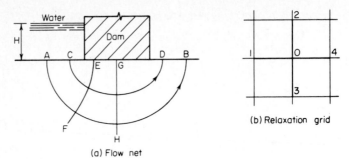

(a) Flow net

(b) Relaxation grid

FIG. 218

3. Write the velocity equation

Let h denote the head on the water at a given point. Equation 158 can be written as

$$v = -k\frac{dh}{dL} \tag{158a}$$

where dL denotes an elemental distance along the flow line.

4. State the particular form of the general Laplace equation

Let x and z denote a horizontal and vertical coordinate axis, respectively. By investigating the two-dimensional flow through an elemental rectangular prism of homogeneous, isentropic soil, and combining Eq. 158a with the equation of continuity, the particular form of the general Laplace equation

$$\frac{\partial^2 h}{\partial x^2} + \frac{\partial^2 h}{\partial z^2} = 0 \tag{160}$$

is obtained.

This equation is analogous to the equation for the flow of an electric current through a conducting sheet of uniform thickness and the equation of the trajectory of principal stress. (This is a curve that is tangent to the direction of a principal stress at each point along the curve. Refer to earlier calculation procedures for a discussion of principal stresses.)

The seepage of moisture through soil may be investigated by analogy with either the flow of an electric current or the stresses in a body. In the latter method, it is merely necessary to load a body in a manner that produces identical boundary conditions and then to ascertain the directions of the principal stresses.

5. Apply the principal-stress analogy

Refer to Fig. 218a. Consider the surface directly below the dam to be subjected to a uniform pressure. Principal-stress trajectories may be readily constructed by applying the principles of elasticity. In the flow net, flow lines correspond to the minor-stress trajectories and equipotential lines correspond to the major-stress trajectories. In this case, the flow lines are ellipses having their foci at the edges of the base of the dam, and the equipotential lines are hyperbolas.

A flow net may also be constructed by an approximate, trial-and-error procedure based on the method of relaxation. Consider that the area through which discharge occurs is covered with a grid of squares, a part of which is shown in Fig. 218b. If it is assumed that the hydraulic gradient is constant within each square, Eq. 160 leads to

$$h_1 + h_2 + h_3 + h_4 - 4h_0 = 0 \tag{161}$$

Trial values are assigned to each node in the grid, and the values are adjusted until a consistent set of values is obtained. With the approximate head at each node thus established, it becomes a simple matter to draw equipotential lines. The flow lines are then drawn normal thereto.

SOIL PRESSURE CAUSED BY POINT LOAD

A concentrated vertical load of 6 kips (26.7 kN) is applied at the ground surface. Compute the vertical pressure caused by this load at a point 3.5 ft (1.07 m) below the surface and 4 ft (1.2 m) from the action line of the force.

Calculation Procedure:

1. Sketch the load conditions

Figure 219 shows the load conditions. In Fig. 219, O denotes the point at which the load is applied, and A denotes the point under consideration. Let R denote the length of OA and r and z denote the length of OA as projected on a horizontal and vertical plane, respectively.

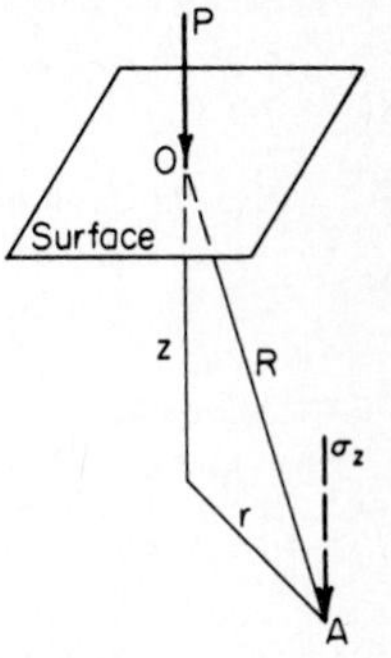

2. Determine the vertical stress σ_z at A

Apply the Boussinesq equation:

$$\sigma_z = \frac{3Pz^3}{2\pi R^5} \qquad (162)$$

FIG. 219

Thus, with $P = 6000$ lb (26,688.0 N), $r = 4$ ft (1.2 m), $z = 3.5$ ft (1.07 m), $R = (4^2 + 3.5^2)^{0.5}$ = 5.32 ft (1.621 m); then $\sigma_z = 3(6000)(3.5)^3/[2\pi(5.32)^5] = 28.8$ lb/ft^2 (1.38 kPa).

Although the Boussinesq equation is derived by assuming an idealized homogeneous mass, its results agree reasonably well with those obtained experimentally.

VERTICAL FORCE ON RECTANGULAR AREA CAUSED BY POINT LOAD

A concentrated vertical load of 20 kips (89.0 kN) is applied at the ground surface. Determine the resultant vertical force caused by this load on a rectangular area 3 × 5 ft (91.4 × 152.4 cm) that lies 2 ft (61.0 cm) below the surface and has one vertex on the action line of the applied force.

Calculation Procedure:

1. State the equation for the total force

Refer to Fig. 220a, where A and B denote the dimensions of the rectangle, H its distance from the surface, and F is the resultant vertical force. Establish rectangular coordinate axes along the

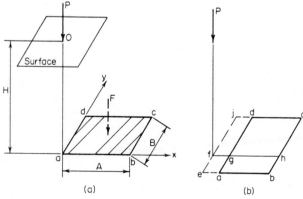

(a)   (b)

FIG. 220

sides of the rectangle, as shown. Let $C = A^2 + H^2$, $D = B^2 + H^2$, $E = A^2 + B^2 + H^2$, $\theta = \sin^{-1} H(E/CD)^{0.5}$ deg.

The force dF on an elemental area dA is given by the Boussinesq equation as $dF = [3Pz^3/(2\pi R^5)]\,dA$, where $z = H$ and $R = (H^2 + x^2 + y^2)^{0.5}$. Integrate this equation to obtain an equation for the total force F. Set $dA = dx\,dy$; then

$$\frac{F}{P} = 0.25 - \frac{\theta}{360°} + \frac{ABH}{2\pi E^{0.5}}\left(\frac{1}{C} + \frac{1}{D}\right) \tag{163}$$

2. Substitute numerical values and solve for F

Thus, $A = 3$ ft (91.4 cm); $B = 5$ ft (152.4 cm); $H = 2$ ft (61.0 cm); $C = 13$; $D = 29$; $E = 38$; $\theta = \sin^{-1} 0.6350 = 39.4°$; $F/P = 0.25 - 0.109 + 0.086 = 0.227$; $F = 20(0.227) = 4.54$ kips (20.194 kN).

The resultant force on an area such as *abcd* (Fig. 220*b*) may be found by expressing the area in this manner: $abcd = ebhf - eagf + fhcj - fgdj$. The forces on the areas on the right side of this equation are superimposed to find the force on *abcd*. Various diagrams and charts have been devised to expedite the calculation of vertical soil pressure.

VERTICAL PRESSURE CAUSED BY RECTANGULAR LOADING

A rectangular concrete footing 6×8 ft (182.9×243.8 cm) carries a total load of 180 kips (800.6 kN), which may be considered to be uniformly distributed. Determine the vertical pressure caused by this load at a point 7 ft (213.4 cm) below the center of the footing.

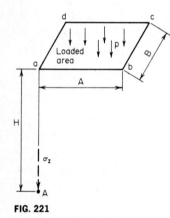

FIG. 221

Calculation Procedure:

1. State the equation for σ_z

Referring to Fig. 221, let p denote the uniform pressure on the rectangle *abcd* and σ_z the resulting vertical pressure at a point A directly below a vertex of the rectangle. Then

$$\frac{\sigma_z}{p} = 0.25 - \frac{\theta}{360} + \frac{ABH}{2\pi E^{0.5}}\left(\frac{1}{C} + \frac{1}{D}\right) \tag{164}$$

2. Substitute given values and solve for σ_z

Resolve the given rectangle into four rectangles having a vertex above the given point. Then $p = 180,000/[6(8)] = 3750$ lb/ft² (179.6 kPa). With $A = 3$ ft (91.4 cm); $B = 4$ ft (121.9 cm); $H = 7$ ft (213.4 cm); $C = 58$; $D = 65$; $E = 74$; $\theta = \sin^{-1} 0.9807 = 78.7°$; $\sigma_z/p = 4(0.25 - 0.218 + 0.051) = 0.332$; $\sigma_z = 3750(0.332) = 1245$ lb/ft² (59.6 kPa).

APPRAISAL OF SHEARING CAPACITY OF SOIL BY UNCONFINED COMPRESSION TEST

In an unconfined compression test on a soil sample, it was found that when the axial stress reached 2040 lb/ft² (97.7 kPa), the soil ruptured along a plane making an angle of 56° with the horizontal. Find the cohesion and angle of internal friction of this soil by constructing Mohr's circle.

Calculation Procedure:

1. Construct Mohr's circle in Fig. 222*b*

Failure of a soil mass is characterized by the sliding of one part past the other; the failure is therefore one of shear. Resistance to sliding occurs from two sources: cohesion of the soil and friction.

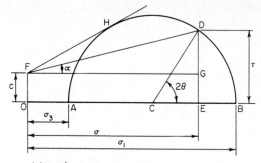

(a) Mohr's diagram for triaxial–stress condition

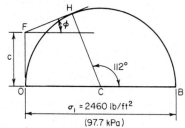

$\sigma_1 = 2460$ lb/ft^2

(97.7 kPa)

(b) Mohr's diagram for unconfined compression test

FIG. 222

Consider that the shearing stress at a given point exceeds the cohesive strength. It is usually assumed that the soil has mobilized its maximum potential cohesive resistance plus whatever frictional resistance is needed to prevent failure. The mass therefore remains in equilibrium if the ratio of the computed frictional stress to the normal stress is below the coefficient of internal friction of the soil.

Consider a soil prism in a state of triaxial stress. Let Q denote a point in this prism and P a plane through Q. Let c = unit cohesive strength of soil; σ = normal stress at Q on plane P; σ_1 and σ_3 = maximum and minimum normal stress at Q, respectively; τ = shearing stress at Q, on plane P; θ = angle between P and the plane on which σ_1 occurs; ϕ = angle of internal friction of the soil.

For an explanation of Mohr's circle of stress, refer to an earlier calculation procedure; then refer to Fig. 222a. The shearing stress ED on plane P may be resolved into the cohesive stress EG and the frictional stress GD. Therefore, $\tau = c + \sigma \tan \alpha$. The maximum value of α associated with point Q is found by drawing the tangent FH.

Assume that failure impends at Q. Two conclusions may be drawn: The angle between FH and the base line OAB equals ϕ, and the angle between the plane of impending rupture and the plane on which σ_1 occurs equals one-half angle BCH. (A soil mass that is on the verge of failure is said to be in *limit equilibrium*.)

In an unconfined compression test, the specimen is subjected to a vertical load without being restrained horizontally. Therefore, σ_1 occurs on a horizontal plane.

Constructing Mohr's circle in Fig. 222b, apply these values: $\sigma_1 = 2040$ lb/ft^2 (97.7 kPa); $\sigma_3 = 0$; angle $BCH = 2(56°) = 112°$.

2. Construct a tangent to the circle

Draw a line through H tangent to the circle. Let F denote the point of intersection of the tangent and the vertical line through O.

3. *Measure OF and the angle of inclination of the tangent*

The results are $OF = c = 688$ lb/ft^2 (32.9 kPa); $\phi = 22°$.

In general, in an unconfined compression test,

$$c = \tfrac{1}{2}\sigma_1 \cot \theta' \qquad \phi = 2\theta' - 90° \tag{165}$$

where θ' denotes the angle between the plane of failure and the plane on which σ_1 occurs. In the special case where frictional resistance is negligible, $\phi = 0$; $c = \tfrac{1}{2}\sigma_1$.

APPRAISAL OF SHEARING CAPACITY OF SOIL BY TRIAXIAL COMPRESSION TEST

Two samples of a soil were subjected to triaxial compression tests, and it was found that failure occurred under the following principal stresses: sample 1, $\sigma_1 = 6960$ lb/ft^2 (333.2 kPa) and $\sigma_3 = 2000$ lb/ft^2 (95.7 kPa); sample 2, $\sigma_1 = 9320$ lb/ft^2 (446.2 kPa) and $\sigma_3 = 3000$ lb/ft^2 (143.6 kPa). Find the cohesion and angle of internal friction of this soil, both trigonometrically and graphically.

Calculation Procedure:

1. *State the equation for the angle ϕ*

Trigonometric method: Let S and D denote the sum and difference, respectively, of the stresses σ_1 and σ_3. By referring to Fig. 222a, develop this equation:

$$D - S \sin \phi = 2c \cos \phi \tag{166}$$

Since the right-hand member represents a constant that is characteristic of the soil, $D_1 - S_1 \sin \phi = D_2 - S_2 \sin \phi$, or

$$\sin \phi = \frac{D_2 - D_1}{S_2 - S_1} \tag{167}$$

where the subscripts correspond to the sample numbers.

2. *Evaluate ϕ and c*

By Eq. 167, $S_1 = 8960$ lb/ft^2 (429.0 kPa); $D_1 = 4960$ lb/ft^2 (237.5 kPa); $S_2 = 12{,}320$ lb/ft^2 (589.9 kPa); $D_2 = 6320$ lb/ft^2 (302.6 kPa); $\sin \phi = (6320 - 4960)/(12{,}320 - 8960)$; $\phi = 23°53'$. Evaluating c, using Eq. 166, gives $c = \tfrac{1}{2}(D \sec \phi - S \tan \phi) = 729$ lb/ft^2 (34.9 kPa).

3. *For the graphical solution, use the Mohr's circle*

Draw the Mohr's circle associated with each set of principal stresses, as shown in Fig. 223.

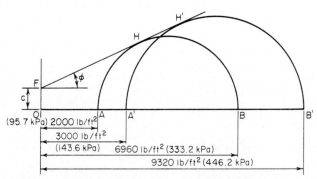

FIG. 223 Composite Mohr's diagram for triaxial compression tests.

4. *Draw the envelope; measure its angle of inclination*

Draw the envelope (common tangent) FHH', and measure OF and the angle of inclination of the envelope. In practice, three of four samples should be tested and the average value of ϕ and c determined.

EARTH THRUST ON RETAINING WALL CALCULATED BY RANKINE'S THEORY

A retaining wall supports sand weighing 100 lb/ft³ (15.71 kN/m³) and having an angle of internal friction of 34°. The back of the wall is vertical, and the surface of the backfill is inclined at an angle of 15° with the horizontal. Applying Rankine's theory, calculate the active earth pressure on the wall at a point 12 ft (3.7 m) below the top.

Calculation Procedure:

1. *Construct the Mohr's circle associated wtih the soil prism*

Rankine's theory of earth pressure applies to a uniform mass of dry cohesionless soil. This theory considers the state of stress at the instant of impending failure caused by a slight yielding of the wall. Let h = vertical distance from soil surface to a given point, ft (m); p = resultant pressure on a vertical plane at the given point, lb/ft² (kPa); ϕ = ratio of shearing stress to normal stress on given plane; θ = angle of inclination of earth surface. The quantity o may also be defined as the tangent of the angle between the resultant stress on a plane and a line normal to this plane; it is accordingly termed the *obliquity* of the resultant stress.

Consider the elemental soil prism *abcd* in Fig. 224*a*, where faces *ab* and *dc* are parallel to the surface of the backfill and faces *bc* and *ad* are vertical. The resultant pressure p_v on *ab* is vertical,

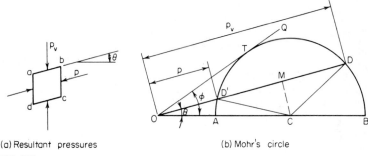

(a) Resultant pressures   (b) Mohr's circle

FIG. 224

and p is parallel to the surface. Thus, the resultant stresses on *ab* and *bc* have the same obliquity, namely, tan θ. (Stresses having equal obliquities are called *conjugate* stresses.) Since failure impends, there is a particular plane for which the obliquity is tan ϕ.

In Fig. 224*b*, construct Mohr's circle associated with this soil prism. Using a suitable scale, draw line OD, making an angle θ with the base line, where OD represents p_v. Draw line OQ, making an angle ϕ with the base line. Draw a circle that has its center C on the base line, passes through D, and is tangent to OQ. Line OD' represents p. Draw CM perpendicular to OD.

2. *Using the Mohr's circle, state the equation for p*

Thus,

$$p = \frac{[\cos \theta - (\cos^2 \theta - \cos^2 \phi)^{0.5}] \, wh}{\cos \theta + (\cos^2 \theta - \cos^2 \phi)^{0.5}} \tag{168}$$

By substituting, w = 100 lb/ft³ (15.71 kN/m³); h = 12 ft (3.7 m); θ = 15°; ϕ = 34°; p = 0.321(100)(12) = 385 lb/ft² (18.4 kPa).

The lateral pressure that accompanies a slight displacement of the wall *away from* the retained soil is termed *active pressure;* that which accompanies a slight displacement of the wall *toward* the retained soil is termed *passive pressure.* By an analogous procedure, the passive pressure is

$$p = \frac{[\cos \theta + (\cos^2 \theta - \cos^2 \phi)^{0.5}]wh}{\cos \theta - (\cos^2\theta - \cos^2\phi)^{0.5}} \tag{169}$$

The equations of active and passive pressure are often written as

$$p_a = C_a wh \qquad p_p = C_p wh \tag{170}$$

where the subscripts identify the type of pressure and C_a and C_p are the coefficients appearing in Eqs. 168 and 169, respectively.

In the special case where $\theta = 0$, these coefficients reduce to

$$C_a = \frac{1 - \sin \phi}{1 + \sin\phi} = \tan^2 (45° - \tfrac{1}{2}\phi) \tag{171}$$

$$C_p = \frac{1 + \sin \phi}{1 - \sin \phi} = \tan^2 (45° + \tfrac{1}{2}\phi) \tag{172}$$

The planes of failure make an angle of $45° + \tfrac{1}{2}\phi$ with the principal planes.

EARTH THRUST ON RETAINING WALL CALCULATED BY COULOMB'S THEORY

A retaining wall 20 ft (6.1 m) high supports sand weighing 100 lb/ft³ (15.71 kN/m³) and having an angle of internal friction of 34°. The back of the wall makes an angle of 8° with the vertical; the surface of the backfill makes an angle of 9° with the horizontal. The angle of friction between the sand and wall is 20°. Applying Coulomb's theory, calculate the total thrust of the earth on a 1-ft (30.5-cm) length of the wall.

Calculation Procedure:

1. *Determine the resultant pressure P of the wall*

Refer to Fig. 225a. Coulomb's theory postulates that as the wall yields slightly, the soil tends to rupture along some plane BC through the heel.

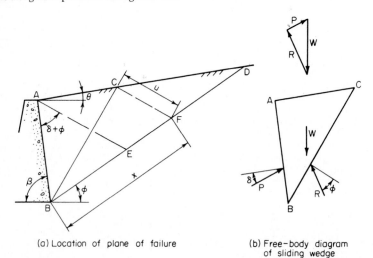

(a) Location of plane of failure

(b) Free–body diagram of sliding wedge

FIG. 225

Let δ denote the angle of friction between the soil and wall. As shown in Fig. 225b, the wedge *ABC* is held in equilibrium by three forces: the weight *W* of the wedge, the resultant pressure *R* of the soil beyond the plane of failure, and the resultant pressure *P* of the wall, which is equal and opposite to the thrust exerted by the earth on the wall. The forces *R* and *P* have the directions indicated in Fig. 225b. By selecting a trial wedge and computing its weight, the value of *P* may be found by drawing the force polygon. The problem is to identify the wedge that yields the maximum value of *P*.

In Fig. 225a, perform this construction: Draw a line through *B* at an angle ϕ with the horizontal, intersecting the surface at *D*. Draw line *AE*, making an angle $\delta + \phi$ with the back of the wall; this line makes an angle $\beta - \delta$ with *BD*. Through an arbitrary point *C* on the surface, draw *CF* parallel to *AE*. Triangle *BCF* is similar to the triangle of forces in Fig. 225b. Then $P = Wu/x$, where $W = w$(area *ABC*).

2. Set dP/dx = 0 and state Rebhann's theorem

This theorem states: The wedge that exerts the maximum thrust on the wall is that for which triangles *ABC* and *BCF* have equal areas.

3. Considering BC as the true plane of failure, develop equations for x², u, and P

Thus,

$$x^2 = BE(BD) \tag{173}$$

$$u = \frac{AE(BD)}{x + BD} \tag{174}$$

$$P = \tfrac{1}{2}wu^2 \sin(\beta - \delta) \tag{175}$$

4. Evaluate P, using the foregoing equations

Thus, $\phi = 34°$; $\delta = 20°$; $\theta = 9°$; $\beta = 82°$; $\angle ABD = 64°$; $\angle BAE = 54°$; $\angle AEB = 62°$; $\angle BAD = 91°$; $\angle ADB = 25°$; $AB = 20 \csc 82° = 20.2$ ft (6.16 m). In triangle *ABD*: $BD = AB \sin 91°/\sin 25° = 47.8$ ft (14.57 m). In triangle *ABE*: $BE = AB \sin 54°/\sin 62° = 18.5$ ft (5.64 m); $AE = AB \sin 64°/\sin 62° = 20.6$ ft (6.28 m); $x^2 = 18.5(47.8)$; $x = 29.7$ ft (9.05 m); $u = 20.6(47.8)/(29.7 + 47.8) = 12.7$ ft (3.87 m); $P = \tfrac{1}{2}(100)(12.7)^2 \sin 62°$; $P = 7120$ lb/ft (103,909 N/m) of wall.

5. Alternatively, determine u graphically

Do this by drawing Fig. 225a to a suitable scale.

Many situations do not lend themselves to analysis by Rebhann's theorem. For instance, the backfill may be nonhomogeneous, the earth surface may not be a plane, a surcharge may be applied over part of the surface, etc. In these situations, graphical analysis gives the simplest solution. Select a trial wedge, compute its weight and the surcharge it carries, and find *P* by constructing the force polygon as shown in Fig. 225b. After several trial wedges have been investigated, the maximum value of *P* will become apparent.

If the backfill is cohesive, the active pressure on the retaining wall is reduced. However, in view of the difficulty of appraising the cohesive capacity of a disturbed soil, most designers prefer to disregard cohesion.

EARTH THRUST ON TIMBERED TRENCH CALCULATED BY GENERAL WEDGE THEORY

A timbered trench of 12-ft (3.7-m) depth retains a cohesionless soil having a horizontal surface. The soil weighs 100 lb/ft³ (15.71 kN/m³), its angle of internal friction is 26°30′, and the angle of friction between the soil and timber is 12°. Applying Terzaghi's general wedge theory, compute the total thrust of the soil on a 1-ft (30.5-cm) length of trench. Assume that the resultant acts at middepth.

Calculation Procedure:

1. Start the graphical construction

Refer to Fig. 226. The soil behind a timbered trench and that behind a cantilever retaining wall tend to fail by dissimilar modes, for in the former case the soil is restrained against horizontal

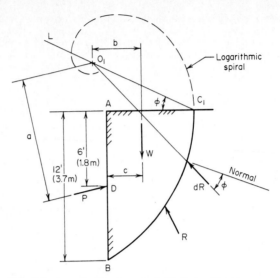

FIG. 226 General wedge theory applied to timbered trench.

movement at the surface by bracing across the trench. Consequently, the soil behind a trench tends to fail along a curved surface that passes through the base and is vertical at its intersection with the ground surface. At impending failure, the resultant force dR acting on any elemental area on the failure surface makes an angle ϕ with the normal to this surface.

The general wedge theory formulated by Terzaghi postulates that the arc of failure is a logarithmic spiral. Let v_o denote a reference radius vector and v denote the radius vector to a given point on the spiral. The equation of the curve is

$$r = r_o e^{\alpha \tan \phi} \tag{176}$$

where r_o = length of v_o; r = length of v; α = angle between v_o and v; e = base of natural logarithms = 2.718. . . .

The property of this curve that commends it for use in this analysis is that at every point the radius vector and the normal to the curve make an angle ϕ with each other. Therefore, if the failure line is defined by Eq. 176, the action line of the resultant force dR at any point is a radius vector or, in other words, the action line passes through the center of the spiral. Consequently, the action line of the total resultant force R also passes through the center.

The pressure distribution on the wall departs radically from a hydrostatic one, and the resultant thrust P is applied at a point considerably above the lower third point of the wall. Terzaghi recommends setting the ratio BD/AB at between 0.5 and 0.6.

Perform the following construction: Using a suitable scale, draw line AB to represent the side of the trench, and draw a line to represent the ground surface. At middepth, draw the action line of P at an angle of 12° with the horizontal.

On a sheet of transparent paper, draw the logarithmic spiral representing Eq. 176, setting ϕ = 26°30′ and assigning any convenient value to r_o. Designate the center of the spiral as O.

Select a point C_1 on the ground surface, and draw a line L through C_1 at an angle ϕ with the horizontal. Superimpose the drawing containing the spiral on the main drawing, orienting it in such a manner that O lies on L and the spiral passes through B and C_1. On the main drawing, indicate the position of the center of the spiral, and designate this point as O_1. Line AC_1 is normal to the spiral at C_1 because it makes an angle ϕ with the radius vector, and the spiral is therefore vertical at C_1.

2. *Compute the total weight W of the soil above the failure line*

Draw the action line of W by applying these approximations:

$$\text{Area of wedge} = \tfrac{2}{3}(AB)AC_1 \qquad c = 0.4AC_1 \tag{177}$$

Scale the lever arms a and b.

3. *Evaluate P by taking moments with respect to O_1*

Since R passes through this point,

$$P = \frac{bW}{a} \tag{178}$$

4. *Select a second point C_2 on the ground surface; repeat the foregoing procedure*

5. *Continue this process until the maximum value of P is obtained*

After investigating this problem intensively, Peckworth concluded that the distance AC to the true failure line varies between $0.4h$ and $0.5h$, where h is the depth of the trench. It is therefore advisable to select some point that lies within this range as the first trial position of C.

THRUST ON A BULKHEAD

The retaining structure in Fig. 227a supports earth that weighs 114 lb/ft³ (17.91 kN/m³) in the dry state, is 42 percent porous, and has an angle of internal friction of 34° in both the dry and submerged state. The backfill carries a surcharge of 320 lb/ft² (15.3 kPa). Applying Rankine's theory, compute the total pressure on this structure between A and C.

Calculation Procedure:

1. *Compute the specific weight of the soil in the submerged state*

The lateral pressure of the soil below the water level consists of two elements: the pressure exerted by the solids and that exerted by the water. The first element is evaluated by applying the appropriate equation with w equal to the weight of the soil in the submerged state. The second element is assumed to be the full hydrostatic pressure, as though the solids were not present. Since there is water on both sides of the structure, the hydrostatic pressures balance one another and may therefore be disregarded.

In calculating the forces on a bulkhead, it is assumed that the pressure distribution is hydro-

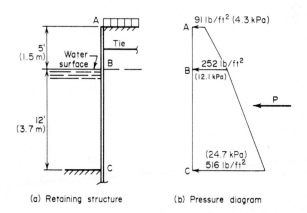

(a) Retaining structure   (b) Pressure diagram

FIG. 227

static (i.e., that the pressure varies linearly with the depth), although this is not strictly true with regard to a flexible wall.

Computing the specific weight of the soil in the submerged state gives $w = 114 - (1 - 0.42)62.4 = 77.8$ lb/ft^3 (12.22 kN/m^3).

2. Compute the vertical pressure at A, B, and C caused by the surcharge and weight of solids

Thus, $p_A = 320$ lb/ft^2 (15.3 kPa); $p_B = 320 + 5(114) = 890$ lb/ft^2 (42.6 kPa); $p_C = 890 + 12(77.8) = 1824$ lb/ft^2 (87.3 kPa).

3. Compute the Rankine coefficient of active earth pressure

Determine the lateral pressure at A, B and C. Since the surface is horizontal, Eq. 171 applies, with $\phi = 34°$. Refer to Fig. 227b. Then $C_a = \tan^2(45° - 17°) = 0.283$; $p_A = 0.283(320) = 91$ lb/ft^2 (4.3 kPa); $p_B = 252$ lb/ft^2 (12.1 kPa); $p_C = 516$ lb/ft^2 (24.7 kPa).

4. Compute the total thrust between A and C

Thus, $P = \frac{1}{2}(5)(91 + 252) + \frac{1}{2}(12)(252 + 516) = 5466$ lb (24,312.7 N).

CANTILEVER BULKHEAD ANALYSIS

Sheet piling is to function as a cantilever retaining wall 5 ft (1.5 m) high. The soil weighs 110 lb/ft^3 (17.28 kN/m^3) and its angle of internal friction is 32°; the backfill has a horizontal surface. Determine the required depth of penetration of the bulkhead.

Calculation Procedure:

1. Take moments with respect to C to obtain an equation for the minimum value of d

Refer to Fig. 228a, and consider a 1-ft (30.5-cm) length of wall. Assume that the pressure distribution is hydrostatic, and apply Rankine's theory.

The wall pivots about some point Z near the bottom. Consequently, passive earth pressure is mobilized to the left of the wall betwen B and Z and to the right of the wall between Z and C.

Let P = resultant active pressure on wall; R_1 and R_2 = resultant passive pressure above and below center of rotation, respectively.

The position of Z may be found by applying statics. But to simplify the calculations, these assumptions are made: The active pressure extends from A to C; the passive pressure to the left of the wall extends from B to C; and R_2 acts at C. Figure 228b illustrates these assumptions.

By taking moments with respect to C and substituting values for R_1 and R_2,

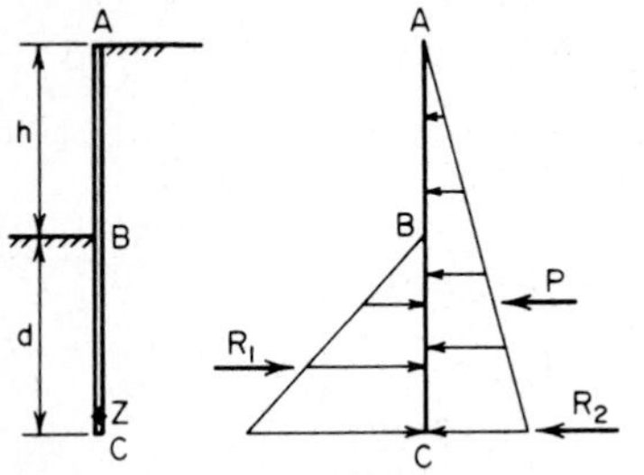

(a) Cantilever bulkhead   (b) Assumed pressures and resultant forces

FIG. 228

$$d = \frac{h}{(C_p/C_a)^{1/3} - 1} \qquad (179)$$

2. Substitute numerical values and solve for d

Thus, $45° + \frac{1}{2}\phi = 61°$; $45° - \frac{1}{2}\phi = 29°$. By Eqs. 171 and 172, $C_p/C_a = (\tan 61°/\tan 29°)^2 = 10.6$; $d = 5/[(10.6)^{1/3} - 1] = 4.2$ ft (1.3 m). Add 20 percent of the computed value to provide a factor of safety and to allow for the development of R_2. Thus, penetration $= 4.2(1.2) = 5.0$ ft (1.5 m).

ANCHORED BULKHEAD ANALYSIS

Sheet piling is to function as a retaining wall 20 ft (6.1 m) high, anchored by tie rods placed 3 ft (0.9 m) from the top at an 8-ft (2.4-m) spacing. The soil weighs 110 lb/ft^3 (17.28 kN/m^3), and its angle of internal friction is 32°. The backfill has a horizontal surface and carries a surcharge of 200 lb/ft² (9.58 kPa). Applying the equivalent-beam method, determine the depth of penetration to secure a fixed earth support, the tension in the tie rod, and the maximum bending moment in the piling.

Calculation Procedure:

1. Locate C and construct the net-pressure diagram for AC

Refer to Fig. 229a. The depth of penetration is readily calculated if stability is the sole criterion. However, when the depth is increased beyond this minimum value, the tension in the rod and

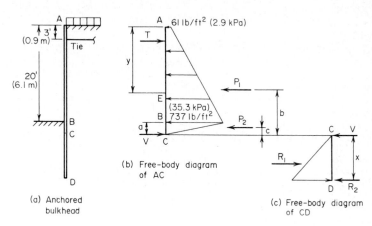

(a) Anchored bulkhead

(b) Free-body diagram of AC

(c) Free-body diagram of CD

FIG. 229

the bending moment in the piling are reduced; the net result is a saving in material despite the increased length.

Investigation of this problem discloses that the most economical depth of penetration is that for which the tangent to the elastic curve at the lower end passes through the anchorage point. If this point is considered as remaining stationary, this condition can be described as one in which the elastic curve is vertical at D, the surrounding soil acting as a fixed support. Whereas an equation can be derived for the depth associated with this condition, such an equation is too cumbersome for rapid solution.

When the elastic curve is vertical at D, the lower point of contraflexure lies close to the point where the net pressure (the difference between active pressure to the right and passive pressure to the left of the wall) is zero. By assuming that the point of contraflexure and the point of zero pressure are in fact coincident, this problem is transformed to one that is statically determinate. The method of analysis based on this assumption is termed the *equivalent-beam* method.

When the piling is driven to a depth greater than the minimum needed for stability, it deflects in such a manner as to mobilize passive pressure to the right of the wall at its lower end. However, the same simplifying assumption concerning the pressure distribution as made in the previous calculation procedure is made here.

Let C denote the point of zero pressure. Consider a 1-ft (30.5-cm) length of wall, and let T = reaction at anchorage point and V = shear at C.

Locate C and construct the net-pressure diagram for AC as shown in Fig. 229b. Thus, $w = 110$ lb/ft³ (17.28 kN/m³) and $\phi = 32°$. Then $C_a = \tan^2 (45° - 16°) = 0.307$; $C_p = \tan^2 (45° + 16°) = 3.26$; $C_p - C_a = 2.953$; $p_A = 0.307(200) = 61$ lb/ft² (2.9 kPa); $p_B = 61 + 0.307(20)(110) = 737$ lb/ft² (35.3 kPa); $a = 737/[2.953(110)] = 2.27$ ft (0.69 m).

2. Calculate the resultant forces P_1 and P_2

Thus, $P_1 = \frac{1}{2}(20)(61 + 737) = 7980$ lb (35,495.0 N); $P_2 = \frac{1}{2}(2.27)(737) = 836$ lb (3718.5 N); $P_1 + P_2 = 8816$ lb (39,213.6 N).

3. *Equate the bending moment at C to zero to find T, V, and the tension in the tie rod*

Thus b = 2.27 + ($\frac{2}{3}$)(737 + 2 × 61)/(737 + 61) = 9.45 ft (2.880 m); c = $\frac{2}{3}$(2.27) = 1.51 ft (0.460 m); ΣM_C = 19.27T − 9.45(7980) − 1.51(836) = 0; T = 3980 lb (17,703.0 N); V = 8816 − 3980 = 4836 lb (21,510.5 N). The tension in the rod = 3980(8) = 31,840 lb (141,624.3 N).

4. *Construct the net-pressure diagram for CD*

Refer to Fig. 229c and calculate the distance x. (For convenience, Fig. 229c is drawn to a different scale from that of Fig. 229b.) Thus p_D = 2953(110x) = 324.8x; R_1 = $\frac{1}{2}$(324.8x^2) = 162.4x^2; ΣM_D = $R_1 x/3$ − Vx = 0; R_1 = 3V; 162.4x^2 = 3(4836); x = 9.45 ft (2.880 m).

5. *Establish the depth of penetration*

To provide a factor of safety and to compensate for the slight inaccuracies inherent in this method of analysis, increase the computed depth by about 20 percent. Thus, penetration = 1.20(a + x) = 14 ft (4.3 m).

6. *Locate the point of zero shear; calculate the piling maximum bending moment*

Refer to Fig. 229b. Locate the point E of zero shear. Thus p_E = 61 + 0.307(110y) = 61 + 33.77y; $\frac{1}{2}y(p_A + p_E)$ = T; or $\frac{1}{2}y(122 + 33.77y)$ = 3980; y = 13.6 ft (4.1 m), and p_E = 520 lb/ft^2 (24.9 kPa); M_{max} = M_E = 3980[10.6 − (13.6/3)(520 + 2 × 61)/(520 + 61)] = 22,300 ft· lb/ft (99.190 N·m/m) of piling. Since the tie rods provide intermittent rather than continuous support, the piling sustains biaxial bending stresses.

STABILITY OF SLOPE BY METHOD OF SLICES

Investigate the stability of the slope in Fig. 230 by the method of slices (also known as the Swedish method). The properties of the upper and lower soil strata, designated as A and B, respectively, are A—w = 110 lb/ft^3 (17.28 kN/m^3); c = 0; ϕ = 28°; B—w = 122 lb/ft^3 (19.16 kN/m^3); c = 650 lb/ft^2 (31.1 kPa); ϕ = 10°. Stratum A is 36 ft (10.9 m) deep. A surcharge of 8000 lb/lin ft (116,751.2 N/m) is applied 20 ft (6.1 m) from the edge.

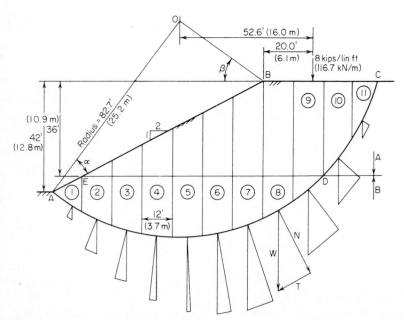

FIG. 230   Analysis of stability of slope by slices.

Calculation Procedure:

1. Locate the center of the trial arc of failure passing through the toe

It is assumed that failure of an embankment occurs along a circular arc, the prism of soil above the failure line tending to rotate about an axis through the center of the arc. However, there is no direct method of identifying the arc along which failure is most likely to occur, and it is necessary to resort to a cut-and-try procedure.

Consider a soil mass having a thickness of 1 ft (30.5 cm) normal to the plane of the drawing; let O denote the center of a trial arc of failure that passes through the toe. For a given inclination of embankment, Fellenius recommends certain values of α and β in locating the first trial arc.

Locate O by setting $\alpha = 25°$ and $\beta = 35°$.

2. Draw the arc AC and the boundary line ED of the two strata

3. Compute the length of arc AD

Scale the radius of the arc and the central angle AOD, and compute the length of the arc AD. Thus, radius = 82.7 ft (25.2 m); arc AD = 120 ft (36.6 m).

4. Determine the distance horizontally from O to the applied load

Scale the horizontal distance from O to the applied load. This distance is 52.6 ft (16.0 m).

5. Divide the soil mass into vertical strips

Starting at the toe, divide the soil mass above AC into vertical strips of 12-ft (3.7-m) width, and number the strips. For simplicity, consider that D lies on the boundary line between strips 9 and 10, although this is not strictly true.

6. Determine the volume and weight of soil in each strip

By scaling the dimensions or using a planimeter, determine the volume of soil in each strip; then compute the weight of soil. For instance, for strip 5: volume of soil A = 252 ft³ (7.13 m³); volume of soil B = 278 ft³ (7.87 m³); weight of soil = $252(110) + 278(122) = 61,600$ lb (273,996.8 N). Record the results in Table 14.

7. Draw a vector below each strip

This vector represents the weight of the soil in the strip. (Theoretically, this vector should lie on the vertical line through the center of gravity of the soil, but such refinement is not warranted in this analysis. For the interior strips, place each vector on the vertical centerline.)

TABLE 14 Stability Analysis of Slope

Strip	Weight, kips (kN)	Normal component, kips (kN)	Tangential component, kips (kN)
1	10.3 (45.81)	8.9 (39.59)	−5.2 (−23.13)
2	28.1 (124.99)	26.0 (115.65)	−10.7 (−47.59)
3	41.9 (186.37)	40.6 (180.59)	−10.4 (−46.26)
4	53.0 (235.74)	52.7 (234.41)	−5.5 (−24.46)
5	61.6 (274.00)	61.5 (273.55)	2.6 (11.56)
6	67.7 (301.13)	66.5 (295.79)	12.8 (56.93)
7	71.0 (315.81)	67.0 (298.02)	23.4 (104.08)
8	67.1 (298.46)	58.8 (261.54)	32.4 (144.12)
9	54.8 (243.75)	43.0 (191.26)	34.0 (151.23)
10	38.3 (170.36)	24.9 (110.76)	29.1 (129.44)
11	14.3 (63.61)	7.0 (31.14)	12.5 (55.60)
Total, 1 to 9		425.0 (1890.40)	
Total, 10 and 11		31.9 (141.89)	
Grand total		456.9 (2032.29)	115.0 (511.52)

8. *Resolve the soil weights vectorially into components normal and tangential to the circular arc*

9. *Scale the normal and tangential vectors; record the results in Table 14*

10. *Total the normal forces acting on soils A and B; total the tangential forces*

Failure of the embankment along arc AC would be characterized by the clockwise rotation of the soil prism above this arc about an axis through O, this rotation being induced by the unbalanced tangential force along the arc and by the external load. Therefore, consider a tangential force as positive if its moment with respect to an axis through O is clockwise and negative if this moment is counterclockwise. In the method of slices, it is assumed that the lateral forces on each soil strip approximately balance each other.

11. *Evaluate the moment tending to cause rotation about O*

In the absence of external loads,

$$DM = r\Sigma T \tag{180}$$

wehre DM = disturbing moment; r = radius of arc; ΣT = algebraic sum of tangential forces.
   In the present instance, DM = 82.7(115) + 52.6(8) = 9930 ft·kips (13,465.1 kN·m).

12. *Sum the frictional and cohesive forces to find the maximum potential resistance to rotation; determine the stabilizing moment*

In general,

$$F = \Sigma N \tan \phi \qquad C = cL \tag{181}$$

$$SM = r(F + C) \tag{182}$$

where F = frictional force; C = cohesive force; ΣN = sum of normal forces; L = length of arc along which cohesion exists; SM = stabilizing moment.
   In the present instance, F = 425 tan 10° + 31.9 tan 28° = 91.9 kips (408.77 kN); C = 0.65(120) = 78.0; total of $F + C$ = 169.9 kips (755.72 kN); SM = 82.7(169.9) = 14,050 ft·kips (19,051.8 kN·m).

13. *Compute the factor of safety against failure*

The factor of safety is FS = SM/DM = 14,050/9930 = 1.41.

14. *Select another trial arc of failure; repeat the foregoing procedure*

15. *Continue this process until the minimum value of FS is obtained*

The minimum allowable factor of safety is generally regarded as 1.5.

STABILITY OF SLOPE BY ϕ-CIRCLE METHOD

Investigate the stability of the slope in Fig. 231 by the ϕ-circle method. The properties of the soil are w = 120 lb/ft³ (18.85 kN/m³); c = 550 lb/ft² (26.3 kPa); ϕ = 4°.

Calculation Procedure:

1. *Locate the first trial position*

The ϕ-circle method of analysis formulated by Krey is useful where standard conditions are encountered. In contrast to the assumption concerning the stabilizing forces stated earlier, the ϕ-circle method assumes that the soil has mobilized its maximum potential *frictional* resistance plus whatever cohesive resistance is needed to prevent failure. A comparison of the maximum available cohesion with the required cohesion serves as an index of the stability of the embankment.
   In Fig. 231, O is the center of an assumed arc of failure AC. Let W = weight of soil mass above arc AC; R = resultant of all normal and frictional forces existing along arc AC; C = resultant cohesive force developed; L_a = length of arc AC; L_c = length of chord AC. The soil above the arc is in equilibrium under the forces W, R, and C. Since W is known in magnitude and direction, the magnitude of C may be readily found if the directions of R and C are determined.

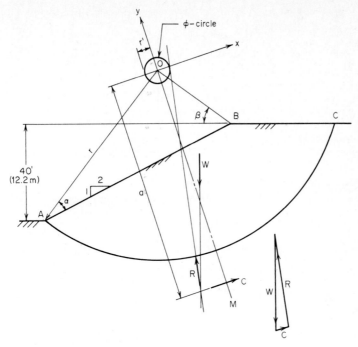

FIG. 231 Analysis of stability of slope by ϕ-circle method.

Locate the first trial position of O by setting $\alpha = 25°$ and $\beta = 35°$.

2. **Draw the arc AC and the radius OM bisecting this arc**

3. **Establish rectangular coordinate axes at O, making OM the y axis**

4. **Obtain the needed basic data**

Scale the drawing or make the necessary calculations. Thus, $r = 78.8$ ft (24.02 m); $L_a = 154.6$ ft (47.12 m); $L_c = 131.0$ ft (39.93 m); area above arc = 4050 ft^2 (376.2 m^2); $W = 4050(120) = 486,000$ lb (2,161,728 N); horizontal distance from A to centroid of area = 66.7 ft (20.33 m).

5. **Draw the vector representing W**

Since the soil is homogeneous, this vector passes through the centroid of the area.

6. **State the equation for C; locate its action line**

Thus,

$$C = C_x = cL_c \tag{183}$$

The action line of C is parallel to the x axis. Determine the distance a by taking moments about O. Thus $M = aC = acL_c$,

$$a = \left(\frac{L_a}{L_c}\right)r \tag{184}$$

Or $a = (154.6/131.0)78.8 = 93.0$ ft (28.35 m). Draw the action line of C.

7. **Locate the action line of R**

For this purpose, consider the resultant force dR acting on an elemental area. Its action line is inclined at an angle ϕ with the radius at that point, and therefore the perpendicular distance r'

from O to this action line is

$$r' = r \sin \phi \tag{185}$$

Thus, r' is a constant for the arc AC. It follows that regardless of the position of dR along this arc, its action line is tangent to a circle centered at O and having a radius r'; this is called the ϕ circle, or *friction circle*. It is plausible to conclude that the action line of the total resultant is also tangent to this circle.

Draw a line tangent to the ϕ circle and passing through the point of intersection of the action lines of W and C. This is the action line of R. (The moment of R about O is counterclockwise, since its frictional component opposes clockwise rotation of the soil mass.)

8. *Using a suitable scale, determine the magnitude of C*

Draw the triangle of forces; obtain the magnitude of C by scaling. Thus, $C = 67{,}000$ lb (298,016 N).

9. *Calculate the maximum potential cohesion*

Apply Eq. 183, equating c to the unit cohesive capacity of the soil. Thus, $C_{max} = 550(131) = 72{,}000$ lb (320,256 N). This result indicates a relatively low factor of safety. Other arcs of failure should be investigated in the same manner.

ANALYSIS OF FOOTING STABILITY BY TERZAGHI'S FORMULA

A wall footing carrying a load of 58 kips/lin ft (846.4 kN/m) rests on the surface of a soil having these properties: $w = 105$ lb/ft^3 (16.49 kN/m^3); $c = 1200$ lb/ft^2 (57.46 kPa); $\phi = 15°$. Applying Terzaghi's formula, determine the minimum width of footing required to ensure stability, and compute the soil pressure associated with this width.

Calculation Procedure:

1. *Equate the total active and passive pressures and state the equation defining conditions at impending failure*

While several methods of analyzing the soil conditions under a footing have been formulated, the one proposed by Terzaghi is gaining wide acceptance.

The soil underlying a footing tends to rupture along a curved surface, but the Terzaghi method postulates that this surface may be approximated by straight-line segments without introducing any significant error. Thus, in Fig. 232, the soil prism OAB tends to heave by sliding downward along OA under active pressure and sliding upward along AB against passive pressure. As stated earlier, these planes of failure make an angle of $\alpha = 45° + \tfrac{1}{2}\phi$ with the principal planes.

Let $b =$ width of footing; $h =$ distance from ground surface to bottom of footing; $p =$ soil

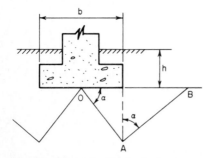

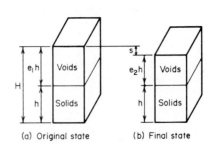

FIG. 232 Failure of soil under footing in accordance with Terzaghi's assumption.

FIG. 233

pressure directly below footing. By equating the total active and passive pressures, state the following equation defining the conditions at impending failure:

$$p = wh \tan^4 \alpha + \frac{wb(\tan^5 \alpha - \tan \alpha)}{4} + 2c \,(\tan \alpha + \tan^3 \alpha) \qquad (186)$$

2. Substitute numerical values; solve for b; evaluate p

Thus, $h = 0$; $p = 58/b$; $\phi = 15°$; $\alpha = 45° + 7°30' = 52°30'$; $58/b = 0.105b(3.759 - 1.303)/4 + 2(1.2)(1.303 + 2.213)$; $b = 6.55$ ft (1.996 m); $p = 58,000/6.55 = 8850$ lb/ft^2 (423.7 kPa).

SOIL CONSOLIDATION AND CHANGE IN VOID RATIO

In a laboratory test, a load was applied to a soil specimen having a height of 30 in (762.0 mm) and a void ratio of 96.0 percent. What was the void ratio when the load settled ½ in (12.7 mm)?

Calculation Procedure:

1. Construct a diagram representing the volumetric composition of the soil in the original and final states

According to the Terzaghi theory of consolidation, the compression of a soil mass under an increase in pressure results primarily from the expulsion of water from the pores. At the instant the load is applied, it is supported entirely by the water, and the hydraulic gradient thus established induces flow. However, the flow in turn causes a continuous transfer of load from the water to the solids.

Equilibrium is ultimately attained when the load is carried entirely by the solids, and the expulsion of the water then ceases. The time rate of expulsion, and therefore of consolidation, is a function of the permeability of the soil, the number of drainage faces, etc. Let H = original height of soil stratum; s = settlement; e_1 = original void ratio; e_2 = final void ratio. Using the given data, construct the diagram in Fig. 233, representing the volumetric composition of the soil in the original and final states.

2. State the equation relating the four defined quantities

Thus,

$$s = \frac{H(e_1 - e_2)}{1 + e_1} \qquad (187)$$

3. Solve for e_2

Thus: $H = 30$ in (762.0 mm); $s = 0.50$ in (12.7 mm); $e_1 = 0.960$; $e_2 = 92.7$ percent.

COMPRESSION INDEX AND VOID RATIO OF A SOIL

A soil specimen under a pressure of 1200 lb/ft^2 (57.46 kPa) is found to have a void ratio of 103 percent. If the compression index is 0.178, what will be the void ratio when the pressure is increased to 5000 lb/ft^2 (239.40 kPa)?

Calculation Procedure:

1. Define the compression index

By testing a soil specimen in a consolidometer, it is possible to determine the void ratio associated with a given compressive stress. When the sets of values thus obtained are plotted on semilogarithmic scales (void ratio vs. logarithm of stress), the resulting diagram is curved initially but becomes virtually a straight line beyond a specific point. The slope of this line is termed the *compression index.*

2. Compute the soil void ratio

Let C_c = compression index; e_1 and e_2 = original and final void ratio, respectively; σ_1 and σ_2 = original and final normal stress, respectively.

Write the equation of the straight-line portion of the diagram:

$$e_1 - e_2 = C_c \log \frac{\sigma_2}{\sigma_1} \tag{188}$$

Substituting and solving give $1.03 - e_2 = 0.178 \log (5000/1200)$; $e_2 = 92.0$ percent. Note that the logarithm is taken to the base 10.

SETTLEMENT OF FOOTING

An 8-ft (2.4-m) square footing carries a load of 150 kips (667.2 kN) that may be considered uniformly distributed, and it is supported by the soil strata shown in Fig. 234. The silty clay has a compression index of 0.274; its void ratio prior to application of the load is 84 percent. Applying the unit weights indicated in Fig. 234, calculate the settlement of the footing caused by consolidation of the silty clay.

Calculation Procedure:

1. Compute the vertical stress at middepth before and after application of the load

To simplify the calculations, assume that the load is transmitted through a truncated pyramid having side slopes of 2 to 1 and that the stress is uniform across a horizontal plane. Take the stress at middepth of the silty-clay stratum as the average for that stratum.

Compute the vertical stress σ_1 and σ_2 at middepth before and after application of the load, respectively. Thus: $\sigma_1 = 6(116) + 12(64) + 7(60) = 1884$ lb/ft^2 (90.21 kPa); $\sigma_2 = 1884 + 150,000/33^2 = 2022$ lb/ft^2 (96.81 kPa).

2. Compute the footing settlement

Combine Eqs. 187 and 188 to obtain

$$s = \frac{H C_c \log (\sigma_2/\sigma_1)}{1 + e_1} \tag{189}$$

Solving gives $s = 14(0.274) \log (2022/1884)/(1 + 0.84) = 0.064$ ft $= 0.77$ in (19.558 mm).

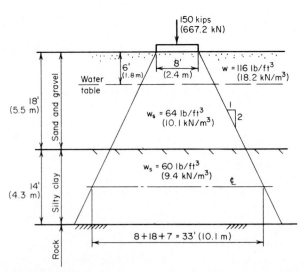

FIG. 234 Settlement of footing.

DETERMINATION OF FOOTING SIZE BY HOUSEL'S METHOD

A square footing is to transmit a load of 80 kips (355.8 kN) to a cohesive soil, the settlement being restricted to ⅝ in (15.9 mm). Two test footings were loaded at the site until the settlement reached this value. The results were

Footing size	Load, lb (N)
1 ft 6 in × 2 ft (45.72 × 60.96 cm)	14,200 (63,161.6)
3 ft × 3 ft (91.44 × 91.44 cm)	34,500 (153,456.0)

Applying Housel's method, determine the size of the footing in plan.

Calculation Procedure:

1. Determine the values of p and s corresponding to the allowable settlement

Housel considers that the ability of a cohesive soil to support a footing stems from two sources: bearing strength and shearing strength. This concept is embodied in

$$W = Ap + Ps \tag{190}$$

where W = total load; A = area of contact surface; P = perimeter of contact surface; p = bearing stress directly below footing; s = shearing stress along perimeter.

Applying the given data for the test footings gives: footing 1, A = 3 ft² (2787 cm²), P = 7 ft (2.1 m); footing 2, A = 9 ft² (8361 cm²), P = 12 ft (3.7 m). Then $3p + 7s$ = 14,200; $9p + 12s$ = 34,500; p = 2630 lb/ft² (125.9 kPa); s = 900 lb/lin ft (13,134.5 N/m).

2. Compute the size of the footing to carry the specified load

Let x denote the side of the footing. Then, $2630x^2 + 900(4x)$ = 80,000; x = 4.9 ft (1.5 m). Make the footing 5 ft (1.524 m) square.

APPLICATION OF PILE-DRIVING FORMULA

A 16 × 16 in (406.4 × 406.4 mm) pile of 3000-lb/in² (20,685-kPa) concrete, 45 ft (13.7 m) long, is reinforced with eight no. 7 bars. The pile is driven by a double-acting steam hammer. The weight of the ram is 4600 lb (20,460.8 N), and the energy delivered is 17,000 ft·lb (23,052 J) per blow. The average penetration caused by the final blows is 0.42 in (10.668 mm). Compute the bearing capacity of the pile by applying Redtenbacker's formula and using a factor of safety of 3.

Calculation Procedure:

1. Find the weight of the pile and the area of the transformed section

The work performed in driving a pile into the soil is a function of the reaction of the soil on the pile and the properties of the pile. Therefore, the soil reaction may be evaluated if the work performed by the hammer is known. Let A = cross-sectional area of pile; E = modulus of elasticity; h = height of fall of ram; L = length of pile; P = allowable load on pile; R = reaction of soil on pile; s = penetration per blow; W = weight of falling ram; w = weight of pile.

Redtenbacker developed the following equation by taking these quantities into consideration: the work performed by the soil in bringing the pile to rest; the work performed in compressing the pile; and the energy delivered to the pile:

$$Rs + \frac{R^2L}{2AE} = \frac{W^2h}{W + w} \tag{191}$$

Finding the weight of the pile and the area of the transformed section, we get w = 16(16)(0.150)(45)/144 = 12 kips (53.4 kN). The area of a no. 7 bar = 0.60 in² (3.871 cm²); n = 9; A = 16(16) + 8(9 − 1)0.60 = 294 in² (1896.9 cm²).

2. Apply Eq. 191 to find R; evaluate P

Thus, s = 0.42 in (10.668 mm); L = 540 in (13,716 mm); E_c = 3160 kips/in^2 (21,788.2 MPa.); W = 4.6 kips (20.46 kN); Wh = 17 ft·kips = 204 in·kips (23,052 J). Substituting gives $0.42R + 540R^2/[2(294)(3160)]$ = 4.6(204)/(4.6 + 12); R = 84.8 kips (377.19 kN); P = $R/3$ = 28.3 kips (125.88 kN).

CAPACITY OF A GROUP OF FRICTION PILES

A structure is to be supported by 12 friction piles of 10-in (254-mm) diameter. These will be arranged in four rows of three piles each at a spacing of 3 ft (91.44 cm) in both directions. A test pile is found to have an allowable load of 32 kips (142.3 kN). Determine the load that may be carried by this pile group.

Calculation Procedure:

1. State a suitable equation for the load

When friction piles are compactly spaced, the area of soil that is needed to support an individual pile overlaps that needed to support the adjacent ones. Consequently, the capacity of the group is less than the capacity obtained by aggregating the capacities of the individual piles. Let P = capacity of group; P_i = capacity of single pile; m = number of rows; n = number of piles per row; d = pile diameter; s = center-to-center spacing of piles; θ = $\tan^{-1} d/s$ deg. A suitable equation using these variables is the Converse-Labarre equation

$$\frac{P}{P_i} = mn - \left(\frac{\theta}{90°}\right)[m(n-1) + n(m-1)] \tag{192}$$

2. Compute the load

Thus P_i = 32 kips (142.3 kN); m = 4; n = 3; d = 10 in (254 mm); s = 36 in (914.4 mm); θ = $\tan^{-1} 10/36$ = 15.5°. Then $P/32$ = 12 − (15.5/90)(4 × 2 + 3 × 3); P = 290 kips (1289.9 kN).

LOAD DISTRIBUTION AMONG HINGED BATTER PILES

Figure 235a shows the relative positions of four steel bearing piles that carry the indicated load. The piles, which may be considered as hinged at top and bottom, have identical cross sections and the following relative effective lengths: A, 1.0; B, 0.95; C, 0.93; D, 1.05. Outline a graphical procedure for determining the load transmitted to each pile.

Calculation Procedure:

1. Subject the structure to a load for purposes of analysis

Steel and timber piles may be considered to be connected to the concrete pier by frictionless hinges, and bearing piles that extend a relatively short distance into compact soil may be considered to be hinge-supported by the soil.

Since four unknown quantities are present, the structure is statically indeterminate. A solution to this problem therefore requires an analysis of the deformation of the structure.

As the load is applied, the pier, assumed to be infinitely rigid, rotates to some new position. This displacement causes each pile to rotate about its base and to undergo an axial strain. The contraction or elongation of each pile is directly proportional to the perpendicular distance p from the axis of rotation to the longitudinal axis of that pile. Let P denote the load induced in the pile. Then $P = \Delta L \, AE/L$. Since ΔL is proportional to p and AE is constant for the group, this equation may be transformed to

$$P = \frac{kp}{L} \tag{193}$$

where k is a constant of proportionality.

If the center of rotation is established, the pile loads may therefore be found by scaling the p distances. Westergaard devised a simple graphical method of locating the center of rotation. This

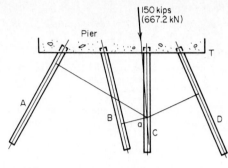

(a) Pile group and load

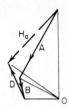

(b) Force polygon
for H_a

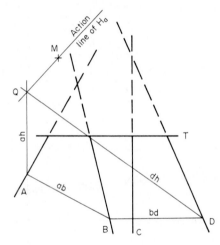

(c) Construction to locate action line of H_a

FIG. 235

method entails the construction of string polygons, described in the first calculation procedure of this handbook.

In Fig. 235a select any convenient point a on the action line of the load. Consider the structure to be subjected to a load H_a that causes the pier to rotate about a as a center. The object is to locate the action line of this hypothetical load.

It is often desirable to visualize that a load is applied to a body at a point that in reality lies outside the body. This condition becomes possible if the designer annexes to the body an infinitely rigid arm containing the given point. Since this arm does not deform, the stresses and strains in the body proper are not modified.

2. Scale the perpendicular distance from a to the longitudinal axis of each pile; divide this distance by the relative length of the pile

In accordance with Eq. 193, the quotient represents the relative magnitude of the load induced in the pile by the load H_a. If rotation is assumed to be counterclockwise, piles A and B are in compression and D is in tension.

3. Using a suitable scale, construct the force polygon

This polygon is shown in Fig. 235b. Construct this polygon by applying the results obtained in step 2. This force polygon yields the direction of the action line of H_a.

4. In Fig. 235b, select a convenient pole O and draw rays to the force polygon

5. Construct the string polygon shown in Fig. 235c

The action line of H_a passes through the intersection point Q of rays ah and dh, and its direction appears in Fig. 235b. Draw this line.

6. Select a second point on the action line of the load

Choose point b on the action line of the 150-kip (667.2-kN) load, and consider the structure to be subjected to a load H_b that causes the pier to rotate about b as center.

7. Locate the action line of H_b

Repeat the foregoing procedure to locate the action line of H_b in Fig. 235c. (The construction has been omitted for clarity.) Study of the diagram shows that the action lines of H_a and H_b intersect at M.

8. Test the accuracy of the construction

Select a third point c on the action line of the 150-kip (667.2-kN) load, and locate the action line of the hypothetical load H_c causing rotation about c. It is found that H_c also passes through M. In summary, these hypothetical loads causing rotation about specific points on the action line of the true load are all concurrent.

Thus, M is the center of rotation of the pier under the 150-kip (667.2-kN) load. This conclusion stems from the following analysis: Load H_a applied at M causes zero deflection at a. Therefore, in accordance with Maxwell's theorem of reciprocal deflections, if the true load is applied at a, it will cause zero deflection at M in the direction of H_a. Similarly, if the true load is applied at b, it will cause zero deflection at M in the direction of H_b. Thus, M remains stationary under the 150-kip (667.2-kN) load; that is, M is the center of rotation of the pier.

9. Scale the perpendicular distance from M to the longitudinal axis of each pile

Divide this distance by the relative length of the pile. The quotient represents the relative magnitude of the load induced in the pile by the 150-kip (667.2-kN) load.

10. Using a suitable scale, construct the force polygon by applying the results from step 9

If the work is accurate, the resultant of these relative loads is parallel to the true load.

11. Scale the resultant; compute the factor needed to correct this value to 150 kips (667.2 kN)

12. Multiply each relative pile load by this correction factor to obtain the true load induced in the pile

LOAD DISTRIBUTION AMONG PILES WITH FIXED BASES

Assume that the piles in Fig. 235a penetrate a considerable distance into a compact soil and may therefore be regarded as restrained against rotation at a certain level. Outline a procedure for determining the axial load and bending moment induced in each pile.

Calculation Procedure:

1. State the equation for the length of a dummy pile

Since the Westergaard construction presented in the previous calculation procedure applies solely to hinged piles, the group of piles now being considered is not directly amenable to analysis by this method.

As shown in Fig. 236a, the pile AB functions in the dual capacity of a column and cantilever beam. In Fig. 236b, let A' denote the position of A following application of the load. If secondary effects are disregarded, the axial force P transmitted to this pile is a function of Δ_y, and the transverse force S is a function of Δ_x.

Consider that the fixed support at B is replaced with a hinged support and a pile AC of identical cross section is added perpendicular to AB, as shown in Fig. 236b. If pile AC deforms an amount Δ_x under an axial force S, the forces transmitted by the pier at each point of support are

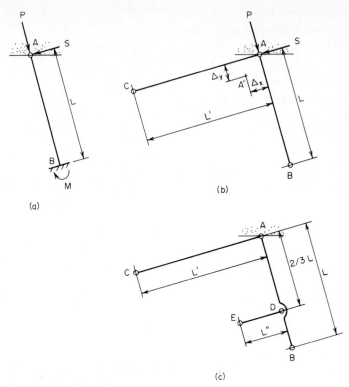

FIG. 236 Real and dummy piles.

not affected by this modification of supports. The added pile is called a *dummy* pile. Thus, the given pile group may be replaced with an equivalent group consisting solely of hinged piles.

Stating the equation for the length L' of the dummy pile, equate the displacement Δ_x in the equivalent pile group to that in the actual group. Or, $\Delta_x = SL'/AE = SL^3/3EI$.

$$L' = \frac{AL^3}{3I} \qquad (194)$$

2.   *Replace all fixed supports in the given pile group with hinged supports*

Add the dummy piles. Compute the lengths of these piles by applying Eq. 194.

3.   *Determine the axial forces induced in the equivalent pile group*

Using the given load, apply Westergaard's construction, as described in the previous calculation procedure.

4.   *Remove the dummy piles; restore the fixed supports*

Compute the bending moments at these supports by applying the equation $M = SL$.

LOAD DISTRIBUTION AMONG PILES FIXED AT TOP AND BOTTOM

Assume that the piles in Fig. 235a may be regarded as having fixed supports both at the pier and at their bases. Outline a procedure for determining the axial load and bending moment induced in each pile.

Calculation Procedure:

1. Describe how dummy piles may be used

A pile made of reinforced concrete and built integrally with the pier is restrained against rotation relative to the pier. As shown in Fig. 236c, the fixed supports of pile AB may be replaced with hinges provided that dummy piles AC and DE are added, the latter being connected to the pier by means of a rigid arm through D.

2. Compute the lengths of the dummy piles

If D is placed at the lower third point as indicated, the lengths to be assigned to the dummy piles are

$$L' = \frac{AL^3}{3I} \quad \text{and} \quad L'' = \frac{AL^3}{9I} \tag{195}$$

Replace the given group of piles with its equivalent group, and follow the method of solution in the previous calculation procedure.

REFERENCES: Crawley and Dillion—*Steel Buildings: Analysis and Design*, Wiley; Bowles—*Structural Steel Design*, McGraw-Hill; ASCE Council on Computer Practices—*Computing in Civil Engineering*, ASCE; American Concrete Institute—*Building Code Requirements for Reinforced Concrete*; American Institute of Steel Construction—*Manual of Steel Construction*; National Forest Products Association—*National Design Specification for Stress-Grade Lumber and Its Fastenings*; Abbett—*American Civil Engineering Practice*, Wiley; Gaylord and Gaylord—*Structural Engineering Handbook*, McGraw-Hill; LaLonde and Janes—*Concrete Engineering Handbook*, McGraw-Hill; Lincoln Electric Co.—*Procedure Handbook of Arc Welding Design and Practice*; Merritt—*Standard Handbook for Civil Engineers*, McGraw-Hill; Timber Engineering Company—*Timber Design and Construction Handbook*, McGraw-Hill; U.S. Department of Agriculture, Forest Products Laboratory—*Wood Handbook (Agriculture Handbook 72)*, GPO; Urquhart—*Civil Engineering Handbook*, McGraw-Hill; Borg and Gennaro—*Advanced Structural Analysis*, Van Nostrand; Gerstle—*Basic Structural Design*, McGraw-Hill; Jensen—*Applied Strength of Materials*, McGraw-Hill; Kurtz—*Comprehensive Structural Design Design Guide*, McGraw-Hill; Roark—*Formulas for Stress and Strain*, McGraw-Hill; Seely—*Resistance of Materials*, Wiley; Shanley—*Mechanics of Materials*, McGraw-Hill; Timoshenko and Young—*Theory of Structures*, McGraw-Hill; Beedle, et al.—*Structural Steel Design*, Ronald; Grinter—*Design of Modern Steel Structures*, Macmillan; Lothers—*Advanced Design in Structural Steel*, Prentice-Hall; Beedle—*Plastic Design of Steel Frames*, Wiley; Canadian Institute of Timber Construction—*Timber Construction*; Scofield and O'Brien—*Modern Timber Engineering*, Southern Pine Association; Dunham—*Theory and Practice of Reinforced Concrete*, McGraw-Hill; Winter, et al.—*Design of Concrete Structures*, McGraw-Hill; Viest, Fountain, and Singleton—*Composite Construction in Steel and Concrete*, McGraw-Hill; Chi and Biberstein—*Theory of Prestressed Concrete*, Prentice-Hall; Connolly—*Design of Prestressed Concrete Beams*, McGraw-Hill; Evans and Bennett—*Pre-stressed Concrete*, Wiley; Libby—*Prestressed Concrete*, Ronald; Magnel—*Prestressed Concrete*, McGraw-Hill; Gennaro—*Computer Methods in Solid Mechanics*, Macmillan; Laursen—*Matrix Analysis of Structures*, McGraw-Hill; Weaver—*Computer Programs for Structural Analysis*, Van Nostrand; Brenkert—*Elementary Theoretical Fluid Mechanics*, Wiley; Daugherty and Franzini—*Fluid Mechanics with Engineering Applications*, McGraw-Hill; King and Brater—*Handbook of Hydraulics*, McGraw-Hill; Li and Lam—*Principles of Fluid Mechanics*, Addison-Wesley; Sabersky and Acosta—*Fluid Flow*, Macmillan; Streeter—*Fluid Mechanics*, McGraw-Hill; Allen—*Railroad Curves and Earthwork*, McGraw-Hill; Davis, Foote, and Kelly—*Surveying: Theory and Practice*, McGraw-Hill; Hickerson—*Route Surveys and Design*, McGraw-Hill; Hosmer and Robbins—*Practical Astronomy*, Wiley; Jones—*Geometric Design of Modern Highways*, Wiley; Meyer—*Route Surveying*, International Textbook; American Association of State Highways Officials—*A Policy on Geometric Design of Rural Highways*; Chellis—*Pile Foundations*, McGraw-Hill; Goodman and Karol—*Theory and Practice of Foundation Engineering*, Macmillan; Huntington—*Earth Pressures and Retaining Walls*, Wiley; Ritter and Paquette—*Highway Engineering*, Ronald; Scott and Schoustra—*Soil: Mechanics and Engineering*, McGraw-Hill; Spangler—*Soil Engineering*, International Textbook; Teng—*Foundation Design*, Prentice-Hall; Terzaghi and Peck—*Soil Mechanics in Engineering Practice*, Wiley; U.S. Department of the Interior, Bureau of Reclamation—*Earth Manual*, GPO.

INDEX

ABOUT THE EDITOR

Tyler G. Hicks, P.E., is a consulting engineer with International Engineering Associates. He has worked in both plant design and operation in a variety of industries, taught at several engineering schools, and lectured both in the United States and abroad on engineering topics. He is a member of ASME and IEEE and holds a bachelor's degree in mechanical engineering from Cooper Union School of Engineering. Mr. Hicks is the author of numerous engineering reference books on equipment and plant design and operation.